Groundwater Quality Sustainability

Selected papers on hydrogeology

17

Series Editor: Dr. Nick S. Robins
Editor-in-Chief IAH Book Series, British Geological Survey, Wallingford, UK

INTERNATIONAL ASSOCIATION OF HYDROGEOLOGISTS

Groundwater Quality Sustainability

Editors:

Piotr Maloszewski

HMGU-Institute of Groundwater Ecology, Neuherberg, Germany

Stanisław Witczak & Grzegorz Malina

AGH University of Science and Technology, Krakow, Poland

CRC Press
Taylor & Francis Group
Boca Raton London New York

CRC Press is an imprint of the
Taylor & Francis Group, an **informa** business
A BALKEMA BOOK

Cover illustrations:
Left: Spring Olczyskie in High Tatra Mountains, Poland, photograph taken by P. Wachniew
Middle: Werenskiold Glacier Spring, Spitsbergen, Svalbard, Norway, photograph taken by P. Wachniew
Right: Wawel Castle and Vistula River in Kraków, photograph taken by B. Maloszewska

Published by:
CRCPress/Balkema
P.O. Box 447, 2300 AK Leiden, The Netherlands
e-mail: Pub.NL@taylorandfrancis.com
www.crcpress.com – www.taylorandfrancis.com

First issued in paperback 2020

CRC Press/Balkema is an imprint of the Taylor & Francis Group, an informa business

Typeset by MPS Limited, Chennai, India

Library of Congress Cataloging-in-Publication Data

Groundwater quality sustainability / editors, Piotr Maloszewski,
Stanislaw Witczak & Grzegorz Malina.
p. cm.
Includes bibliographical references and index.
ISBN 978-0-415-69841-2 (hardback : alk. paper)
1. Groundwater—Quality. 2. Groundwater—Pollution. 3. Groundwater recharge. I. Maloszewski, Piotr. II. Witczak, Stanislaw. III. Malina, Grzegorz. IV. International Association of Hydrogeologists. Congress (38th : 2010 : Kraków, Poland)
TD403.G74 2012
628.1'14—dc23

2012016910

ISBN 13: 978-0-367-57673-8 (pbk)
ISBN 13: 978-0-415-69841-2 (hbk)

Table of contents

Foreword

Sustainable groundwater development requires knowledge of the appropriate recharge and transport processes. This is a prerequisite to understanding: (i) groundwater resources and their availability, and (ii) the dependence between groundwater and the environment. Conceptual understanding of groundwater flow at both temporal and spatial scales (local and regional) is essential for management that will support engineering, industry, agriculture, ecology, and all environmentally related issues.

All sustainable groundwater development programmes require knowledge of the system. This is implied as a prerequisite to understand: (i) groundwater resources and their availability, and (ii) the dependence between groundwater and the environment. Such awareness could be achieved through understanding of groundwater flow: from local to regional scale. Understanding of groundwater flow at its relevant temporal and spatial scales is essential in studies involving engineering, geography, agriculture, ecology, and in a broader sense, in environmentally related issues.

There are five sections to this book that reflect the sub-sections within the 38th IAH Congress held in Kraków, Poland between 12 and 17 September, 2010, where these selected papers were first presented: (i) Groundwater Quality, (ii) Management of Groundwater, (iii) Groundwater-Surface Water Interactions, (iv) Regional Groundwater Problems and (v) Data Processing & Modelling. The overall theme is groundwater quality sustainability, which in Europe is interwoven with the EU Water Framework Directive, but is equally important the world over.

This book has been prepared for scientists, researchers, students, engineers, water resources specialists, groundwater consultants, government administrators and teachers. It is of direct and applied interest to practitioners in hydrogeology and groundwater (resources, quality, pollution, protection and clean-up), geochemistry and hydrogeochemical modelling, and investigators into environmental hydrology, groundwater dependent ecosystems and other practical environmental issues.

Finally, we would like to dedicate this volume to our good friend and the Chair of the Scientific Committee of the IAH 2010 Congress in Kraków, Professor Andrzej Zuber who unexpectedly passed away on 28 February 2011. Professor Zuber was a most distinguished and world renowned scientists working in the field of isotope hydrology. He will be sorely missed not only in Poland but across the globe.

Editors:
Piotr Maloszewski
Stanisław Witczak
Grzegorz Malina

March 2012

About the editors

Prof. Dr. Piotr Maloszewski, PhD, DSc, is a Deputy Director and Head of Hydrological Modelling Group at the Institute for Groundwater Ecology of the Helmholtz Centre Munich – German Research Centre for Environmental Health. He is Professor of hydrology at the University Freiburg, Germany, and a President of the International Commission on Tracers of IAHS/IUGG. He is author and co-author of above 170 publications and one book ("Tracers in Hydrology") in areas of application of artificial and environmental tracers in hydrology and in mathematical modelling of experimental data. He was three years in the Editorial Board of the Journal of Hydrology and five years associated editor of the Hydrological Science Journal. He was also reviewer for several journals like e.g.: Water Resources Research, Contaminant Hydrology, or Hydrogeology Journal. He is a member of the IAH, IAHS and AGU.

Prof. Dr. Stanisław Witczak, PhD, DSc, is a Professor of hydrogeology at the Department of Hydrogeology and Engineering Geology of the AGH-University of Science and Technology in Krakow, Poland. He is author and co-author of above 160 publications in the areas of hydrogeology, hydrogeochemistry, groundwater protection and modelling of regional groundwater flow and transport. Among others, he is co-author of the Map of the Critical Protection Areas (CPA) of the Major Groundwater Basins (MGWB) in Poland and Groundwater Vulnerability Map of Poland. Professional affiliations: IAH – International Association of Hydrogeologists; NGWA – National Ground Water Association; IWA – International Water Association; IAGC – International Association of GeoChemistry; AGU – American Geophysical Union.

Prof. Dr. Grzegorz Malina, PhD, DSc, EurIng, is a Professor of Earth sciences at the Department of Hydrogeology and Engineering Geology of the AGH-University of Science and Technology in Krakow, Poland. He is author and co-author of above 160 publications and technical notes in the areas of environmental protection/engineering and hydrogeology, and 4 books; co-editor of "Soil and Sediment Remediation: Mechanisms, technologies and applications" Integrated Environmental Technology Series, IWA Publishing, 2005; and editor of 5 books and monographs. He was reviewer for such journals like: Ecology and Ecohydrology, Water Practice and Technology, the Open Environmental Engineering Journal, Journal of Environmental Management, Polish Journal of Environmental Sciences, Archives of Environmental Protection. Since 2008 he is member of the Editorial Advisory Board of the Benthon Open Environmental Engineering Journal.

List of all contributors

AHRNS Johannes J.	Department of Civil Engineering/Architecture, University of Applied Sciences Dresden, Friedrich-List-Platz 1 01069 Dresden, Germany
ALEXEEV Sergey V.	Institute of the Earth's Crust SB RAS, Lermontov str., 128, Irkutsk, Russia
ALEXEEVA Ludmila P.	Institute of the Earth's Crust SB RAS, Lermontov str., 128, Irkutsk, Russia
AL-SIBAI Mahmoud	Arab Center for the Studies of Arid Zones and Dry Lands (ACSAD), Damascus, Syria
ANDREO Bartolomé	Department of Geology and Centre of Hydrogeology of University of Malaga, Spain
BABA Alper	Izmir Institute of Technology Department of Civil Engineering Izmir, Turkey
BARBERÁ Juan A.	Department of Geology and Centre of Hydrogeology of University of Malaga, Spain
BEER Antje	Landesamt für Geologie und Bergwesen Sachsen-Anhalt, Köthener Str. 34, 06118 Halle, Germany
BIONDIĆ Božidar	University of Zagreb, Faculty of Geotechnical Engineering, Hallerova aleja 7, 42 000 Varaždin, Croatia
BIONDIĆ Ranko	University of Zagreb, Faculty of Geotechnical Engineering, Hallerova aleja 7, 42 000 Varaždin, Croatia
BLOKHIN Maksim G.	Far East Geological Institute, Russian Academy of Sciences, 100-let Vladivostoku – 159, Vladivostok, Russian Federation
BRAGIN Ivan V.	Far East Geological Institute, Russian Academy of Sciences, 100-let Vladivostoku – 159, Vladivostok, Russian Federation
BROSE D.	Landesamt für Bergbau, Geologie und Rohstoffe Brandenburg, Inselstr. 26, D-03046 Cottbus, Germany
BUDZIAK Dörte	Landesamt für Bergbau, Energie und Geologie, Stilleweg 2, 30655 Hannover, Germany
CABARET Olivier	Institut EGID, Université Bordeaux-3, 1, allée F. Daguin, 33607 Pessac, France

CARRASCO CANTOS Francisco	Centre of Hydrogeology, Department of Geology, Faculty of Science, University of Málaga, Campus de Teatinos s.n., 29071 Málaga, Spain
CASTALDELLI Giuseppe	Department of Biology and Evolution, University of Ferrara, Via L. Borsari, 46, 44100 Ferrara, Italy
CHARLÉ Christoph	PROTEKUM Umweltinstitut GmbH, Lehnitzstrasse 73, 16151 Oranienburg, Germany
CHELNOKOV George A.	Far East Geological Institute, Russian Academy of Sciences, 100-let Vladivostoku – 159, Vladivostok, Russian Federation
CLEMENTINE Cyprien	Institute of Environmental Engineering, ETH Zurich, Wolfgang-Pauli-Strasse 15, CH-8093 Zürich, Switzerland
CLOS Patrick	Bundesanstalt für Geowissenschaften und Rohstoffe, Stilleweg 2, 30655 Hannover, Germany
COLOMBANI Nicolò	Department of Earth Sciences, University of Ferrara, Via Saragat, 1, 44100 Ferrara, Italy
CSOMA Anita É.	ConocoPhillips, Subsurface Technology, Basin and Sedimentary Systems, 600 N Dairy Ashford, PR3064, Houston, 77079 TX, USA
CUSTODIO Emilio	Departament of Geotechnical Engineering and Geosciences, Technical University of Catalonia (UPC), Jordi Girona, 1–3. E08034 Barcelona, Spain
DORCA Helena	Fundación Centro Interacional de Hidrología Subterránea (FCIHS), C/Provença 102, Barcelona 08029, Spain
DREHER Thomas	Landesamt für Geologie und Bergbau Rheinland-Pfalz, Emy-Roeder-Str. 5, 55129 Mainz, Germany
DROUBI Abdallah	Arab Center for the Studies of Arid Zones and Dry Lands (ACSAD), Damascus, Syria
DUPUY Alain	Institut EGID, Université Bordeaux-3, 1, allée F. Daguin, 33607 Pessac, France
EBERMANN Jakob P.,	Department of Civil Engineering/Architecture, University of Applied Sciences Dresden, Friedrich-List-Platz 1 01069 Dresden, Germany
ECKERT Paul	Waterworks, Stadtwerke Duesseldorf AG, Himmelgeister Landstraße 1, 40589 Duesseldorf, Germany
EICHHORN Dieter	Dr.-Ing. Dieter Eichhorn, Possendorfer Straße 5, 01217 Dresden, Germany
ELPIT Handan	Dokuz Eylul University Department of Environmental Engineering Izmir, Turkey
ERŐSS Anita	Department of Physical and Applied Geology, Institute of Geography and Earth Sciences, Eötvös Loránd University, 1/c Pázmány Péter sétány, 1117 Budapest, Hungary

FARNLEITNER Andreas H.	Vienna University of Technology, Institute of Chemical Engineering, Department of Applied Biochemistry and Gene Technology, Gumpendorferstrasse 1a, 1060 Vienna, Austria
FRAILE Josep	Agencia Catalana de l'Aigua (ACA), C/Provença 204-208, E08036-Barcelona, Spain
FRITSCHE Hans-Gerhard	Hess. Landesamt für Umwelt und Geologie, Abt. Geologie, Dezernat Hydrogeologie, Postfach 3209, 65022 Wiesbaden, Germany
GANJI KHORRAMDEL Naser	Department of Water Engineering, Arak University, Arak, Iran
GARRIDO Teresa	Agencia Catalana de l'Aigua (ACA), C/Provença 204-208, E08036-Barcelona, Spain
GRISCHEK Thomas	Department of Civil Engineering/Architecture, University of Applied Sciences Dresden, Friedrich-List-Platz 1 01069 Dresden, Germany
GUNDUZ Orhan	Dokuz Eylul University Department of Environmental Engineering Izmir, Turkey
GUNKEL Günter	Berlin University of Technology, Dept. Water Quality Control, Strasse des 17. Juni 135, 10623 Berlin, Germany
HENNINGS Volker	Federal Institute for Geosciences and Natural Resources (BGR), Stilleweg 2, 30655 Hannover, Germany
HOWARD Ken	Department of Physical and Environmental Sciences, University of Toronto Scarborough, 1265 Military Trail Toronto, M1C 1A4 Canada
HUBER Markus	Geo: Tools, Brünnsteinstr. 10, 81541 München, Germany
HUGMAN Rui	Geo-Systems Centre/CVRM – Universidade do Algarve, Campus de Gambelas, 8005-139 Faro, Portugal
HÜBSCHMANN Matthias	Sächsisches Landesamt für Umwelt, Landwirtschaft und Geologie, Postfach 540137, 01311 Dresden, Germany
JAVADI Saman	Water Research Institute, Ministry of Energy, Tehran, Iran
JUAREZ Iker	Amphos XXI Consulting S.L., Passeig de García i Faria 49-51, E08019- Barcelona, Spain
KAYZER Dariusz	Department of Mathematical and Statistical Methods, Poznan University of Life Sciences, Poznan, Poland
KĘDZIORA Andrzej	Institute for Agricultural and Forest Environment, Polish Academy of Sciences, Poznan, Poland
KINZELBACH Wolfgang	ETH Zurich, Institute of Environmental Engineering, Wolfgang-Pauli-Strasse 15, 8093 Zurich, Switzerland

KONONOV Alexander M.	Institute of the Earth's Crust SB RAS, Lermontov str., 128, Irkutsk, Russia
KOZLOWSKI Cezary	Institute of Chemistry, Environmental Protection and Biotechnology, Jan Dlugosz University of Czestochowa, Armii Krajowej 13/15, 42-200 Czestochowa, Poland
KROM Thomas D.	Mosevej 18; 8600 Silkeborg; Denmark
KUEHN Stephan	Struppe & Dr. Kühn Umweltberatung, Hochspannungsweg 22, 12359 Berlin, Germany
LANE Richard	Christchurch; New Zealand
LARROQUE François	Institut EGID, Université Bordeaux-3, 1, allée F. Daguin, 33607 Pessac, France
LEIS Albrecht	Joanneum Research, Institute of Water, Energy and Sustainability, Dept. for Water Resources Management, Elisabethstrasse 16, 8010 Graz, Austria
MACHELEIDT Wolfgang	Department of Civil Engineering/Architecture, University of Applied Sciences Dresden, Friedrich-List-Platz 1 01069 Dresden, Germany
MÁDL-SZŐNYI Judit	Department of Physical and Applied Geology, Institute of Geography and Earth Sciences, Eötvös Loránd University, 1/c Pázmány Péter sétány, 1117 Budapest, Hungary
MALINA Grzegorz	Department of Hydrogeology and Engineering Geology, AGH University of Science and Technology, Mickiewicza 30, 30-059 Cracow, Poland
MARCZINEK Silke	Bayerisches Landesamt für Umwelt, Hans-Högn-Str. 12, 95030 Hof, Germany
MATROCICCO Micòl	Department of Earth Sciences, University of Ferrara, Via Saragat, 1, 44100 Ferrara, Italy
MAßMANN Jobst	Federal Institute for Geosciences and Natural Resources (BGR), Stilleweg 2, 30655 Hannover, Germany
MEAŠKI Hrvoje	University of Zagreb, Faculty of Geotechnical Engineering, Hallerova aleja 7, 42 000 Varaždin, Croatia
MOHAMMADI Kourosh	Tarbiat Modares University, Tehran, Iran
MOLINERO Jorge	Amphos XXI Consulting S.L., Passeig de García i Faria 49-51, E08019- Barcelona, Spain
MONEM Mohammad J.	Tarbiat Modares University, Tehran, Iran
MONTEIRO José Paulo	Geo-Systems Centre/CVRM – Universidade do Algarve, Campus de Gambelas, 8005-139 Faro, Portugal
NUNES Luís	Geo-Systems Centre/CVRM – Universidade do Algarve, Campus de Gambelas, 8005-139 Faro, Portugal
OLICHWER Tomasz	Institute of Geological Sciences, University of Wroclaw, ul. Cybulskiego 30, 50-205 Wroclaw, Poland
ORTUÑO Felip	Agencia Catalana de l'Aigua (ACA), C/Provença 204-208, E08036-Barcelona, Spain

OTERO Neus	Grup de Mineralogia Aplicada i Medi Ambient. Facultat de Geologia-Universitat de Barcelona. Departament de Cristallografia, Mineralogia i Dipòsits Minerals. C/Martí i Franqués s/n 08028 Barcelona
PALAU Jordi	Grup de Mineralogia Aplicada i Medi Ambient. Facultat de Geologia-Universitat de Barcelona. Departament de Cristallografia, Mineralogia i Dipòsits Minerals. C/Martí i Franqués s/n 08028 Barcelona
PEÑA-HARO Salvador	Institute of Environmental Engineering, ETH Zurich, Wolfgang-Pauli-Strasse 15, CH-8093 Zürich, Switzerland
PETERS Anett	Thüringer Landesanstalt für Umwelt und Geologie, Carl-August-Allee 8/10, 99423 Weimar, Germany
PLIESCHNEGGER Markus	Joanneum Research, Institute of Water, Energy and Sustainability, Dept. for Water Resources Management, Elisabethstrasse 16, 8010 Graz, Austria
POESER Heidrun	Sächsisches Landesamt für Umwelt, Landwirtschaft und Geologie, Postfach 540137, 01311 Dresden, Germany
PULIDO-VELAZQUEZ Manuel	Department of Hydraulics and Environmental Engineering, Universidad Politécnica de Valencia, Camino de Vera s/n, 46022, Valencia, Spain
RAY Chittaranjan	Department of Civil and Environmental Engineering & Water Resources Research Center, University of Hawaii at Manoa, 2540 Dole Street, Holmes 383, Honolulu, HI 96822, USA
RIBEIRO Luís	Geo-Systems Centre/CVRM – Instituto Superior Técnico, Av. Rovisco Pais, 1049-001 Lisbon, Portugal
RIBERA Fidel	Fundación Centro Interacional de Hidrología Subterránea (FCIHS). C/Provença 102, Barcelona 08029. Spain
RUBINIĆ Josip	University of Rijeka, Faculty of Civil Engineering, V. C. Emina 5, 51 000 Rijeka, Croatia
SALEMI Enzo	Department of Earth Sciences, University of Ferrara, Via Saragat, 1, 44100 Ferrara, Italy
SÁNCHEZ GARCÍA Damián	Centre of Hydrogeology, Department of Geology, Faculty of Science, University of Málaga, Campus de Teatinos s.n., 29071 Málaga, Spain
SCHELKES Klaus	Federal Institute for Geosciences and Natural Resources (BGR), Stilleweg 2, 30655 Hannover, Germany
SCHOENHEINZ Dagmar	Division of Water Sciences, Department of Civil Engineering & Architecture, University of Applied Sciences Dresden, Friedrich-List-Platz 1, 01069 Dresden, Germany

SCHUSTER Hans-Jörg	Geologischer Dienst NRW, De-Greiff-Str. 195, 47803 Krefeld, Germany
SKRITEK Paul	University of Applied Sciences – Technikum Wien, Inst. of Telecommunication and Internet Technologies, Hoechstaedtplatz 5 1200 Vienna, Austria
SOARES Marcus	Berlin University of Technology, Dept. Water Quality Control, Strasse des 17. Juni 135, 10623 Berlin, Germany
SOLER Albert	Grup de Mineralogia Aplicada i Medi Ambient. Facultat de Geologia-Universitat de Barcelona. Departament de Cristallografia, Mineralogia i Dipòsits Minerals. C/Martí i Franqués s/n 08028 Barcelona
STADLER Hermann	Joanneum Research, Institute of Water, Energy and Sustainability, Dept. for Water Resources Management, Elisabethstrasse 16, 8010 Graz, Austria
STAŚKO Stanisław	Institute of Geological Sciences, Wroclaw University, ul. Cybulskiego 30, 50-205 Wroclaw, Poland
STAUFFER Fritz	Institute of Environmental Engineering, ETH Zurich, Wolfgang-Pauli-Strasse 15, CH-8093 Zürich, Switzerland
STIGTER Tibor	Geo-Systems Centre/CVRM – Instituto Superior Técnico, Av. Rovisco Pais, 1049-001 Lisbon, Portugal
STRUPPE Thomas	Struppe & Dr. Kühn Umweltberatung, Hochspannungsweg 22, 12359 Berlin, Germany
TARKA Robert	Institute of Geological Sciences, University of Wroclaw, ul. Cybulskiego 30, 50-205 Wroclaw, Poland
WAGNER Bernhard	Bayerisches Landesamt für Umwelt, Hans-Högn-Str. 12, 95030 Hof, Germany
WAGNER Frank	Bundesanstalt für Geowissenschaften und Rohstoffe, Stilleweg 2, 30655 Hannover, Germany
WALTER Thomas	Landesamt für Umwelt- und Arbeitsschutz, Don-Bosco-Str. 1, 66119 Saarbrücken, Germany
WIRSING Günther	Regierungspräsidium Freiburg, Landesamt für Geologie, Rohstoffe und Bergbau, Albertstraße 5, 79104 Freiburg i. Br., Germany
WOLFER Johannes	Federal Institute for Geosciences and Natural Resources (BGR), Stilleweg 2, 30655 Hannover, Germany
WOLTER Rüdiger	Umweltbundesamt (UBA), Wörlitzer Platz 1, 06844 Dessau, Germany
VINCENZI Fabio	Department of Biology and Evolution, University of Ferrara, Via L. Borsari, 46, 44100 Ferrara, Italy
ZAWIERUCHA Iwona	Institute of Chemistry, Environmental Protection and Biotechnology, Jan Dlugosz University of Czestochowa, Armii Krajowej 13/15, 42-200 Czestochowa, Poland

Part I

Groundwater quality

Chapter 1

The role of the unsaturated zone in determining nitrate leaching to groundwater

Micòl Matrocicco[1,2], Nicolò Colombani[1,2], Enzo Salemi[1], Fabio Vincenzi[3] & Giuseppe Castaldelli[2,3]

[1]*Department of Earth Sciences, University of Ferrara, Ferrara, Italy*
[2]*LT Terra & Acqua Tech, HTN Emilia-Romagna, Ferrara, Italy*
[3]*Department of Biology and Evolution, University of Ferrara, Ferrara, Italy*

ABSTRACT

In order to identify the dominant processes affecting nitrate leaching in the Po River Delta area, field tests were performed to determine the fate and transport of nitrogen species. Nitrogen (urea) was applied at a rate of 300 kg-N/ha/y, in both a sandy and a silty loamy sites cultivated with maize; the sandy soil was fertilised with chicken manure (7000 kg/ha), the silty loamy soil was not fertilised. Each field site was equipped with soil moisture probes, suction cups and piezometers to quantify the presence of nitrogen and carbon dissolved species in the subsurface. Nitrate leaching was observed in the silty loamy soil, while in the sandy soil the elevated dissolved organic matter, resulting from chicken manure decomposition, prevented the nitrate migration towards the aquifer. Results highlight the reliability of increasing the labile organic matter in the more permeable and intrinsically vulnerable sandy soil to prevent nitrate leaking.

1.1 INTRODUCTION

Nitrate (NO_3^-) is a pervasive inorganic pollutant often found in shallow aquifers (Galloway *et al.*, 2008; Rivett *et al.*, 2008). NO_3^- contamination can be due to agricultural fertilisers or industrial discharges and municipal sewer systems (Wakida & Lerner, 2005). NO_3^- concentrations are spatially and temporally variable in aquifers (Böhlke *et al.*, 2002; Thayalakumaran *et al.*, 2008), this is usually related to variations in groundwater flow direction and nitrate attenuation (Tesoriero *et al.*, 2000; Almasri & Kaluarachchi, 2007). In Italy, the Po River valley is the largest and more intensively farmed alluvial plain and is heavily impacted by NO_3^- groundwater contamination (Mastrocicco *et al.*, 2010a; Onorati *et al.*, 2006; Cinnirella *et al.*, 2005) and surface water eutrophication (Provini *et al.*, 1992; Palmieri *et al.*, 2005). However, in agricultural practices, the types of soils and soil tillage, different crops and irrigation techniques and different nitrogen fertilisers, form a variety of terms depending on the nitrogen load and subsequent denitrification (Seitzinger *et al.*, 2006). A generally well understood and quantified process of nitrogen attenuation from surface and groundwater systems is heterotrophic denitrification which uses NO_3^- as electron acceptor and a carbon source as electron donor, producing nitrogen gases (Coyne, 2008; Schipper *et al.*, 2008).

The purpose of this research was to investigate the fluxes of NO_3^- from the top soil to the groundwater in cultivated plots and to determine if the addition of chicken manure from organic farming is a viable alternative to diminish the NO_3^- concentration in shallow groundwater bodies. This was tested in two different sites with the same rates of application but with different soil textures, sandy and silty loamy. In the sandy soil, chicken manure was employed to increase its low intrinsic fertility by augmenting the soil labile organic matter (Whitmore, 2007).

1.2 MATERIAL AND METHODS

1.2.1 Field sites

The entire Po delta area is an intensively farmed region due to its flat topography and abundance of surface water for irrigation; the primary agricultural land use is maize cropping. In the study area, located in Ferrara province (Italy) at an altitude ranging from 5 to −3 m above sea level (asl.), two sites (named CCR and MON) were selected to monitor the water and nitrogen transport in the unsaturated/saturated zone. Both sites are cultivated under a rotation of cereals, mainly maize and wheat, using urea as nitrogen fertilizer at an average rate of 300 kg-N/ha/y. The surface area of the plot in each site was 1 ha, its slope was less than 0.5% (and mostly less than 0.05%). For this reason, it was assumed surface runoff has been minimal and water movement in the unsaturated zone has been dominantly vertical. Meteorological stations recording rainfall, wind speed, solar radiation, temperature and humidity are located from 0.5 to 5 km far from the field sites. Data are available on-line from meteorological regional service (www.dexter.it) and from local web service (www.meteoveneto.com). A rain gauge was installed in each site to record daily rainfall. The linear correlation of daily data between the rain gauges and the nearest regional weather stations ($R^2 > 0.9$ for all sites), consented to extend the data measured by the on line weather stations to the field sites.

To preserve natural conditions no irrigation was applied. As shown in Table 1.1 the average meteorological parameters recorded in the two sites are typical of coastal plain environments, with a sub-coastal climate characterised by cold winters and warm summers, with moderate precipitations, elevated humidity, low wind speed, moderate daily and seasonal temperature variations.

The predominant soil textures in Ferrara province are silty loam and silty clay (68% of the territory), while sandy soils are less common (11% of the territory). The CCR soils are in general moderately alkaline, with the upper horizons characterised by

Table 1.1 Average meteorological parameters for the two field sites during the two monitored years.

	Cumulative rainfall (mm)	*Average air temperature (°C)*	*Solar radiation ($MJ/m^2/d$)*	*Humidity (%)*	*Wind speed (km/d)*
CCR	1813	13.2	1.60E−04	76	5.5
MON	999	13.6	1.66E−04	70	8.6

silty clay loam texture and moderate carbonates content; the lower horizons exhibit silty loam texture and are highly calcareous (Tab. 1.2). Briefly, the hydrogeological units present in the CCR site are the unconfined aquifer composed of recent fluvial sandy deposits with clay and silt lenses, from 0 to around 4 m below ground level (bgl.), and the underlying aquiclude constituted of fluvial clay and silty sediments, from 4 to almost 14 m bgl.

The representative MON soil profile shows upper horizons of approximately 40–60 cm thickness characterized by fine sand texture, with moderate carbonates content and slightly alkaline pH; while the lower horizons exhibit alkaline pH and medium sand texture (Tab. 1.2). The hydrogeological units present in the MON site are the unconfined aquifer composed of coastal plain medium and fine sandy deposits, from 0 to around 12 m (bgl.), and the underlying aquiclude constituted of prodelta silt and clay sediments, from 12 to almost 15 m bgl. (Stefani & Vincenzi, 2005).

1.2.2 Analytical and field methods

To better define the site stratigraphy core logs were drilled manually with an Ejielkamp Agrisearch auger equipment down to 2 m (bgl.). The soil stratification was divided in two distinct layers: the upper one stressed by tillage, roots growth and weathering and the lower undisturbed one (Tab. 1.3). In CCR site the upper layer was 0.75 m thick and in MON was 0.65 m, while the lower layer extended until 2 m bgl. at both sites. From collected core samples at 0.25, 0.50, 0.75, 1 and 2 m bgl., particle size curves were obtained using a settling tube for the sandy fraction and an X-ray Micromeritics Sedigraph 5100 for the finer one. Organic matter content was measured by loss of ignition method (Tiessen & Moir, 1993), while bulk density was determined gravimetrically.

Two arrays of Watermark soil moisture probes were vertically inserted into augered holes, at the same depths of 0.25, 0.50, 0.75 and 1 m bgl. in each field site. Watermark

Table 1.2 Soil characterization for the two field sites.

	Pedological classification	*Sedimentological environment*	*Textural classification*
CCR	Haplic Calcisols	Fluvial	Silty loam
MON	Calcaric Arenosols	Coastal plain	Sand

Table 1.3 Average grain size distribution, bulk density and organic matter content, measured for the upper and lower layers.

	CCR		*MON*	
Parameter	*Upper layer*	*Lower layer*	*Upper layer*	*Lower layer*
Grain size (%)				
Sand	7.7	23.9	95.6	98.1
Silt	63.2	58.1	3.0	1.9
Clay	29.1	18.0	1.4	0.0
Bulk density (g/cm^3)	1.5	1.6	1.4	1.7
Organic matter (%)	2.3	1.3	2.0	1.1

soil moisture probes were used to monitor the soil water potential (measurement range 0–250 cbar). A copper-constantan thermocouple was inserted adjacent to each soil moisture probe to compensate for soil temperature. Standard Irrometer tensiometers (measurement range 0–80 cbar) were installed at 0.25 and 0.50 m depths to monitor and correct any deviance of soil moisture probes readings.

A series of nested piezometer (2.5 cm inner diameter) screened from 1.5 to 4 m asl., were installed near the soil moisture arrays to monitor the level and quality of groundwater. Monitoring started on 27 March 2008 and is still going on. LTC M10 Levelogger Solinst dataloggers were placed in piezometers to monitor hourly groundwater level, electrical conductivity and temperature. All the piezometers were sampled at variable intervals, via low flow purging, for major ions and TOC/TIC determination. Two arrays of soil solution suction samplers were installed at 0.25, 0.50, 0.75 and 1 m bgl. in each site to analyze soil water in the unsaturated zone. In addition to suction sampler, the unsaturated zone was sampled every four months by means of auger coring (from 0 to 2 m bgl.), and sediments were analyzed for major anions and cations. Unsaturated zone sediment analysis consisted of a batch with a sediment/water ratio of 1:10, using 10 g of air dried sample dispersed in 100 ml of Milly-Q water (Millipore, US). A biological inhibitor (1 g/l phenylmercuric acetate) was added to prevent microbial activity and the solution was stirred for 1 hour and then allowed to stand for one day. The insoluble residue was removed by filtration and analyzed for major cations and anions.

In-well parameters were determined with the HANNA Multi 340i instrument which includes a HIcell-31 pH combined electrode with a built-in temperature sensor for pH measurements, a CellOx 325 galvanic oxygen sensor for DO measurements, a combined AgCl-Pt electrode for Eh measurement and a HIcell-21 electrode conductivity cell for EC measurements. The major cations, anions and oxianions (acetate and formate) were determined with isocratic dual pump ion chromatography ICS-1000 Dionex, equipped with an AS9-HC 4 × 250 mm high capacity column and an ASRS-ULTRA 4mm self-suppressor for anions, and a CS12A 4 × 250 mm high capacity column and a CSRS-ULTRA 4 mm self-suppressor for cations. Samples were filtered through 0.22 μm Dionex vial caps. An AS-40 Dionex auto-sampler was employed to run the analyses, Quality Control (QC) samples were run every 10 samples. The standard deviation for all QC samples run was better than 4%. Charge balance errors in all analyses were less than 5% and predominantly less than 3%. Total organic carbon (TOC) and total inorganic carbon (TIC) were determined with a carbon analyzer (Carbon Analyzer Shimadzu TOC-V-CSM) after acidification with one drop of 2 M HCl to remove dissolved carbonate.

1.3 RESULTS AND DISCUSSION

1.3.1 Unsaturated zone monitoring

The matric potential measured at different depth in CCR (Fig. 1.1) shows that during the autumn/winter seasons the upper and the lower horizons are near the saturation state. This implies that recharge is taking place especially during the late winter season, where the saturation state is reached in all the measuring points. From the sowing to

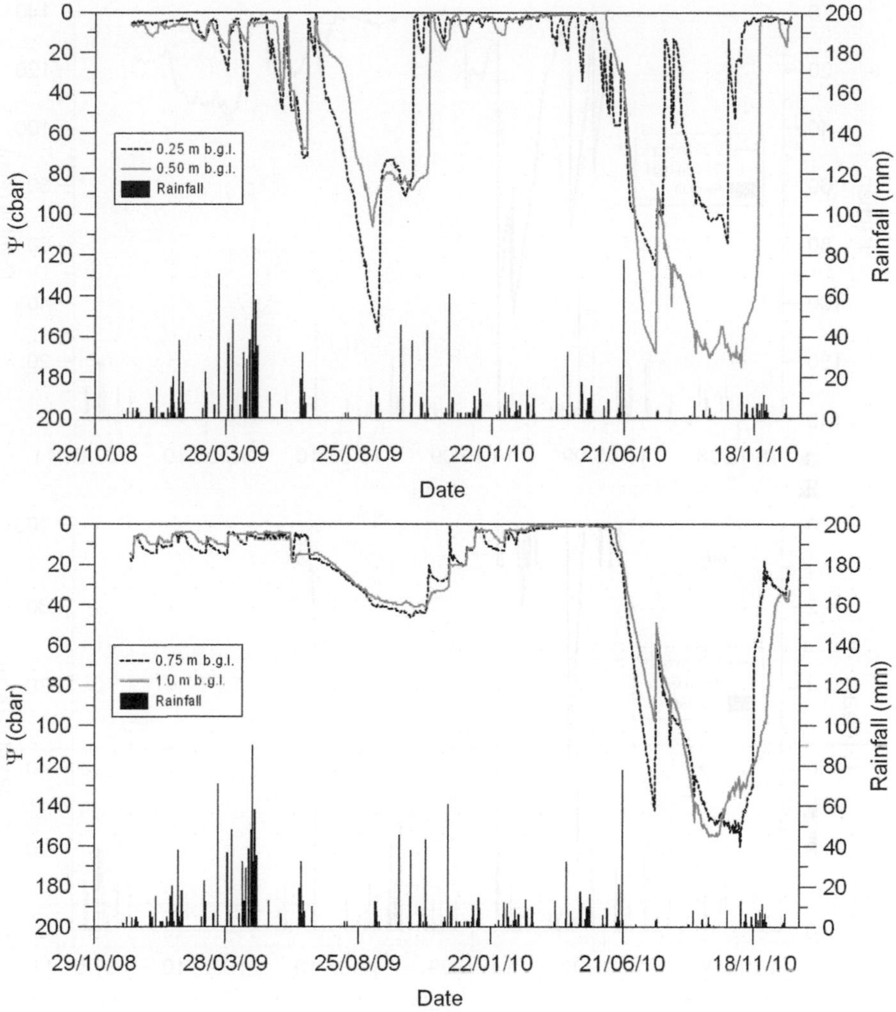

Figure 1.1 Rainfall and matrix potential (in cbar) recorded at CCR site during the monitoring period at different depth 0.25, 0.5, 0.75 and 1.0 m b.g.l.

the harvest of maize (May to September 2009) the soil became dry in the upper horizon as evapotranspiration was higher. (Mastrocicco *et al.*, 2010b).

For crop rotation needs, in May 2010 beetroot was sowed instead of maize at CCR site and the soil became dry also in the lower horizon, since the rooting system of the beetroot is deeper than in maize (Christiansen *et al.*, 2006). In addition Figure 1.1 shows a clear temporal shift during the wetting cycle, from the sensor located at 0.25 m bgl. and the one located at 0.5 m bgl.; this is due to the low permeability of these soils, which do not allow fast vertical transfer of water.

Figure 1.2 depicts the matric potential measured at different depths in MON, here the upper soil did not reach the values recorded in CCR because the water table was

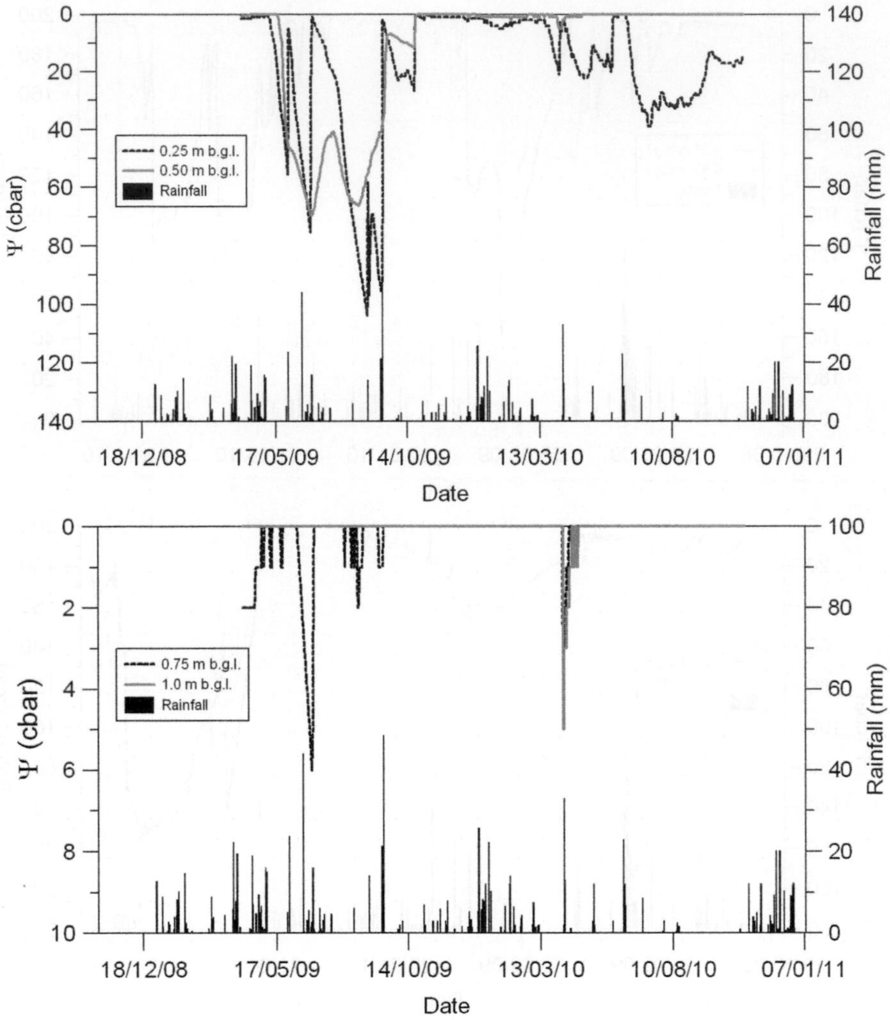

Figure 1.2 Rainfall and matric potential (in cbar) recorded at MON site during the monitoring period at different depth 0.25, 0.5, 0.75 and 1.0 m b.g.l.

on average at 0.8 m bgl. This condition provided a continuous source of water for the maize rooting system, which did not necessitate of irrigation during the cropping season. As a consequence, the lower horizon was always saturated with groundwater and the elevated permeability of these sandy soils allowed fast travel times of recharge water towards the shallow unconfined aquifer.

It is also noticeable that during the summer 2009 the soil became dry to 0.5 m bgl., while during the summer 2010 only the first 0.25 m were unsaturated. This was due to the water table that was slightly higher during summer 2010 (see Figure 1.6 in section 1.3.3).

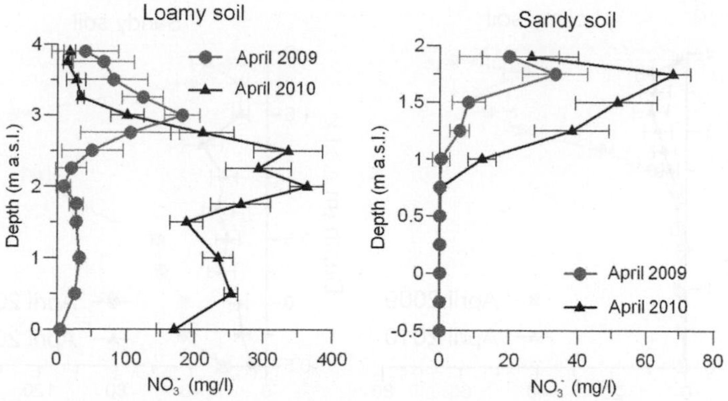

Figure 1.3 NO_3^- depth profiles at CCR (on the left) and at MON (on the right) collected in April 2009 and April 2010 (concentrations measured in solids corrected for bulk density and water content; error bars show the standard deviation of three replicates).

1.3.2 Unsaturated/saturated zone profiles

Since the most intense recharge occurs during spring time, Figure 1.3 shows a series of NO_3^- profile collected in CCR and in MON sites in April 2009 and 2010 before the fertilisation. In CCR the NO_3^- concentration observed in April 2009 showed a peak at 3 m asl. in correspondence of the water table, while in the saturated zone NO_3^- decreased rapidly to concentrations below 50 mg/l.

In April 2010 the observed NO_3^- profile showed very high concentration with a maximum peak below the water table. The elevated NO_3^- concentration was due to a combination of factors: (i) the water table was higher than in 2009, (ii) the temperature was lower than in 2009 and (iii) precipitation was less intense.

The higher water table suggests that more NO_3^- trapped in the vadose zone can be dissolved in groundwater, the lower temperature may decrease denitrification rate and lower rainfall contributes to concentrate NO_3^-. The same trend is visible for the MON site, although NO_3^- concentrations were about five times lower and the profile appears quite different from CCR. In fact, NO_3^- was found only in the upper part of the profile, while below 1 m a.s.l. NO_3^- was always below detection limits. Nitrate disappearance below 1 m a.s.l. was attributed to denitrification supported by the addition of chicken manure, in April 2008, which provided labile organic matter used as an electron donor. The process is microbially catalyzed and can be written as an overall reaction:

$$5CH_2O + 4NO_3^- \rightarrow 4HCO_3^- + CO_2 + 2N_2 + 3H_2O$$

The process involves also nitrite as intermediate compounds that are transiently produced and then reduced to nitrogen gas. When denitrification is not limited by organic substrate, nitrite remains at low concentration. Otherwise, when organic substrates become limiting nitrite tends to accumulate (Nair *et al.*, 2007; Israel *et al.*, 2009). In fact, in MON site nitrite remained at very low level, always below 1 mg/l, while in CCR site concentrations up to 100 mg/l were occasionally recorded in suction

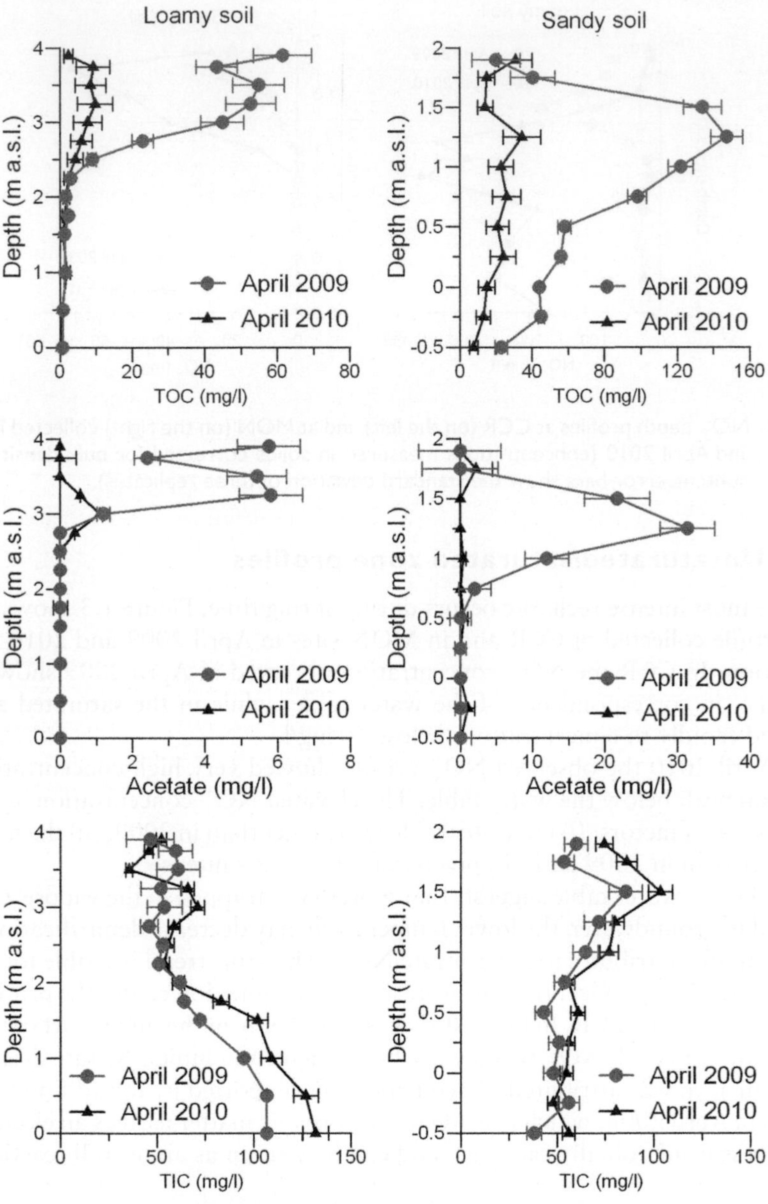

Figure 1.4 TOC, acetate and TIC depth profiles at CCR (on the left) and at MON (on the right) collected in April 2009 and April 2010 (concentrations measured in solids corrected for bulk density and water content; error bars show the standard deviation of three replicates).

cups. This provided evidence of incomplete denitrification due to a lack of organic substrate. Ammonium was always observed below 2 mg/l and thus is not shown.

Figure 1.4 shows the TOC, acetate and TIC in both sites collected in April 2009 and April 2010. The TOC observed in CCR site was always lower than the TOC in MON

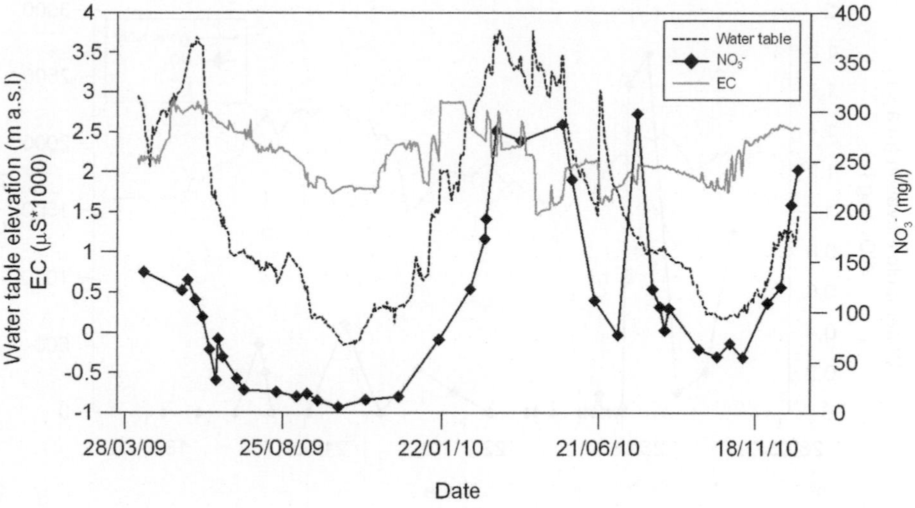

Figure 1.5 Groundwater level fluctuations of NO_3^- and EC trends in groundwater at CCR (loamy soil) throughout the monitoring period.

site, this can be directly linked to chicken manure addition at MON site. Acetate concentration followed the same trend of TOC in both sites: in fact TOC decreased with depth and also acetate, but the latter became exhausted at shallower depths since this organic acid is readily available as a carbon source (Mastrocicco *et al.*, 2011). Acetate is one of the most reactive organic acids, produced as intermediate during organic matter degradation and thus is a good proxy to evaluate the reactivity of organic matter (Strobel, 2001). The elevated abundance of TOC after the addition of chicken manure in MON is another evidence of the excess of electron donors compared to nitrate, this led to a full nitrate reduction within the first meter of soil. On the contrary, at CCR site the TOC and acetate concentrations were not sufficient to support denitrification, leading to accumulation of nitrite. This conceptual model was also supported by the measured dissolved oxygen at MON and CCR sites in the piezometers: at MON oxygen was always below detection limits, while in CCR oxygen was present between 3 to 5 mg/l. Finally, the TIC vertical distribution supports the postulated denitrification reaction, because in MON the TIC concentration was elevated in the vadose zone, where the NO_3^- was consumed. While, in CCR the TIC increased towards the bottom of the aquifer, simply following the accumulation trend of total dissolved species (not shown).

1.3.3 Saturated zone monitoring

Figure 1.5 shows a variable groundwater table with large seasonal variations in the loamy soil, where the groundwater flow is linked with canal levels. In particular, the

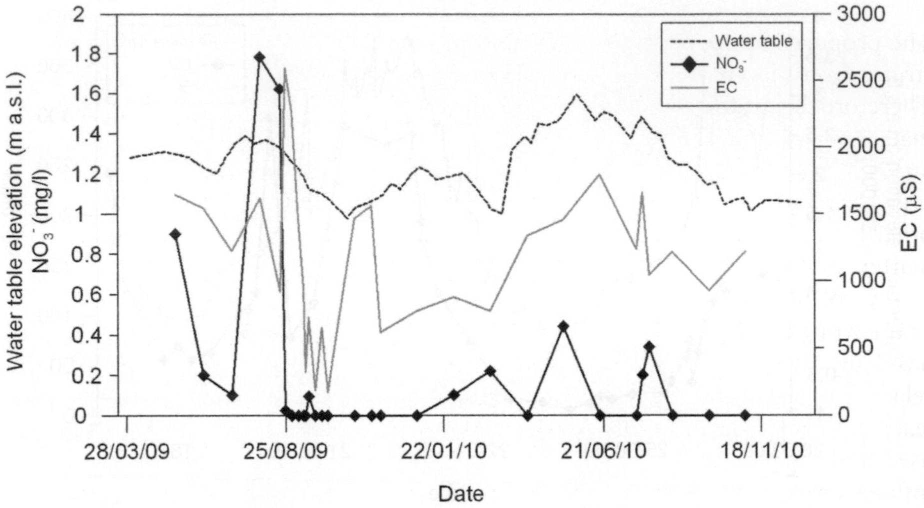

Figure 1.6 **Groundwater level fluctuations of NO_3^- and EC trends in groundwater at MON site (sandy soil) throughout the monitoring period.**

sharp peak recorded on 21 June 2010 was due to a flood event that increased the level of the nearby canal by 2.5 m, but the soil moisture sensor placed at 1 m bgl. was not in a saturation condition (Figure 1.1). This proves that the peak was due to groundwater fluctuation induced by the canal and not by recharge.

At the CCR site nitrate mass transfer to the unconfined aquifer was slow as shown in Figure 1.5 and concentrated at the end of the winter season, the water table rose and brought in solution the available NO_3^-. For comparison with NO_3^- (Figure 1.5) is plotted also the EC, but is evident that there is not a direct relation between these two parameters. EC was generally increasing when the water table was rising, since the latter dissolved the salts accumulated in the vadose zone. While during the summer seasons EC decreased because groundwater was replaced by canal water that had a lower EC.

Figure 1.6 shows that in the MON site groundwater fluctuations were less pronounced than in CCR, since MON site is located near the coast and the water table fluctuations are moderated. At the MON site, despite the fast transfer of mass induced by the elevated permeability of the soil, NO_3^- concentration was very low. In fact, NO_3^- concentration never exceeded 2 mg/l confirming that the denitrification process efficiently removed all the NO_3^-. The EC monitoring was less continuous due to malfunction of the EC probe, thus only the EC measured during sampling campaigns is shown. In general the EC seems to be related to the rainfall, exhibiting decreasing values after prolonged rainy periods although is not possible to infer clear trends with these sparse data. At the CCR site, the EC trend cannot be directly related to NO_3^- concentration in groundwater.

1.4 CONCLUSIONS

The proposed approach has shown that in the sandy soil, the release of organic substrates from the mineralization of manure was sufficient to prevent nitrate leaching. Therefore this study emphasizes the importance of considering the role of labile organic matter in buffering NO_3^- excess via denitrification reactions. In particular, the augmentation of the soil labile organic matter with chicken manure, provided a source of labile organic acids, like acetate, used as substrates for the denitrification process. On the other hand, in the unamended silty loamy soil the chronic deficiency of labile organic matter, could not prevent nitrogen leaking towards the shallow unconfined aquifer.

According to the sustainable agricultural practices recommended by the Water Framework Directive (2000/60 EU) to prevent groundwater pollution, this approach has demonstrated that even in the intrinsically more vulnerable area (the sandy soil), a relatively low but constant release of organic substrates was sufficient to prevent nitrate leaching. Nevertheless, further research should pay attention to the kind of manure used and to the farming practices, organic or industrial. In fact, this last term may influence organic matter transformation kinetics and nitrogen mineralisation rates, because of the presence of hormones, antibiotics and other undesired chemicals that could impact the groundwater quality and could interfere with bacterial activities.

There is an increasing interest in using ^{15}N and ^{18}O nitrate isotopes to distinguish the origin of nitrate, and numerical modelling to quantify the nitrate leaching mass transport. The present study was done on relatively simple sites, where depth dependent monitoring allowed discriminating between different flow and reactions pathways. Nevertheless, in our opinion the isotopic and numerical modelling approaches are key tools to track the fate of nitrogen species in the subsurface.

ACKNOWLEDGEMENTS

Dr. Umberto Tessari and Dott. Sandro Bolognesi are acknowledged for the technical and scientific support. The Emilia–Romagna ARPA SIMC is acknowledge for the meteorological data. The work presented in this paper was made possible and financially supported by PARC-AGRI under "Contratto di Programma (Delib.CIPE n.202)" and by the Province of Ferrara within the EU-Water Project of the South-East Europe Program.

REFERENCES

Almasri M. N., Kaluarachchi J. J. (2007) Modeling nitrate contamination of groundwater in agricultural watersheds. *J Hydrol* 343: 211–229.

Böhlke J. K., Wanty R., Tuttle M., Delin G., Landon M. (2002) Denitrification in the recharge area and discharge area of a transient agricultural nitrate plume in a glacial outwash sand aquifer, Minnesota. *Wat Resour Res* 38(7): 10.1–10.26.

Christiansen J. S., Thorup-Kristensen K. and Kristensen H. L. (2006) Root development of beetroot, sweet corn and celeriac, and soil N content after incorporation of green manure. *J Horticult Sci and Biotech* 81(5): 831–838.

Cinnirella S., Buttafuoco G., Pirronea N. (2005) Stochastic analysis to assess the spatial distribution of groundwater NO_3^- concentrations in the Po catchment (Italy). *Environmental Pollution* 133: 569–580.

Coyne M. S. (2008) Biological denitrification. In: Schepers J. S., Raun W. (eds) Nitrogen in Agricultural Systems, ASA-CSSSA-SSSA Agronomy Monograph 49, Madison, WI, pp. 197–249.

Galloway J. N., Townsend A. R., Erisman J. W., Bekunda M., Cai Z., Freney J. R., Martinelli L. A., Seitzinger S. P., Sutton M. A. (2008) Transformation of the nitrogen cycle: recent trends, questions, and potential solutions. *Science* 320: 889–892.

Giuliano G. (1995) Ground water in the Po basin: some problems relating to its use and protection. *Sci Total Environ* 171: 17–27.

Israel S., Engelbrecht P., Tredoux G., Fey M. V. (2009) In situ batch denitrification of nitrate-rich groundwater using sawdust as a carbon source—Marydale, South Africa. *Water Air and Soil Poll* 204: 177–194.

Mastrocicco M., Colombani N., Castaldelli G., Jovanovic N. (2010a) Monitoring and modeling nitrate persistence in a shallow aquifer. *Water Air and Soil Poll* 217: 83–93.

Mastrocicco M., Colombani N., Salemi E., Castaldelli G. (2010b) Numerical assessment of effective evapotranspiration from maize plots to estimate groundwater recharge in lowlands. *Agricult Wat Manag* 97(9): 1389–1398.

Mastrocicco M., Colombani N., Salemi E., Castaldelli G. (2011) Reactive Modeling of Denitrification in Soils with Natural and Depleted Organic Matter. *Water Air and Soil Poll* doi: 10.1007/s11270-011-0817-6.

Onorati G., Di Meo T., Bussettini M., Fabiani C., Farrace M. G., Fava A., Ferronato A., Mion F., Marchetti G., Martinelli A., Mazzoni M. (2006) Groundwater quality monitoring in Italy for the implementation of the EU water framework directive. *Phys and Chem of the Earth* 31: 1004–1014.

Nair R. R., Dhamole P. B., Lele S. S., D'Souza S. F. (2007) Biological denitrification of high strength nitrate waste using preadapted denitrifying sludge. *Chemosphere* 67: 1612–1617.

Palmeri L., Bendoricchio G., Artioli Y. (2005) Modelling nutrient emissions from river systems and loads to the coastal zone: Po River case study, Italy. *Ecolog Modelling* 184: 37–53.

Provini A., Crosa G., Marchetti R. (1992) Nutrient export from the Po and Adige river basin over the last 20 years. *Sci Total Environ* Suppl: 291–313.

Rivett M. O., Buss S. R., Morgan P., Smith J. W. N., Bemment C. D. (2008) Nitrate attenuation in groundwater: A review of biogeochemical controlling processes. *Water Res* 42: 421–4232.

Seitzinger S., Harrison J. A., Böhlke J. K., Bouwman A. F., Lowrance R., Peterson B., Tobias C., Van Drecht G. (2006) Denitrification across landscapes and waterscapes: A synthesis. *Ecolog Appl* 16(6): 2064–2090.

Schipper L. A., Robertson W. D., Gold A. J., Jaynes, D. B., Cameron S. C. (2010) Denitrifying bioreactors-An approach for reducing nitrate loads to receiving waters. *Ecolog Eng* doi:10.1016/j.ecoleng.2010.04.008.

Stefani M., Vincenzi S. (2005) The interplay of eustasy, climate and human activity in the late Quaternary depositional evolution and sedimentary architecture of the Po Delta system. *Marine Geol* 222–223: 19–48.

Strobel B. W. (2001) Influence of vegetation on low-molecular-weight carboxylic acids in soil solution—a review. *Geoderma* 99: 169–198.

Tesoriero A. J., Liebscher H., Cox S. E. (2000) Mechanism and rate of denitrification in an agricultural watershed: Electron and mass balance along groundwater flow paths. *Wat Resour Res* 36(6): 1545–1559.

Thayalakumaran T., Bristow K. L., Charlesworth P. B., Fass T. (2008) Geochemical conditions in groundwater systems: Implications for the attenuation of agricultural nitrate. *Agricult Wat Manag* 95: 103–115.

Tiessen H., Moir J. O. (1993) Total and organic carbon. In: Soil Sampling and Methods of Analysis, M.E. Carter, (Ed) Lewis Publishers: Ann Arbor, MI, pp. 187–211.

Youngs E. G. (1995) Developments in the physics of infiltration. *Soil Sci Soc Am J* 59(2): 307–313.

Wakida F. T., Lerner D. N. (2005) Non-agricultural sources of groundwater nitrate: A review and case study. *Water Res* 39(1): 3–16.

Whitmore A. (2007) Determination of the mineralization of nitrogen from composted chicken manure as affected by temperature. *Nutr Cycl Agroecosyt* 77(3): 225–232.

Chapter 2

Coupling an unsaturated model with a hydro-economic framework for deriving optimal fertilizer application to control nitrate pollution in groundwater

Salvador Peña-Haro[1], Fritz Stauffer[1], Cyprien Clementine[1] & Manuel Pulido-Velazquez[2]

[1] *Institute of Environmental Engineering, ETH Zurich, Zürich, Switzerland*
[2] *Department of Hydraulics and Environmental Engineering, Universidad Politécnica de Valencia, Valencia, Spain*

ABSTRACT

In deriving efficient management policies to control groundwater nitrate pollution, it is important to conduct integrated modeling which takes into consideration soil, unsaturated and saturated zones. In groundwater management regional studies the influence of the unsaturated zone is often neglected. The unsaturated zone can have an important influence on the time delay of nitrate transport and therefore on accomplishing the good groundwater chemical status required by the EU Water Framework Directive. In this paper a simple unsaturated zone model is coupled with a hydro-economic framework that obtains the spatial and temporal fertilizer application rate that maximizes the net benefits in agriculture constrained by the quality requirements in groundwater at various control sites. The integrated model was applied to El Salobral-Los Llanos aquifer in Spain. The fertilizer allocation that accomplishes with the nitrate concentration in groundwater was obtained. The results show that a delay of about 7 years can be expected on accomplishing the quality requirements when considering the unsaturated zone, while the total benefits do not have a considerable change.

2.1 INTRODUCTION

The EU Water Framework Directive (Directive 2000/60/EC) (WFD), proclaims an integrated management framework for sustainable water use, and requires that all water bodies reach a good status by 2015. According to article 4 of the WFD, this deadline can be extended if the "Member States determine that all necessary improvements in the status of the bodies of water cannot reasonably be achieved within the timescale". Therefore, the correct estimation of the timescale to achieve the good quality of the groundwater bodies is very important.

To control groundwater diffuse pollution it is necessary to analyze and implement management decisions. The efficiency of these decisions depends on the inertia of the soil-unsaturated-groundwater system. In order to predict future groundwater-quality values, especially after implementation of environmental measures such as reduction in fertilizer use, the response time has to be determined. The response time between the fertilizer application and the contamination at a particular site depends on the soil parameters, on the degree of saturation water which in turn is controlled by the variable water input at the land surface (Mattern & Vanclooster, 2010) and on the distance between the source area and the control site where water quality is analysed (Gutierrez & Baran, 2009). Denitrification occurs mainly in the root zone while advective-dispersive transport will continue to transport nitrate through the unsaturated and saturated zones (Geyer *et al.*, 1992). Peña-Haro *et al.* (2009) developed a hydro-economic model to obtain the optimal allocation of fertilizer application in order to maximize the benefits in agriculture while maintaining the nitrate concentrations in groundwater below a predefined standard. In this methodology the unsaturated zone was not explicitly taken into account. In this chapter the delay time that nitrate suffers while going through the unsaturated zone was coupled with the hydro-economic model. A crude salute transport model based on water flux calculation was used. The methodology was applied to El Salobral-Los Llanos aquifer (Mancha Oriental) in Spain.

2.2 HYDROECONOMIC MODEL

The hydro-economic model (Peña-Haro *et al.*, 2009; 2010) determines the spatial and temporal fertilizer application rate that maximizes the net benefits in agriculture constrained by the groundwater quality requirements at various control sites. The original formulation was modified to take into consideration the unsaturated zone as follows:

$$\text{Max} \prod = \sum_{s}\sum_{y} \frac{1}{(1+r)^{y}} A_s(p_s \cdot Y_{s,y} - p_n \cdot N_{s,y} - p_w \cdot W_{s,y} - C_s + S_s)$$

subject to:

$$\sum_{s} \mathbf{RM}_{c,txs,y} \cdot \mathbf{cr}_{s,y} + \mathbf{NRM}_{c,txs,y} \cdot \mathbf{ncr}_{s,y} + \mathbf{DL}_{c,t} + \mathbf{IC}_{c,t} \leq \mathbf{q}_{c,t} \quad \forall c,t,y$$

where: $\prod$ is the objective function to be maximized and represents the present value of the net benefit from agricultural production (€) defined as crop revenues minus fertilizer and water variable costs; A_s is the area cultivated for crop located at source s; p_s is the crop price (€/kg); $Y_{s,y}$ is the production yield of crop located at source s at planning year y (kg/ha), that depends on the nitrogen fertilizer and irrigation water applied; p_n is the nitrogen price (€/kg); $N_{s,y}$ is the fertilizer applied to crop located at source s at year y (kg/ha), p_w is the price of water (€/m^3), and $W_{s,y}$ is the water applied to crop located at source s at each planning year y (m^3); C_s is the aggregate of the remaining per hectare costs for crop located at source s (€/ha); S_s are the subsidies for the crop located at source s (€/ha); r is the annual discount rate; $\mathbf{RM}_{c,txs,y}$ is the unitary pollutant concentration response matrix due to the fertilizer inputs before fertilizer reduction; $\mathbf{NRM}_{c,txs,y}$ is the unitary pollutant concentration response matrix due to fertilizer inputs after fertilizer reduction; $\mathbf{q}_{c,t}$ is a vector of water quality standard imposed at the control sites over the simulation time (kg/m^3); $\mathbf{cr}_{s,y}$ is a vector representing the nitrate concentration recharge (kg/m^3) reaching groundwater from a crop located at source s, which is obtained by dividing the nitrate leached over the water that recharges the aquifer; $\mathbf{ncr}_{s,y}$ is a vector representing the nitrate concentration recharge (kg/m^3) due to new fertilizer application. $\mathbf{DL}_{c,t}$ is the nitrate concentration at the control site c and the planning horizon y due to fertilizer application in dryland; $\mathbf{IC}_{c,t}$ is the nitrate concentration at the control site c and the planning horizon y due to the initial nitrate concentrations in the aquifer.

The hydro-economic model couples agronomic, flow and transport models in an optimization framework. From agronomic simulations quadratic functions are derived that relate the nitrate leached and the crop yield with the water and fertilizer use. These equations can be obtained from the results of agronomic simulations models such as EPIC (Williams, 1995; Liu *et al.*, 2007). The groundwater flow and nitrate transport models are used to obtain the concentration response matrices, which show the influence of a pollutant source upon nitrate concentrations at different control sites over time. The integration of the response matrix in the constraints of the management model allows for simulation by superposition of the evolution of groundwater nitrate concentration at different control sites over time, resulting from multiple pollutant sources distributed over time and space. The benefits in agriculture were determined through crop prices and crop production functions.

A simple unsaturated model, which only considers the gravity flow, was coupled into the hydro-economic model (Figure 2.1). With this model the nitrate travel time from the root zone to the saturated zone was estimated, it was considered as a time

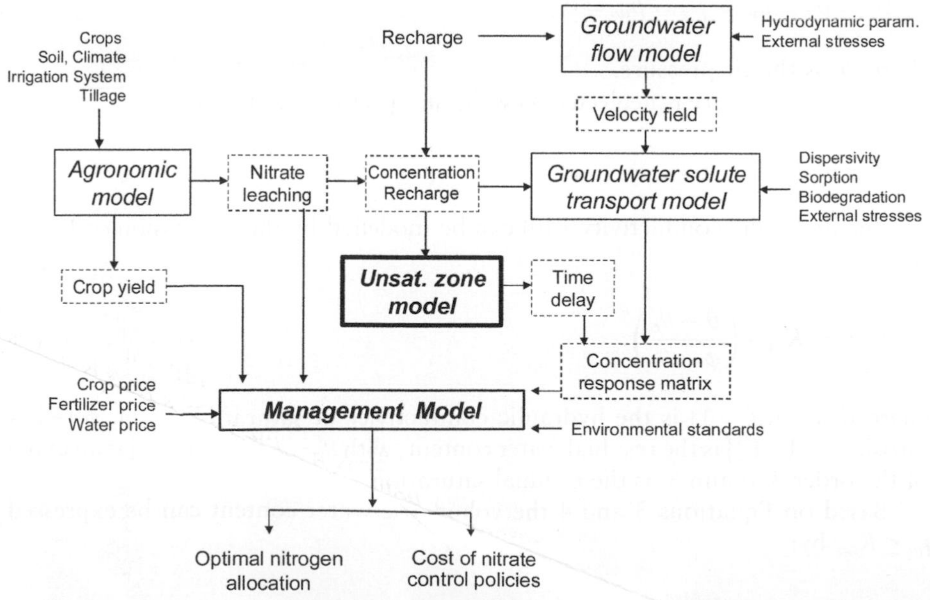

Figure 2.1 Modeling framework.

delay, which was included into $\mathbf{NRM}_{c,txs,y}$. The $\mathbf{NRM}_{c,txs,y}$ is the unitary pollutant concentration response matrix due to the application of the fertilizer subjected to optimization, where each column is the time series of nitrate concentration for each crop area, s, times the number of years within the planning horizon, y, the number of rows equals the number of control sites, c, times the number of simulated time steps, t, in the frame of the problem. The time delay due to the nitrate transport trough the unsaturated zone was included into the concentration response matrix as a shift on the time series of nitrate concentration at each control site.

2.3 UNSATURATED ZONE MODEL

For the simulation of the water flow through the unsaturated zone, a simple model considering a simplification of Richards' equation with kinematic wave was used. One-dimensional vertical unsaturated flow can be described by the Darcy's law, as follows:

$$q_w(z,t) = -K(\theta(z,t)) \cdot \left[1 + \frac{1}{\rho g}\frac{\partial p_w(z,t)}{\partial z}\right] \tag{2.1}$$

where: q_w [L/T] is the specific flux, θ [L^3/L^3] is the volumetric water content, $K(\theta)$ [L/T] is the hydraulic conductivity, z [L] is the vertical coordinate, p_w [M/(LT^2)] is the water pressure, ρ [M/L^3] is the water density, g [L/T] is the acceleration due to gravity.

If capillarity in unsaturated flow is neglected, then the capillary pressure p_c [M/(LT^2)] is:

$$p_c = p_a - p_w = p_a = p_w = 0 \tag{2.2}$$

where: p_a is the air pressure.

If the air pressure is neglected as well, the specific flux q gets:

$$q_w = -K(\theta) \tag{2.3}$$

The hydraulic conductivity $K(\theta)$ can be modelled by the expression of Brooks & Corey (1964):

$$K(\theta) = K_{sat} \cdot \left(\frac{\theta - \theta_r}{\phi - \theta_r}\right)^{\varepsilon} \tag{2.4}$$

where $K_{sat} = K(\theta = 1)$ is the hydraulic conductivity at saturation [L/T], ϕ [−] is the porosity, θ_r [L^3/L^3] is the residual water content, with $\theta_r = \phi S_r$, and ε [−] is an exponent (of the order 3.5) and S_r is the residual saturation.

Based on Equations 3 and 4 the volumetric water content can be expressed for $q_w \leq K_{sat}$ by:

$$\theta = \theta_r + \left(\frac{q_w}{K_{sat}}\right)^{1/s} \cdot (\phi - \theta_r) \tag{2.5}$$

For steady-state flow conditions and constant K_{sat} the volumetric water content profile is homogeneous. With transient flow conditions and a change in the infiltration rate from q_1 to q_2, a shock front with velocity w will be developed. Flow continuity at the moving shock front requires:

$$\theta_1 \cdot \left(\frac{K_1(\theta_1)}{\theta_1} - w \right) = \theta_2 \cdot \left(\frac{K_2(\theta_2)}{\theta_2} - w \right) \tag{2.6}$$

The shock front velocity w can be expressed by:

$$w = \frac{K_1(\theta_1) - K_2(\theta_2)}{\theta_1 - \theta_2} \tag{2.7}$$

A crude solute transport model (without diffusion, dispersion, retardation and decay effects) can be obtained by assigning solute concentrations to the inflow rates. In this approximation only the mobile region of the porous medium is taken into account, thus neglecting immobile solute concentration. Irregular outflow concentrations can be obtained by integrating the solute mass over time leaving the soil column, and the solute mass at the corresponding outflow times, and by calculating the solute mass and volume increments in order to get outflow concentrations. The transport of nitrate in the saturated zone is assumed to be conservative; most nitrogen cycle processes take place within the root-zone (Ledoux *et al.*, 2007).

2.4 EL SALOBRAL LOS LLANOS AQUIFER

The methodology was applied to El Salobral-Los Llanos Domain (SLD), which is located in the southeast of the Mancha Oriental System and extends over about 420 km^2. 80% of the land is agriculture from which 100 km^2 are irrigated crops (CHJ, 2004). The climate is Mediterranean. The mean summer temperature is about 22°C and the mean winter temperature about 6°C. The mean annual precipitation is about 360 mm. The average groundwater recharge of 165 mm/year is estimated (CHJ, 2008). The irrigated area has increased considerably, from about 29 km^2 in 1961 (Spanish Geological Survey, IGME, 1976) to 100 km^2 in 2004. This has provoked a decline of the groundwater levels of between 60 and 80 m as well as high nitrate concentrations. The highest nitrate concentration of about 54 mg/l was recorded in the well El Salobral (Moratalla *et al.*, 2009) (Figure 2.2), exceeding the allowed concentration for human consumption of 50 mg/l (Drinking Water Directive, 80/778/EEC). All these facts ended up with the declaration of the aquifer as a Nitrate Vulnerable Area by the Castilla-La Mancha regional government (DOCM, 1998).

The El Salobral-Los Llanos aquifer is formed mainly by 2 units. The deepest one is constituted by mid-Jurassic dolostones and limestones that can reach a thickness of 250 m with a mean hydraulic conductivity of 250 m/day. A detrital aquitard overlies it and reaches a maximum thickness of about 75 m, with a mean hydraulic conductivity of 10 m/day (Sanz *et al.*, 2011).

2.5 RESULTS

The main inputs to the hydro-economic model are the crop yield, leaching functions and the response matrix. The agronomic functions were taken from Peña-Haro *et al.* (2010), where an agronomic model was applied to the main crops in the area in order to obtain the coefficients of the quadratic function representing the crop yield and nitrate leaching as a function of water and fertilizer use.

The influence of unitary fertilizer application depends on management areas (pollution sources). These pollutant areas represent administrative zones where the fertilizer application will be subjected to standards. Two criteria were taken into account: the type of crop; and the administrative distribution of the crop irrigation districts. The number of resulting areas was 23 (Figure 2.2).

The groundwater flow was modelled with MODFLOW (McDonald & Harbough, 1988). According to the geology two layers were considered. The model was discretized into a finite-difference grid of 500 × 500 m with 60 rows and 82 columns (Peña-Haro *et al.*, 2010). The recharge and pumping values were taken from CHJ (2008) and Sanz *et al.* (2011). The water extraction from the aquifer in 2001 was about 58 Mm^3/year and 30 Mm^3/year in 2005. The groundwater recharge due to rainfall has an average value of 7 Mm^3/year.

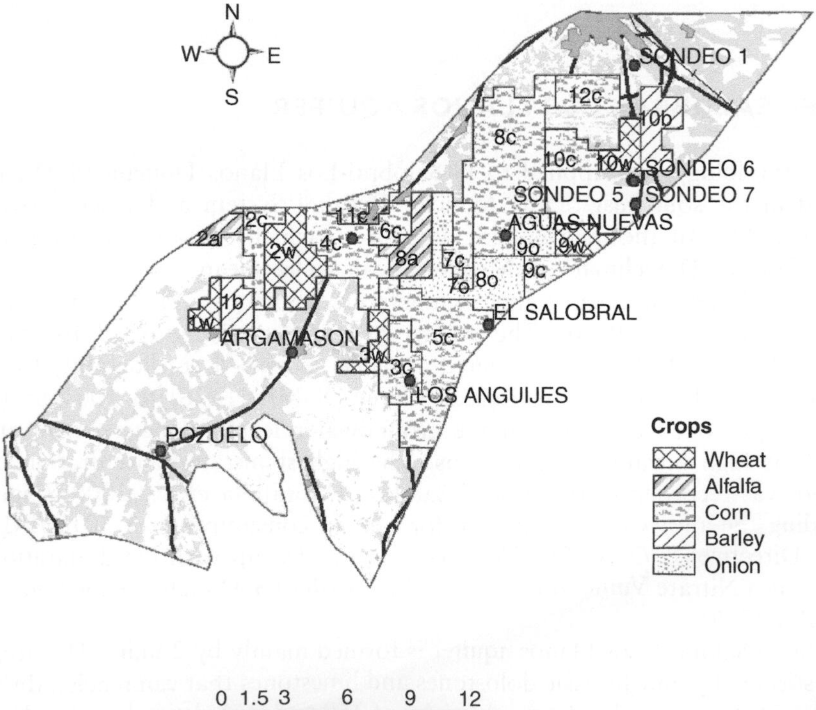

Figure 2.2 Control sites (dots), agricultural management areas and dry-land areas.

Nitrate fate and transport in groundwater was simulated using the MT3DMS model (Zheng & Wang, 1999). The nitrate loading areas were defined using remote sensing images. In order to estimate the nitrate input loads, the leaching functions were used. The highest nitrate concentrations are in the middle part of the aquifer where the irrigated agriculture is located, while the lowest ones are located in forested areas (Peña-Haro *et al.*, 2010). The highest value of 54.1 mg/l was recorded in the well named El Salobral.

The simulation of the unsaturated zone flow starts with an estimation of the groundwater recharge, since it is an agricultural area; the recharge is due to both precipitation and irrigation. Groundwater recharge due to precipitation was taken from Sanz *et al.* (2011); it was estimated using the distributed SIMPA model (Ruiz, 1999; Font, 2004) and the results of the model Hydromore (UCLM, 2008). For estimating the time delay the recharge values of 2005 were applied. The average recharge was 160 mm/year. The recharge from irrigation water use was estimated using the irrigation from 2005 reported by the ITAP (ITAP, 2005) and the efficiency reported by Castaño, *et al.* (2005) (Table 2.1). The unsaturated zone thickness was calculated by considering the difference between the average ground level and the average groundwater level in 2005 for each management area (Figure 2.3). The unsaturated zone thickness varies from 123 m in the south to 73 m in the north. Other parameters to the unsaturated zone model were: $K_s = 1.5$ m/day, $\varepsilon = 3.5$ and $n = 0.2$.

The results from the unsaturated zone model shows a delay time between 4 and 13 years, with a mean value of 7 years (Figure 2.4). The time delay was estimated as time difference between the centre of gravity of the input nitrate concentration and the output. The highest delay times are located in areas were the unsaturated zone thickness is big and the recharge by irrigation is low. More comprehensive models are available; however they require more data for calibration.

With the values of the delay time, the **NRM** was applied and the management model was run. The results of the optimal fertilizer allocation were obtained as a reduction from the actual fertilizer use. The areas that have to undergo a higher fertilizer reduction are 5c and 9c, with a 93 and 86% respectively. Comparing these results with the ones obtained if the unsaturated zone is neglected, areas 5c and 9c have to further reduce the fertilizer application by 6 and 26% more than when the unsaturated zone is not considered. However, the total benefits were almost the same, they were only reduced from 95.4 to 95.2 M€/year. On the other hand, when taking into consideration the unsaturated zone, the results from the optimization shows that the target value of 50 mg/l cannot be reached before 2023, while without taking it into

Table 2.1 Irrigation and efficiency (2005).

Crop	*Efficiency (%) (*Castaño et al., 2005)*	*Irrigation (m³/ha) (*ITAP, 2005)*
Wheat	63.6	4,170
Corn	91.9	6,020
Barley	63.6	3,300
Alfalfa	75.7	9,011
Onion	91.9	5,990

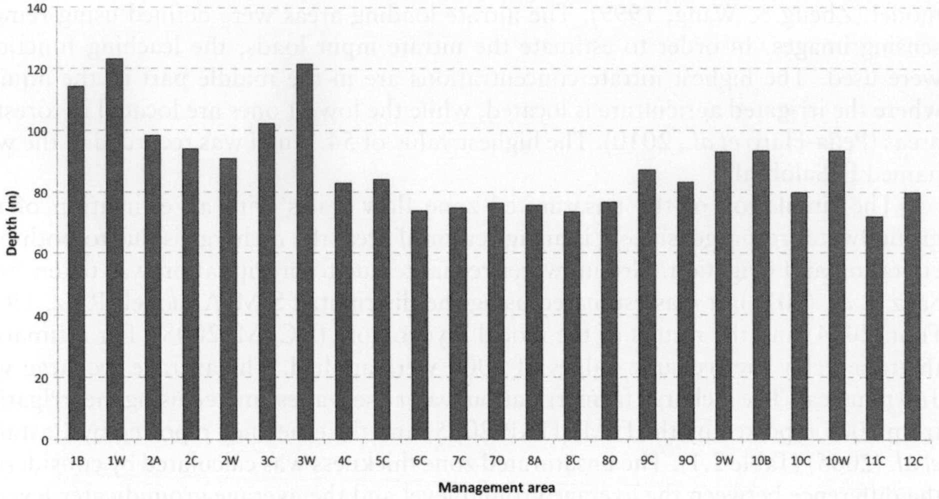

Figure 2.3 Depth to groundwater table (2005).

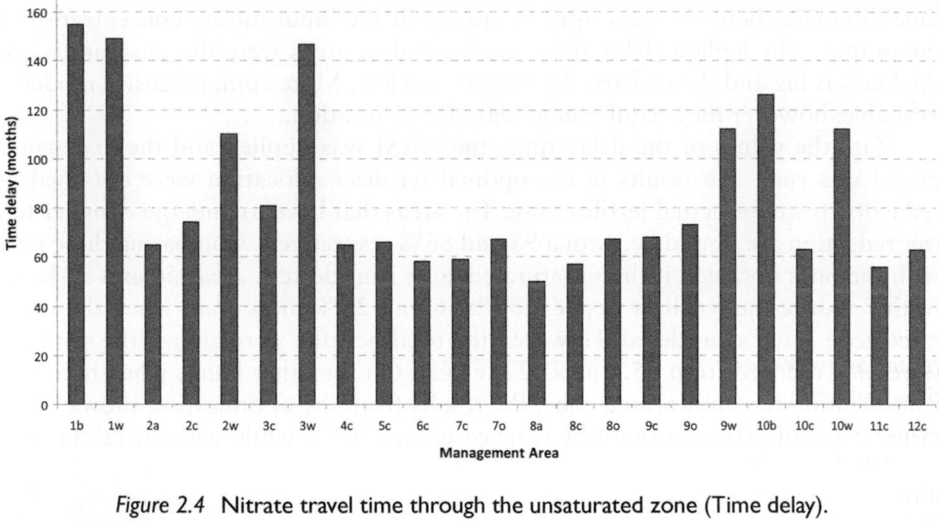

Figure 2.4 Nitrate travel time through the unsaturated zone (Time delay).

consideration the groundwater nitrate concentration limit can be achieved by 2015. This result is significant when analyzing the timescale to achieve the good quality status of the groundwater bodies.

The difference in time to achieve the target limits is not only influenced by the time that reduced leaching takes to arrive to the control sites, but also because there is high nitrate concentrations already travelling through the unsaturated zone, which

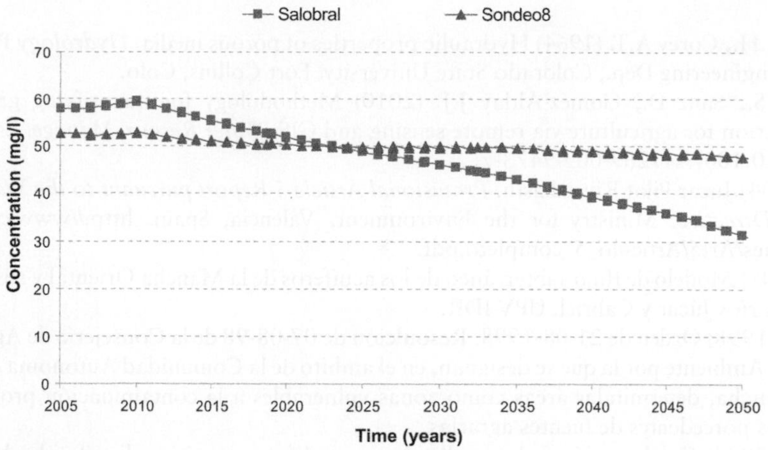

Figure 2.5 Nitrate concentrations for the optimal fertilizer allocation at two different control sites (see Fig. 2.2).

explains the upward trend in nitrate concentrations even after the application of the fertilizer standards (Figure 2.5).

2.6 CONCLUSIONS

The WFD established the year 2015 as a deadline for the achievement of good chemical status and also the possibility to extend it. The average travel time of the nitrate through the unsaturated zone, considering the conditions of 2005, was 7 years, and the good chemical status cannot be achieved before 2023.

Even though, when taking into consideration the nitrate travel time through the unsaturated zone, the fertilizer has to be further reduced, this reduction did not represent an important decrease in the total benefits. The major influence is on the travel time, which has an important effect on the time when the good quality status can be achieved.

The travel time trough the unsaturated zone is mostly influenced by the magnitude of the recharge, and the wet-dry periods. If changes in precipitation and/or in land use patterns, different time delays can be expected. This stresses the role of the unsaturated zone in deriving management policies to reduce groundwater nitrate pollution.

ACKNOWLEDGMENTS

The study was supported by the European Community 7th Framework Project GENESIS (226536) on groundwater systems.

REFERENCES

Brooks R.H., Corey A.T. (1964) Hydraulic properties of porous media. *Hydrology* Paper no. 3, Civil Engineering Dep., Colorado State University, Fort Collins, Colo.

Castaño S., Sanz D., Goméz-Alday J.J. (2010) Methodology for quantifying groundwater abstraction for agriculture via remote sensing and GIS. *Water Resour. Manage.* 24:795–814. DOI: 10.1007/s11269-009-9473-7.

CHJ (2004) Jucar Pilot River Basin. *Provisional Article 5 Report pursuant to the Water Framework Directive.* Ministry for the Environment, Valencia, Spain. http://www.chj.es/CPJ3/imagenes/Art5/Articulo_5_completo.pdf.

CHJ (2008) Modelo de flujo subterráneo de los acuíferos de la Mancha Oriental y sus relaciones con los ríos Júcar y Cabriel. UPV-IDR.

DOCM (1998) Orden de 21-08-1998. Resoulción de 07-08-98 de la Consejería de Agriculutra y Medio Ambiente por la que se designan, en el ambito de la Comunidad Autonoma de Castilla-La Mancha, determindas áreas como zonas vulnerables a la contaminación producida por nitratos porcedentes de fuentes agrarias.

Font E. (2004) Colaboración y desarrollo de un modelo matemático distribuido de flujo subterráneo de la Unidad Hidrogeológica 08.29 Mancha Oriental en las provincias de Albacete, Cuenca y Valencia. BSc Thesis, Polytechnical University of Valencia, Spain.

Geyer D.J., Keller C.K., Smith J.L., Johnstone D.L. (1992) Subsurface fate of nitrate as a function of depth and landscape position in Missouri Flat Creek watershed, U.S.A. *J Cont Hydrol*, 11 (1992) 127–147.

Gutierrez A. and Baran N. (2009) Long-term transfer of diffuse pollution at catchment scale: Respectives roles of soil, and the unsaturated and saturated zones (Brèvilles, France). *J Hydrol* 369. 391–391.

IGME (1976) Informe sobre la zona regable de Los Llanos. Madrid, España: Instituto Geológico y Minero de España.

ITAP (2005) Anuario Tecnico 2005. Instituto Tecnico Agronomico Provincial de Albacete.

Ledoux E., Gomez E., Monget J.M., Viavattene, Viennot P., Ducharne A., Benoit M., Mignolet C., Schott C., Mary B. (2007) Agriculture and groundwater nitrate contamination in the Seine basin. The STICS–MODCOU modelling chain. *Sci. Total Environ.* 375 (2007) 33–47.

Liu J., Williams J.R., Zehnder A.J.B., Hong Y. (2007) GEPIC – modelling wheat yield and crop water productivity with high resolution on a global scale. *Agricultural Systems* 94 (2), 478–493.

Mattern S., Vanclooster M. (2010) Estimating travel time of recharge water through a deep vadose zone using a transfer function model. *Environ. Fluid. Mech.* 10:121–135. DOI: 10.1007/s10652-009-9148-1.

McDonald M.G., Harbough A.W. (1988) A Modular Three-Dimensional Finite-Difference Groundwater Flow Model, *US Geological Survey Technical Manual of Water Resources Investigation*, Book 6, US Geological Survey, Reston, Va, 586 p.

Moratalla A., Gómez-Alday J.J., De las Heras J., Sanz D., Castaño S. (2009) Nitrate in the water-supply well in the Mancha Oriental hydrogeological system (SE Spain). *Water Resour. Manage.* 23. 1621–1640. DOI 10.1007/s11269-008-9344-7.

Peña-Haro S., Pulido-Velazquez M., Sahuquillo A. (2009) A hydro-economic modeling framework for optimal management of groundwater nitrate pollution from agriculture. *J Hydrol* doi:10.1016/j.jhydrol.2009.04.024.

Peña-Haro S., Llopis-Albert C., Pulido-Velazquez M., Pulido-Velazquez D. (2010) Fertilizer standards for controlling groundwater nitrate pollution from agriculture: El Salobral-Los Llanos case study, Spain. *J Hydrol.* doi:10.1016/j.jhydrol.2010.08.006.

Ruiz J.M. (1999) Modelo distribuido para la evaluación de recursos hídricos. *CEDEX*, Madrid, 245 pp.

Sanz D., Castaño S., Casiraga E., Sauquillo A., Gomey-Alday J.J., Peña-Haro S., Calera A. (2011) Modeling aquifer-river interactions under the influence of groundwater abstraction in the Mancha Oriental System (SE Spain). *Hydrogeol J*, doi:10.1007/s10040-010-0694-x.

UCLM (2008) Hydrological model for estimation recharge and evapotranspiration by remote sensing and GIS. UCLM, Albacete, Spain. http://www.hidromore.es/.

Williams J.R. (1995) The EPIC model. In: Singh, V.P. (Ed.), Computer Models of Watershed Hydrology. *Water Resources Publisher*, pp. 909–1000.

Zheng C., Wang P. (1999) MT3DMS: A Modular Three-Dimensional Multispecies Transport Model for Simulation of Advection, Dispertion and Chemical Reactions of Contaminants in Groundwater Systems, Documentation and User's Guide.

Chapter 3

Field tests for subsurface iron removal at a dairy farm in Saxony, Germany

Jakob P. Ebermann[1]*, Dieter Eichhorn*[2]*, Wolfgang Macheleidt*[1]*, Johannes J. Ahrns*[1] *& Thomas Grischek*[1]

[1]*Department of Civil Engineering/Architecture, University of Applied Sciences Dresden, Dresden, Germany*

[2]*Dr.-Ing. Dieter Eichhorn, Dresden, Germany*

ABSTRACT

A field test approach for subsurface iron removal, based on field investigations at a dairy farm in Saxony, Germany was carried out. The water supply there is based on two groundwater abstraction wells. The thickness of the sandy aquifer is 10 to 14 m. The groundwater has a mean concentration of 10 mg/l Fe(II), 0.5 mg/l Mn, 0.1 mg/l ammonium, and a mean pH of 6.5. By oxidation and adsorption in the subsurface reaction zone, it is possible to remove iron. At the field site, a tracer test to determine the dimensions of the reaction zone was conducted by measuring the electrical conductivity and dissolved oxygen concentration. Results of the tests are used to achieve an optimal operation of the wells and aeration units, especially in planning long-term subsurface treatment using aeration or technical oxygen.

3.1 INTRODUCTION

Iron and manganese are commonly present in anoxic groundwater worldwide. High iron concentrations are not harmful to human and animal health, but can result in technical problems such as clogging of production wells, precipitation and incrustation in the water supply distribution systems, orange/brown colour and bad taste of drinking water (Sharma, 2001).

Iron removal from groundwater can be accomplished in two different ways. The quality of raw water, the available financial resources and the philosophy of the water company are the main criteria in deciding on either of the groundwater treatment processes. The most common process is the treatment of groundwater on the surface (after abstraction). Another treatment process is the in-situ subsurface removal of iron from groundwater, using the Vyredox method (Hallberg & Martinell, 1976), Subterra (Rott & Friedle, 2000; Rott *et al.*, 2002; Herlitzius *et al.*, 2008) or Uneis methods (Eichhorn, 1985).

Under certain boundary conditions subsurface iron removal can be a sustainable and cost-effective technique for groundwater treatment. The oxidation and filtration/adsorption processes necessary for the removal of dissolved iron (also manganese and arsenic) are transferred from conventional treatment facilities into the aquifer. By infiltrating oxygen-enriched water through an abstraction well iron hydroxides are formed in the aquifer around the well providing adsorption sites for Fe(II) and other cations. Periodic alternation between infiltration and abstraction eventually results in the production of iron-free water.

Subsurface iron removal is also an ecologically sound in-situ oxidation treatment without chemicals. For this technology the subsurface soil matrix is used as a natural bio-chemical reactor.

Nowadays more than 10 large waterworks and more than 10 000 single well operators in Europe apply the technique of subsurface iron removal successfully. The volume of drinking water produced varies from 1 m^3/d to more than 30 000 m^3/d. The dissolved iron and manganese concentrations in the groundwater also range widely from 0.5 to 55 mg/l for iron and 0.1 to 3.0 mg/l for manganese.

Despite the fact that subsurface iron removal is proven in theory and practice to be a cost-effective and sustainable technique in the process of drinking water treatment, the awareness amongst waterworks is not widespread and therefore its potential is not fully exploited. Subsurface iron removal can be applied on both small and large scales and combined with other treatment techniques (Herlitzius *et al.*, 2010).

The mobile field test unit has been used to study subsurface iron removal at a 30 years old shaft well and a recently built vertical filter well. Promising results were obtained after 1 month of operation.

3.2 TECHNICAL ASPECTS OF SUBSURFACE IRON REMOVAL

Figure 3.1 shows a scheme of a subsurface iron removal facility with two wells, aeration unit, pipes, and degassing tank. The technique includes the following steps:

1 Abstraction of groundwater from the first well, aeration of a portion of the pumped water and re-infiltration of the aerated groundwater containing dissolved oxygen into the second well (Fig. 3.1).
2 In the aquifer: Oxidation of the dissolved and adsorbed Fe(II), transformation of the soluble Fe(II) to its less soluble form Fe(III), formation and precipitation of iron(hydr)oxides providing further adsorption sites for Fe(II).
3 Abstraction of anoxic groundwater and removal of Fe(II) by adsorption onto iron(hydr)oxide in the reaction zone around the pumped well. The second well is used for abstraction and the first well is used for infiltration of aerated/oxygenated water.

With technical or atmospheric oxygen and concentrations ranging from 10 mg/l to 50 mg/l a portion of the abstracted groundwater from well 1 will be infiltrated into well 2. A typical tool for the aeration unit is a water jet air pump. The selection of the oxygenation technique is based on the oxygen consumption during the infiltration in the reaction zone. The main oxygen consumers are:

1 Fe(II)
2 Mn(II)
3 Ammonium
4 Easily degradable organic compounds

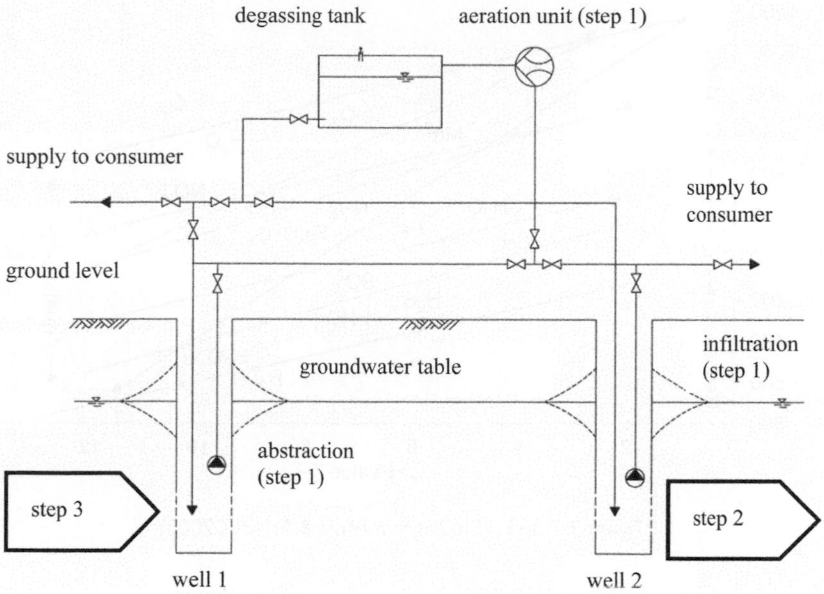

Figure 3.1 Illustration of subsurface iron removal with two vertical wells (DVGW 2005, modified).

Additionally, nitrite and sulphide consume oxygen depending on their concentrations. Conditions that hinder subsurface iron removal are low pH, low hardness, and high Fe(II), Mn(II), and ammonium concentrations. The oxidation of Fe(II) to Fe(III) requires a pH value greater than 5.5 (Eichhorn, 1985). During iron oxidation the pH decreases, depending on the buffering capacity of the water-soil system.

Before planning and building an in-situ treatment plant for iron removal it is useful to carry out a subsurface iron removal field test and groundwater sampling to calculate the functionality principle with the chemical composition of the natural groundwater. For both investigations at least one monitoring well at the site chosen for the in-situ treatment is necessary.

The efficiency of the subsurface iron removal technique is determined as the volumetric ratio of the water abstracted to the water infiltrated, which commonly is between 3 and 5, but could reach 10. This ratio is called "efficiency coefficient" (Rott & Friedle, 2000). A high efficiency coefficient is the aim of optimising design and operation of the technique.

Through adsorption in the reaction zone it is also possible to remove arsenic, manganese, and dissolved organic carbon. During groundwater abstraction in the first month the iron adsorption on soil grains occurs, but before manganese adsorption and oxidation can take place the iron oxidation in this zone around the well has to be complete. A pH-Eh diagram (Fig. 3.2) shows that the immobilisation of manganese can start only if iron and ammonium have been oxidised. Given sufficient room for iron and manganese precipitation around the well after some period of time the manganese will be adsorbed onto the soil grains (Fig. 3.3). The acceleration of manganese oxidation also depends on autocatalytic effects of microorganisms.

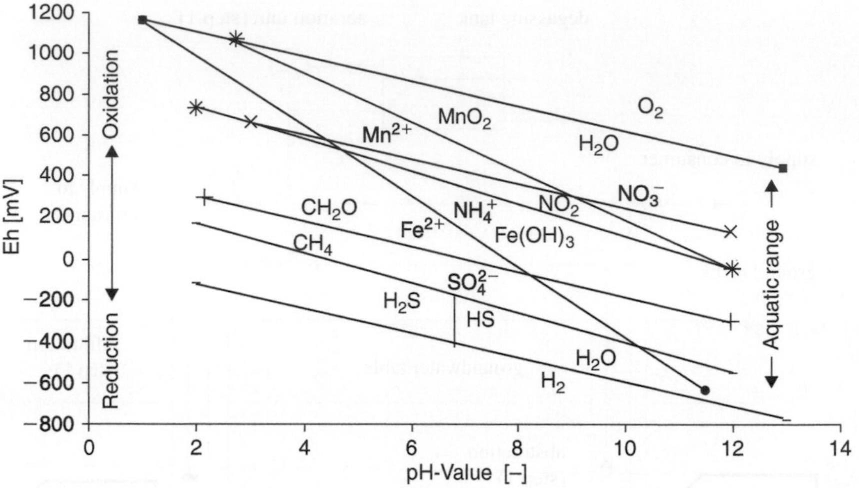

Figure 3.2 pH – Eh diagram (Rott & Friedle, 2000).

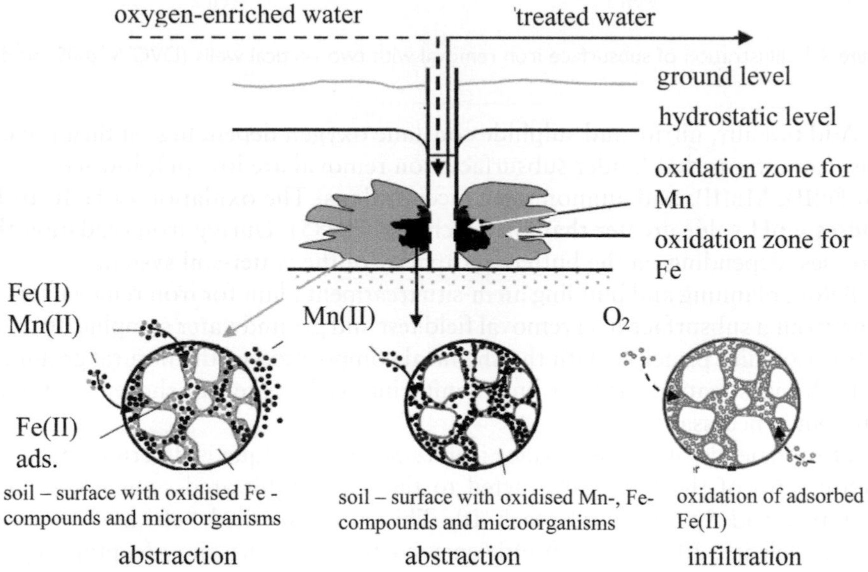

Figure 3.3 Precipitation zoning for iron and manganese (Meyerhoff, 1996 modified).

3.3 STUDY SITE AND METHODOLOGY

The research site at the farm in Dobra is located in northern Saxony 40 km north of the city of Dresden. Pleistocene sediments are underlain by greywacke with a thickness of 10–14 m. The sediments are fine to medium sands with a hydraulic conductivity of about 1×10^{-3} m/s. At the time of the experiment the water table was 1.5 m beneath the ground surface, and the groundwater had a concentration of 5–10 mg/l Fe(II), 0.3–0.5 mg/l Mn, 0.1 mg/l ammonium, and a mean pH of 6.5.

A mobile unit on a vehicle trailer was developed for pilot investigations to test the applicability of subsurface iron removal at field scale. It contains an aeration unit (Red-y smart) to regulate and measure the technical oxygen flow (Vögtlin, Aesch, CH), a static mixing unit (SULZER, Winterthur, CH), a degassing unit, valves for discharge control and devices (Multi 350i) for continuous measurement of discharge, pH, temperature, and electrical conductivity (WTW, Weilheim, Germany). The field test unit was developed for site investigation using abstraction/ infiltration wells or observation wells with infiltration rates between 5 and 10 m^3/h. The trailer can be brought to all sites and allows self-sustaining operation. All measurement equipment is protected in the closed trailer and can be operated automatically to reduce staff hours on-site.

To conduct the field test for subsurface iron removal in Dobra, an existing old shaft well (PW 2) with a diameter of 2 m and a depth of 5 m, and a new vertical well (PW 1) having a diameter of 325 mm and a depth of 14 m were used. Mean iron concentration in PW 1 and PW 2 were of 10 mg/l and 5 mg/l, respectively. For measuring dissolved oxygen concentration (O_2), electrical conductivity (EC) and pH in the reaction zone, two monitoring wells (MW 1 & MW 2) having a depth of 12 m were installed at a distance of 5 m and 7.5 m from the new vertical well PW 1 (Fig. 3.4 & 3.5).

The potential interaction of the aquifer and a nearby creek was investigated and the determination of the creek bed infiltration rate was found to be negligible.

According to Ahrns *et al.* (2009) bank filtration could have an effect on the efficiency of subsurface iron removal. Table 3.1 provides data on water quality and design of the wells in Dobra.

The field tests were conducted from September to December 2009, with a total abstraction rate of 4.5 m^3/hour of which 1.2 m^3/hour was infiltrated. One test period consisted of a cycle of an abstraction from one well and infiltration into the other well. During the first testing period it was necessary to create a reaction zone with a radius of approximately 6 m. Due to an iron concentration of about 10 mg/l in PW 1 the O_2 concentration was set to 15–25 mg/l. In the first period the water was abstracted from PW 2 and infiltrated in PW 1. Using data loggers in the production well and monitoring well MW 1 it was possible to measure O_2, pH, temperature and EC automatically at regular intervals. Samples of abstracted water were taken during all 7 test periods.

A tracer test to determine the size of the reaction zone around one well was conducted by measuring the EC and O_2 concentration. The infiltrate containing 15–25 mg/l O_2 and a chloride tracer was injected in PW 1, during the first period. Dissolved oxygen and EC were measured in PW 1 and MW 1.

3.4 RESULTS

The results (Fig. 3.6) show a breakthrough of the injected tracer in MW 1 after 3.5 days, peaking at 1300 μS/cm after about 4 days. A breakthrough of oxygen (from the oxygen-rich infiltrated water) was observed in MW 1 after about 7 days with a peak of 22 mg/l at 8.5 days. During the infiltration in period 3 (P3) there was a breakthrough of oxygen

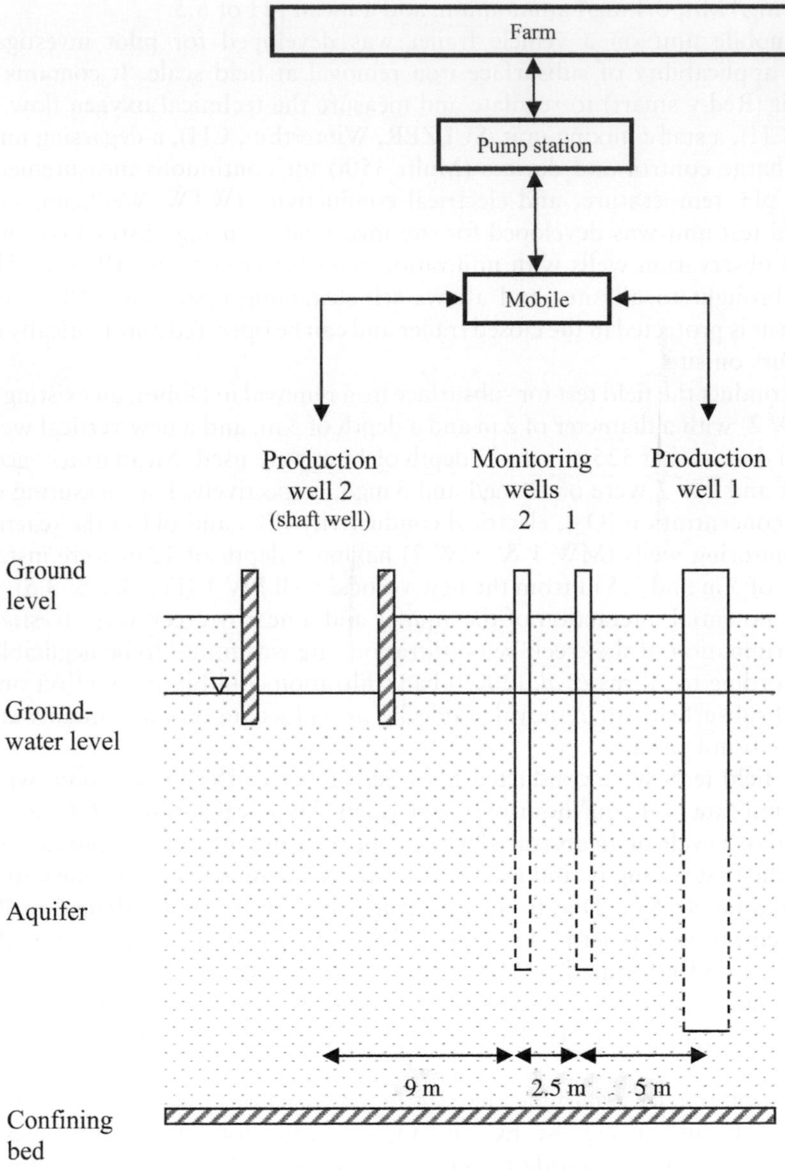

Figure 3.4 Configuration of production and monitoring wells (not drawn to scale).

after only about 4 days peaking at 15 mg/l and matching the time of breakthrough of the tracer in period 1. The earlier breakthrough of oxygen in the third period indicates that during the first period a lot of oxygen consumers use more oxygen than in the following periods. The size of the reaction zone has a diameter of about 14 m and a height of about 6 m.

Figure 3.5 Mobile trailer based unit for subsurface iron removal tests, production wells (PW), monitoring wells (MW) (© Ebermann, 2009).

Table 3.1 Parameters at the site in Dobra and technical configuration of the wells.

Parameter	*Unit*	*PW 1* *Vertical well*	*PW 2* *Shaft well*	*MW 1* *Monitoring well*
calcium	mg/l	50.5	49.9	56.3
potassium	mg/l	6.0	12.3	5.2
magnesium	mg/l	16.4	16.9	18.0
sodium	mg/l	19.7	18.7	17.9
chloride	mg/l	34.8	33.6	32.5
nitrate	mg/l	<5	24.5	<5
nitrite	mg/l	<0.01	–	–
ammonium	mg/l	0.09	–	–
sulphate	mg/l	123	118	111
aluminium	mg/l	<0.1	<0.1	<0.1
arsenic	mg/l	0.011	<0.01	0.014
iron	**mg/l**	**9.8**	**4.5**	**11.0**
manganese	mg/l	0.43	0.29	0.56
carbonate hardness	°dH	3.4	2.4	4.3
$K_{S4.3}$	mmol/l	1.2	0.8	1.6
$K_{B8.2}$	mmol/l	1.8	1.6	2.0
pH	–	6.9	6.1	6.5
electrical conductivity	μS/cm	485	508	505
abstraction depth	**m bgl**	**13**	**3**	**12**
aquifer thickness	m	14	14	14
hydrostatic level	m bgl	1.5	1.5	1.5
hydraulic conductivity	m/s	1×10^{-3}	1×10^{-3}	1×10^{-3}
aquifer material	–	fine and medium sand	fine and medium sand	fine and medium sand
filter length	m bgl	7.5–13.5	5	10–12
well diameter	mm	325	2000	171

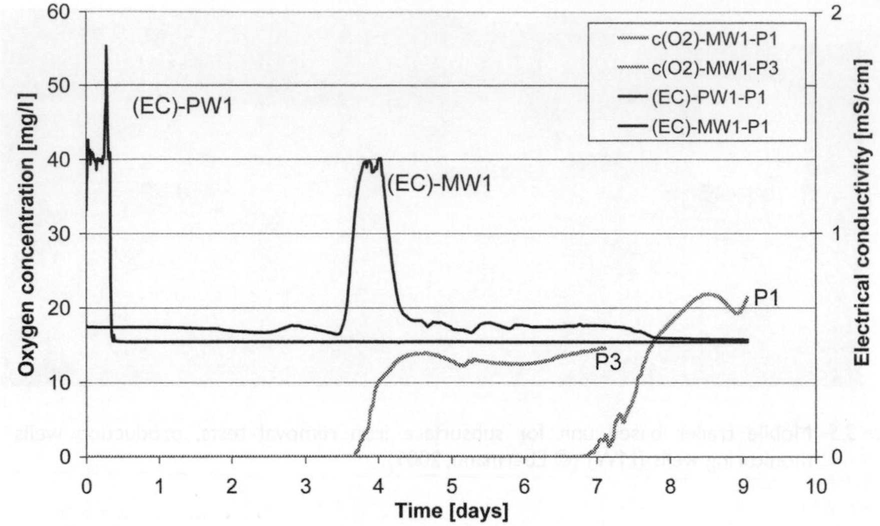

Figure 3.6 Oxygen concentration and electrical conductivity during the tracer test in the production well 1 (PW 1) and monitoring well 1 (MW 1) during periods 1 and 3 (P 1, P 3).

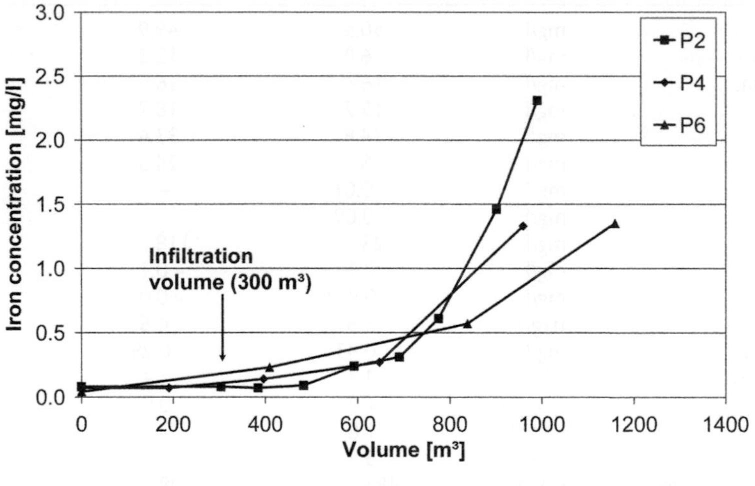

Figure 3.7 Iron concentration during abstraction in periods 2, 4, and 6 at production well 1 (PW 1).

The infiltrated volume before every abstraction was around 300 m^3 in both of the wells. The iron concentration of 10 mg/l at a depth of 14 m in PW 1 was reduced to less than 0.5 mg/l in three periods after an abstraction volume of 700 m^3 (Fig. 3.7). The iron removal in PW 1 was improving after subsequent periods of infiltration and abstraction. The faster increase of iron concentration during the abstraction in the first period is due to the higher oxygen consumption. In the following periods the increase

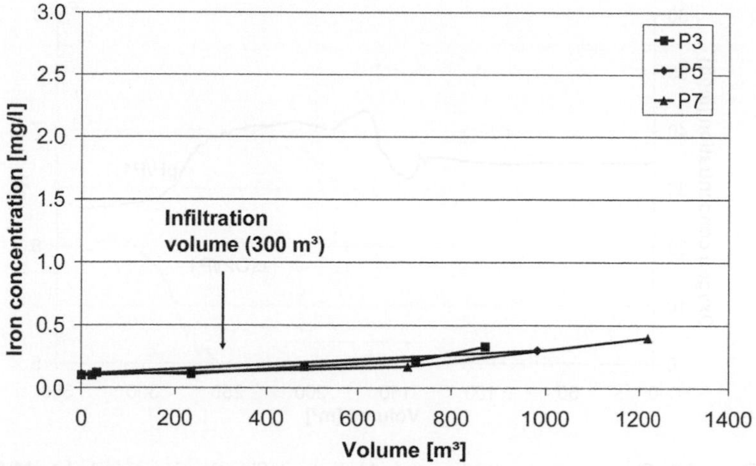

Figure 3.8 Iron concentration during abstraction in periods 3, 5, and 7 at production well 2 (PW 2).

of iron concentration was gradually lowered. In the first 3 periods the "efficiency coefficient" (maximum iron concentration of 0.5 mg/l in abstracted water) was 2.3 but increased to 4 after a few months. This is caused by additional adsorption sites provided by the precipitated iron(hydr)oxides.

In PW 2 it was possible to lower the iron concentration from around 5 mg/l to less than 0.5 mg/l for an abstraction volume of up to 1200 m^3 (Fig. 3.8). The iron concentration at the filter depth is important for the efficiency of subsurface iron removal. The natural iron concentration at the PW 2 of 5 mg/l was half of the concentration at the PW 1 and the "efficiency coefficient" quickly reached a value of 4. This means that the adsorption capacity around both wells is similar.

During the infiltration into PW 1, pH and O_2 were measured in MW 1. Figures 3.9, 3.10 and 3.11 show the dissolved oxygen concentration in comparison with pH. In period 1 the pH increases from 6.6 to 7.0 after about 150 m^3 with the infiltration front. When the oxygen front in period 1 reaches MW 1 the pH decreases from 7.0 to 6.4, because the oxidation process results in a release of H^+ (Fig. 3.9). In period 3 the pH decreases from 6.5 to 5.9 as the O_2 front reaches the monitoring well (Fig. 3.10). As soon as the oxidation of the consumers is finished, the pH of the water around MW 1 increases to 6.4. Figure 3.11 shows the changing pH during period 5, where the decrease in pH is lower.

3.5 CONCLUSIONS

After application of the subsurface iron removal technique, an iron concentration of less than 0.2 mg/l (a 98% change from background) was achieved in the abstracted groundwater within one week. Results from a tracer test were used to determine the size of the reaction zone. The experiment was continued to provide data for different

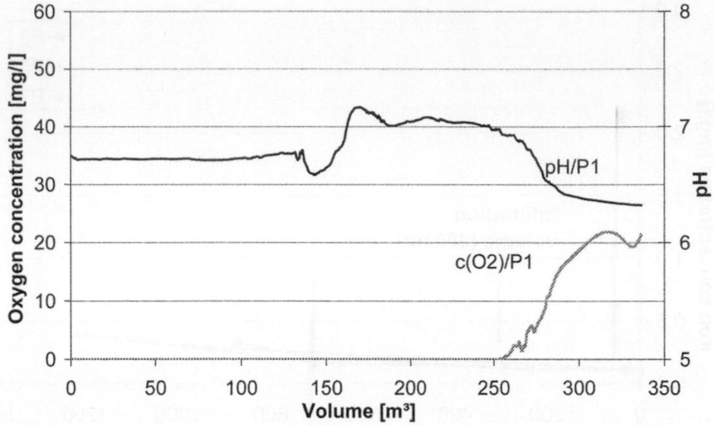

Figure 3.9 Oxygen concentration and pH during infiltration in periods 1 at MW 1.

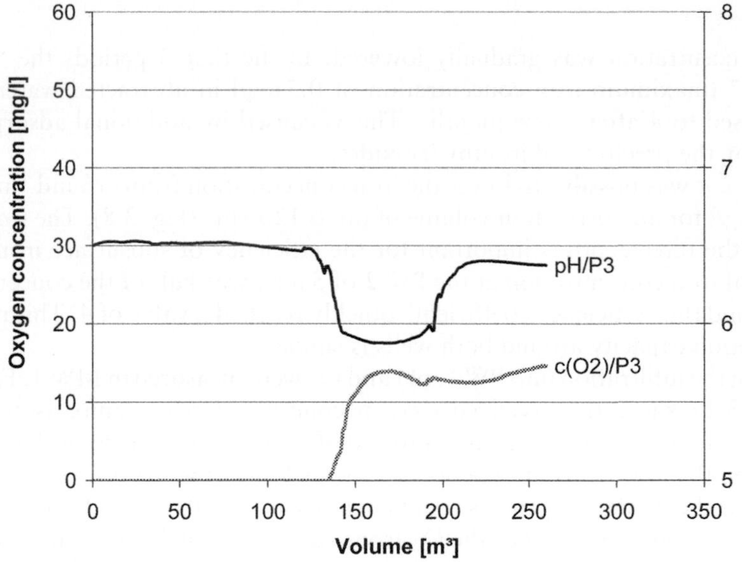

Figure 3.10 Oxygen concentration and pH during infiltration in period 3 at MW 1.

oxygen concentrations in the infiltrate, the resulting efficiency of iron removal and oxygen consumption by other processes. Results were used to design an optimal operation of the wells and the oxygenation unit. To quantify oxygen consumption processes in the aquifer and to predict the changes in pH at sites with high Fe(II) concentration, it is important to decide between aeration with air or technical oxygen. The mobile unit can be used to characterise site-specific aquifer conditions within one week and to test, whether subsurface iron removal can be successfully applied at a certain site.

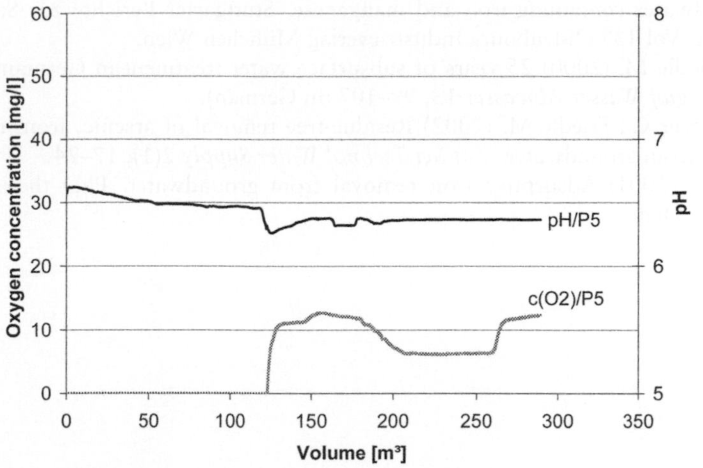

Figure 3.11 Oxygen concentration and pH during infiltration in period 5 at MW 1.

ACKNOWLEDGEMENTS

The authors are grateful to the financial support of the German Federal Ministry of Education and Research (BMBF) to the project "Development of mobile modules for automated subsurface iron removal and well conservation" (BMBF 1735X07). The logistic support of the agricultural company Agrargenossenschaft Dobra is also gratefully acknowledged.

REFERENCES

Ahrns J., Klügel S., Schoenheinz D., Eichhorn D., Grischek T. (2009) Subsurface iron removal at river bank filtration sites. *Proc. IWA Eastern European Regional Young Water Professionals Conf.*, 21.–22.05.09, Minsk, 290–297.

DVGW (2005) Arbeitsblatt W 223-3 Removal of iron and manganese. Part 3: Planning and operation of facilities for subsurface water treatment. DVGW, Bonn (in German).

Eichhorn D. (1985) Contribution to the theory of iron removal through subsurface water treatment. PhD thesis, *Faculty of Water Sciences*, Dresden University of Technology (in German).

Hallberg R.O., Martinell R. (1976) Vyredox – In situ purification of ground water. *Ground Water* 14(2), 88–93.

Herlitzius J., Sumpf H., Grischek T., Rothe S. (2008) Chances for in-situ treatment of groundwater in Russia. *DVGW energie wasser praxis* 9, 8–13 (in German).

Herlitzius J., Sumpf H., Kulakov V.V., Grischek T., Pillukeit W. (2010) Further development of subsurface iron removal in combination with other techniques. *Proc. IWA Specialist Conference on Water and Wastewater Treatment Plants in Towns and Communities of the XXI Century: Technologies, Design and Operation*, 2–4 June 2010, Moscow, 123–129.

Meyerhoff R. (1996) Development of criteria for planning and application of in-situ treatment of groundwater containing iron and manganese. Stuttgarter Berichte zur Siedlungswasserwirtschaft. Vol 139 Oldenbourg Industrieverlag München Wien.

Rott U., Friedle M. (2000) 25 years of subsurface water treatment in Germany – review and prospects. *gwf Wasser Abwasser* 13, 99–107 (in German).

Rott U., Meyer C., Friedle M. (2002) Residue-free removal of arsenic, iron, manganese and ammonia from groundwater. *Wat Sci Technol Water Supply* 2(1), 17–24.

Sharma S.K. (2001) Adsorptive iron removal from groundwater. PhD thesis, Wageningen University. Delft.

Chapter 4

DNA-microarrays for monitoring natural attenuation of emissions from abandoned landfill sites in contaminated groundwater plumes

Christoph Charlé[1], Stephan Kühn[2] & Thomas Struppe[2]

[1] *Protekum Umweltinstitut GmbH, Oranienburg, Germany*

[2] *Struppe & Dr. Kühn Umweltberatung, Berlin, Germany*

ABSTRACT

A DNA-microarray for the detection and characterisation of microbial communities involved in degradation of organic pollutants in contaminated groundwater plumes was developed and tested. Universal PCR primers were used to amplify the 5′ region of *Bacteria* and *Archaea* 16S rRNA-genes from groundwater total DNA. Amplification products were fluorescently labeled and hybridized to the prototype microarray, which consisted of 78 16S rRNA gene-targeted oligonucleotide probes for simultaneous detection of *Bacteria*- and *Archaea*-families. The results of this study demonstrate that DNA microarray technology facilitates the fast and efficient analysis of complex microbial communities in methodological challenging habitats such as contaminated groundwater. Hence, DNA-microarray technology is applicable for monitoring natural attenuation (NA) in contaminated groundwater.

4.1 INTRODUCTION

Degradation of organic contaminants by microbial communities in soil and groundwater is the main activity in natural attenuation (Weiss and Cozzarelli, 2008). The effects of contaminants on growth characteristics and composition of microbial communities have been confirmed by DGGE of PCR-amplified marker-gene fragments such as the 16S rRNA gene (Röling *et al.*, 2001). Until now, PCR-DGGE is the method of choice to generate genetic fingerprints for comparison of microbial communities involved in degradation of most varying contaminants (Hendrickx *et al.*, 2005; Nozawa-Inoue *et al.*, 2005; Ferguson *et al.*, 2007; Tas *et al.*, 2009). Although a standard technique in microbial ecology, PCR-DGGE has some major disadvantages. It is strictly site dependent and does not facilitate the direct identification of microbial communities involved in degradation of contaminants and, therefore, the precise monitoring of natural attenuation. Over the past years developments in the field of microbial ecology improved considerably. The application of DNA-microarrays in the area of marine biology as well as soil- and water ecology facilitated the analysis of microbial communities on species-level by specific detection of microbial rRNA-genes (Brodie *et al.*, 2007; Koizumi *et al.*, 2002; Loy *et al.*, 2002, 2004, 2005; Peplies *et al.*, 2003, 2004, 2006). These findings led to the assumption that DNA-microarray-technology is applicable for monitoring NA in contaminated groundwater. Compared to other methods, which are state of the art in molecular biology, the potential to analyse a large number of parameters in parallel is a major advantage of DNA-microarrays. Based on known probe-sequences the composition of microbial communities can be analyzed on any taxonomic level. To overcome the limitations of PCR-DGGE and to apply the benefits of DNA-microarray technology to environmental analytics, a 16S rRNA directed DNA-microarray for the analysis of microorganisms, which have been detected in soil and groundwater (Gliesche, 1999), was designed. The main objective in microarray-design was to detect as many microorganisms as possible with a minimal number of probes. Therefore, a higher taxonomic level was selected to detect bacterial families instead of single species. Since more than 90% of the existing prokaryotes still remain unknown (Curtis *et al.*, 2002; Schloss and Handelsmann, 2004), a microarray for the detection of microorganisms on species level is not applicable for routine monitoring of natural attenuation. The DNA-microarray developed in this study facilitated the simultaneous detection of *Bacteria* and *Archaea*-families in groundwater for the first time ever and confirmed results obtained with PCR-DGGE (Kerndorff *et al.*, 2008). Furthermore, the composition of microbial communities in tested groundwater samples could be assessed.

4.2 MATERIALS AND METHODS

4.2.1 Area of investigation

The area of investigation containing abandoned landfill sites is situated in the southwestern part of Berlin/Germany at Kladow/Gatow (Fig. 4.1) in sands, gravels and

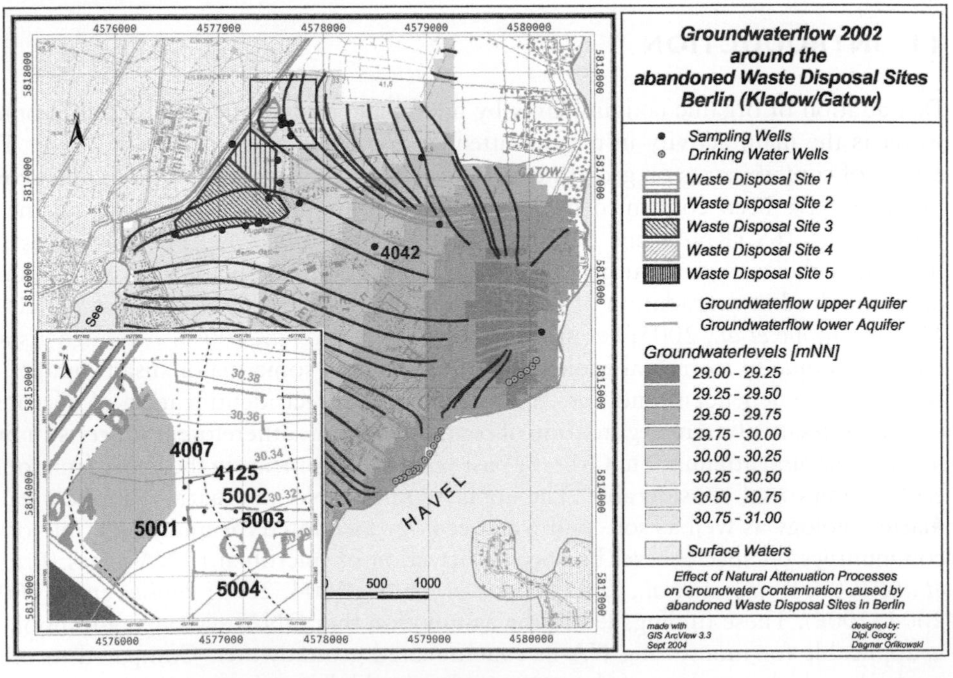

Figure 4.1 Area of investigation.

minor mudstones of the Quaternary system. The main aquifer shows k_f values between 2.6×10^{-4} and 5.6×10^{-4} m/s, has an average thickness of more than 100 m and is divided by a 10 m layer of mudstone in the southwest. The *Havel River* forms the eastern and southern boundaries of the investigation area. The western and northern boundaries of the investigation area are formed by the border to the *Land Brandenburg* and the trenches along the southern edge of the irrigation fields, respectively. A water divide and the drainage into the Havel River in the east and in the west characterise groundwater flow from WNW to ESE to the Havel River respectively. The detailed geological and hydrogeological conditions of the study area have been described previously (Kerndorff *et al.*, 2008). Filling pits, which originate from sand quarrying, created the abandoned landfill sites in the study area. Landfill site 1 was mainly used for disposal of household waste while site 2 and 3 were used primarily to dispose of construction waste. Groundwater from groundwater observation wells situated downstream from waste site 1, has been analysed (Fig. 4.1). Borehole 4010 is situated in the upstream area in the NW of landfill site 2 and serves as a general source for low contaminated upstream samples. Boreholes 4007 and 5001 are situated in the direct vicinity of landfill site 1 between 0–20 m in the downstream direction. Boreholes 5002, 5003 and 5004 are situated along the downstream flow in a distance of about 20–60 m, 60–120 m and 120–200 m, respectively. Borehole 4024 is situated more than 1000 m away in the downstream direction of the landfill sites.

4.2.2 Geochemical analysis

Groundwater samples were analyzed by Gas Chromatography/Mass Spectrometry and Liquid Chromatography/Mass Spectrometry (GC/MS and LC/MS) as described previously (Kerndorff *et al.*, 2008). Briefly, parameters indicating microbial activity, such as ammonia (NH_3), carbonate (HCO_3^-/CO_2), iron (FeII), sulfate (SO_4^{2-}) and total organic carbon (TOC) were employed for the characterisation of NA.

4.2.3 Sample preparation

Groundwater samples were collected according to the protocol by Kerndorff *et al.* (2008). Groundwater from observation wells in the area of investigation were transferred to N_2 filled sterile 1000 ml gas collection tubes under anaerobic conditions and kept at 4°C until further processing. Microbial cells were collected on a 0.22 μm Durapore PVDF-Membrane (Millipore, Billerica, MA, USA) by vacuum-filtration under anaerobic conditions to avoid precipitation of ferrihydrite, which can interfere with the analysis of the samples. After complete extraction of the groundwater sample, the membranes were transferred to a sterile 15 ml centrifuge tube and stored at −20°C.

4.2.4 Epifluorescent direct cell counting

Concentration of microorganisms in groundwater samples was determined by staining of formaldehyde fixed samples (10% v/v 37% formaldehyde) with DAPI (4′,6-Diamidino-2-phenyindole). 1/100 vol of a 1 mg/ml DAPI-Solution (Sigma-Aldrich, Germany) was added to 3 ml–10 ml formaldehyde fixed sample followed by 20 min incubation at room temperature in the dark. Stained bacterial cells were collected on a black Millipore GTTPO2500 polycarbonate filter (Millipore, Billerica, MA, USA) by vacuum filtration. The sample-membranes were washed with 0.1% IGEPAL CA-630 (Sigma-Aldrich, Germany) twice to reduce background fluorescence and mounted with Citifluor AF2 (Citifluor Ltd., UK). Stained cells were enumerated via fluorescence microscopy using a Zeiss Axiovert 100 fluorescence microscope (Carl Zeiss Microimaging GmbH, Germany) at 1000x magnification (lamp: HBO 50, filter Set: Zeiss 02, Ex 365, FT 395, Em LP 420).

4.2.5 DNA-extraction

DNA was extracted from groundwater-samples according to a slightly modified standard protocol (Tsai and Olson, 1991, 1992; Kilb, 1999). PVDF-filters carrying microbial cells and sediment from groundwater samples were cut into small pieces, transferred to 2 ml reaction tubes and 480 μl of lysis solution (0.15 M NaCl, 0.1 M EDTA [pH 8.0]) containing 15 mg of lysozyme/ml was added. After vigorous vortexing and incubation for 1 h at 37°C, 120 μl of 0.1 M NaCl-0.5 M Tris-HCl (pH 8.0) and 10% w/v sodium dodecyl sulfate (SDS) was added to each sample followed by five cycles of freezing in a −70°C dry ice-ethanol bath and thawing in a 70°C water bath. Subsequently 3 μl RNase (29 mg/ml) were added, followed by incubation at 37°C for 30 min. Microbial DNA was extracted from the crude extracts with a QIAGEN DNA-extraction kit (Qiamp Tissue Kit, QIAGEN GmbH, Germany). DNA was concentrated by ethanol precipitation (Maniatis *et al.*, 1982) and resuspended in 30 μl TE-buffer (10 mM Tris-HCl [pH = 8.0], 1 mM EDTA).

4.2.6 PCR amplification of 16S rRNA genes

Previously described primers (Baker & Cowan, 2004, Wilson *et al.*, 1990 & Murray *et al.*, 1996) were used for the amplification of 16S rRNA Genes from *Archaea* (T7_ARC_333_fw: $^{5'}$TCC AGG CCCTAC GGG$^{3'}$; T7_ARC_344_fw: $^{5'}$ACG GGG TGC AGC AGG CGC GA$^{3'}$ & SP6_ARC_1115_rv: $^{5'}$GGG TCT CGC GTT G$^{3'}$) and *Bacteria* (T7_27f: $^{5'}$AGA GTT TGA TCM TGG CTC AG$^{3'}$ & SP6_517r: $^{5'}$ATT ACC GCG GCT GCT GG C$^{3'}$). PCRs were performed in a final volume of 50 μl and contained 2.5 U DFS Taq DNA Polymerase (Bioron GmbH, Germany), 1/10 vol 10 × DFS Taq Reaction buffer "complete" (160 mM $(NH_4)_2SO_4$, 670 mM Tris-HCl pH 8.8, 0.1% Tween-20, 25 mM $MgCl_2$), 500 μM primers, 330 μM of each deoxynucleotide triphosphate, 0.5 μl of DNA extract, and sterile H_2O upto 100 μl. Thermal cycling was performed using an iCycler (Bio-Rad, Hercules, Calif.) with initial denaturation at 94°C (5 min), followed by 35 cycles of 94°C (1 min), 60°C (1 min), and 72°C (1 min) using maximum temperature ramp rates, and a final extension at 72°C (7 min). During the first 10 cycles of the protocol the annealing temperature was decreased from 70°C−>60°C at a rate of 1°C/cycle. PCR products (Amplicons) were analysed by agarose gel electrophoresis using 3% agarose gels containing 12 μg/ml ethidiumbromide. Bands on the gel were visualized by fluorescence imaging using a BioRad Molecular Imager FX fluorescence laser scanner (Bio-Rad, Hercules, USA). Finally, amplicons were purified with the aid of the QiaQuick® PCR-clean up kit (QIAGEN

GmbH, Germany). The concentration of purified ampules was determined by UV spectrometry using a NanoDrop® UV/vis spectrophotometer (Nandrop/Thermo Fisher Inc.) or by densitometry.

4.2.7 Preparation of fluorescently labeled single-stranded target DNA

Fluorescently labeled single-stranded sample DNA (ssDNA) was prepared based on a modified standard procedure (Eberwine *et al.*, 1992, Schena, 2003, Bittner, 2003). Labeled target molecules were prepared from sample DNA by amplification of microbial 16S rRNA genes via PCR with modified 16S rRNA specific primers. *Bacterial* and *Archaea* 16S-rRNA genes were amplified with primer pair T7_27f/SP6_517r and primer pair T7_arc333/344fw/SP6_arc1115rv, respectively. cRNA was synthesized by in vitro transcription of 0.5 μg–3 μg purified amplicon for 4 h at 37°C with SP6 RNA-polymerase using a RiboMAXTM Large Scale RNA Production System-SP6 (Promega GmbH, Germany). Template DNA (amplicons) was removed from the reaction by DNase digestion and subsequent purification using an RNeasy®-mini kit (QIAGEN GmbH, Germany). Purified cRNA was analyzed by denaturing agarose gel electrophoresis using 2% agarose gels containing 0.41 mol/l formaldehyde and 15 μg/ml ethidiumbromide. Bands on the gel were visualised by fluorescence imaging using a BioRad Molecular Imager FX fluorescence laser scanner (Bio-Rad, Hercules, USA). Concentration of purified cRNA was determined by UV spectrometry using a NanoDrop® UV/vis spectrophotometer (Nandrop/Thermo Fisher Inc.). Fluorescently labeled ssDNA was synthesized by reverse transcription of 0.5 μg–1 μg purified cRNA. The reactions were performed for 90 min at 41°C in a final volume of 40 μl and contained 400 U Reverase^{TM} (M-MuLV reverse transcriptase RNaseH^{-}; Bioron GmbH, Germany), 8 μl 5× RT-buffer "complete" (250 mM Tris-HCl, pH 8.3; 500 mM KCl, 15 mM $MgCl_2$, 50 mM DTT), 40 nmol each dNTP and 300 pmol 5′-Cy3 or 5′-Cy5 labeled T7-Primer (5′GGT AAT ACG ACT CAC TAT AGG G3′; Metabion, Germany). The reaction was stopped by heating to 70°C for 10 min and subsequent degradation of cRNA-template with 20 U RNaseH for 60 min at 15°C. Labeled ssDNA was purified using a QiaQuick® PCR-clean up kit (QIAGEN GmbH, Germany). Finally, the concentration of the labeled ssDNA and the labeling efficiency was determined by UV spectrometry using a NanoDrop® UV/vis spectrophotometer (Nandrop/Thermo Fisher Inc.).

4.2.8 Oligonucleotide probe set

Based on published studies (e.g. Gliesche, 1999), set of 78 non-redundant 16S rRNA-targeted oligonucleotide probes, targeting phylogenetic groups that are expected to be present in groundwater, was designed. The targeted phylogenetic families from the domains *Bacteria* and *Archaea* are listed in table 4.1a and 4.1b as well as the probes and their sequences. The general probes ARC_333_C, ARC_344_C, ARC_1115_C, 27F_C and 517R_RC served as positive control and aditionally for normalization. Probe dT served as general background control since each probe contains a $p(dT)_{15}$ spacer module. The probes GAPDH and ß2M which target the human genes glyceraldehyde-3-phosphate dehydrogenase (GAPDH) and ß2-microglobulin respectively, served as general negative control (sequences not listed in table 4.1b). All probes were designed to have a mi0nimum GC content of 50%, a length of approximately 20 nucleotides (not including the p(dT)$_{15}$-spacer), a uniform melting temperature (Tm = 68°C) as predicted by the nearest neighbor model (Breslauer *et al.*, 1986). 16S rRNA-sequences from the phylogenetic groups of interest were retrieved from public databases (EMBL, RDP release 9). Oligonucleotide-probes were designed by using the multiple alignment tools of the EMBOSS package (Rice *et al.*, 2000) to generate consensus sequences for the phylogenetic groups of interest. Probe sequences were selected by using the eprimer3 tool of EMBOSS. The probe-check module of the ribosomal Database Project II (Cole *et al.*, 2005) and BLAST searches (Altschul *et al.*, 1990) were applied to evaluate probe specificity.

4.2.9 Microarray matrix, probe configuration, and spotting

3′-amino-modified DNA-oligonucleotide probes (Metabion, Martinsried, Germany) were spotted in 6 replicates onto epoxy-coated glass slides (Genetix Ltd, UK) using a SpotBot^{TM} microarraying device (ArrayIt Corporation, Sunnyvale, CA) equipped with ArrayIt SMP3 Stealth Pins (ArrayIt Corporation, Sunnyvale, CA) at 18°C and 40% relative humidity. The 3′-amino group of the oligonucleotide probes was linked via a C7-spacer to the 3′-end of the probes oligonucleotide-part. To avoid surface-mediated steric effects affecting the hybridization efficiency, the oligonucleotide itself was composed of a dT$_{15}$ spacer module and a 20 nt–30 nt 16S rRNA specific probe sequence in 3′–>5′ direction to ensure a distance of at least 5 nm between array surface and probe-sequence (see table 4.1). Concentration of oligonucleotide probes in Nexterion® spotting solution (SCHOTT Technical Glass Solutions GmbH, Germany) was 20 μM. Normal spot diameter was approximately 100 μm and 120 μm under the spotting conditions applied.

4.2.10 Pretreatment-, hybridization- and washing of DNA microarrays

Prior to hybridisation, spotted slides were blocked in 50 ml of Nexterion® blocking solution (SCHOTT Technical Glass Solutions GmbH, Germany) at room temperature (RT) for 1 h followed by an incubation in 25% ethanolamine for 30 min at RT. Blocked slides were dried by centrifugation at 6,000 × g in a microarray high speed centrifuge (ArrayIt Corporation, Sunnyvale CA, USA) for 10 sec without additional washing. DNA-microarrays were hybridised and washed based on a modified standard protocol (Schena *et al.*, 1996). Briefly, 60 μl of hybridization solution was made of 250–500 ng Cy3/Cy5-labeled ssDNA in hybridization buffer (25% formamide, 3 × SSPE, 5 × Denhardt's solution, 1% SDS and 5% dextrane sulfate). Prior to hybridisation, the solution was denatured at 95°C for 5 min, and cooled to 37°C in a BioRad iCycler. The hybridisation solution was applied to a pretreated and prewarmed microarray under a 22-by 40-mm LifterSlip (Erie Scientific, Portsmouth, NH). Hybridisation was performed in a HC4 hybridisation chamber (Quantifoil Instruments GmbH, Germany) for

14 h at 42°C. Hybridised microarray slides were washed in wash buffer A (2 × SSPE, 0.1% SDS) for 5 min at room temperature followed by a washing step in wash buffer B (2 × SSPE) for 5 min using a high throughput microarray wash station (ArrayIt Corporation, Sunnyvale CA, USA). Subsequently, the microarray slides were dipped into 0.2 × SSPE to remove traces of salt, dried by centrifugation at 6000 × g rpm for 3 min and stored in the dark until signal detection.

4.2.11 Signal detection and data analysis

All slides were scanned at a resolution of 5 nm using a DNAScope LM microarray scanner (Biomedical Photometrics, Waterloo, Ontario, Canada) at a general sensitivity setting of the photomultiplier at 60% and a laser power of 90%. The exact sensitivity settings depended on the overall signal intensity of the microarrays. The microarray data analysis software tool MACROView™ (Biomedical Photometrics, Waterloo, Ontario, Canada) was employed for raw data acquisition and the microarray analysis software suite AIDA® (Raytest GmbH, Germany) was utilised for spot detection, quantification of spot signals (mean pixel intensity) and background subtraction. Each spot signal was tested for statistical significance, and series of probe replicates were tested for outliers using the *t* test. Each data point represents the arithmetic mean of the normalised background-corrected mean pixel intensity of the positive replicates for the corresponding probe. To account for variation from array to array, hybridisation signals of probes targeting the sequences of the PCR-primes used for the amplification of *Bacteria*- and *Archaea*-16S rRNA-genes were utilized as standards for normalisation. The background-corrected intensity value of each hybridisation signal on a given array was divided by the background-corrected intensity value of the normalisation standards.

4.3 RESULTS AND DISCUSSION

4.3.1 Geochemical conditions and microorganism concentration

Due to natural attenuation, a drastic reduction of pollutant concentrations can be observed down gradient from investigated landfills (Fig. 4.2). Primary pollutants seeping from landfills are usually organic compounds with a relatively high molecular mass, which are represented as the sum parameter – total organic carbon (TOC). Elevated levels of NH_3, H_2CO_3 and CH_4 originating from metabolization of organic compounds constitute an indicator for bacterial degradation of contaminants. Since the groundwater in the investigation area is anaerobic, the major source of electron acceptors for anaerobic respiration is SO_4^{2-}, which presents a substantial prerequisite for anoxic carbon metabolism. The bacterial concentration in groundwater samples from the area of investigation is directly proportional to the degree of contamination. This provides an indication that the pollutants leaving the landfills do not inhibit microbial growth. The concentration of microorganisms in contaminated samples is about 5 times higher than in neutral samples and decreases significantly along the groundwater contamination plume. Therefore, elevated bacterial concentrations represent a high level of pollutant degradation.

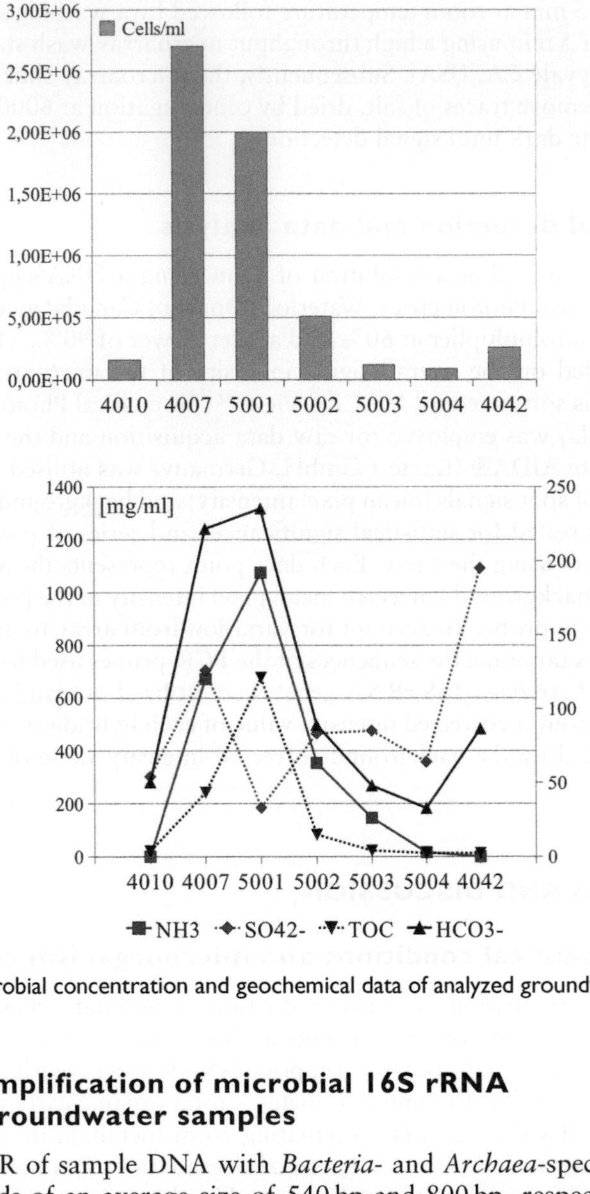

Figure 4.2 Microbial concentration and geochemical data of analyzed groundwater samples.

4.3.2 PCR-amplification of microbial 16S rRNA from groundwater samples

As expected, PCR of sample DNA with *Bacteria*- and *Archaea*-specific primers corresponds to bands of an average size of 540 bp and 800 bp, respectively (Fig. 4.3). PCR amplification yields of DNA from analysed groundwater samples were reciprocally proportional to microbial concentrations quantified by epifluorescent counting in the corresponding samples (Fig. 4.3). Highly polluted samples showed a significant decrease of PCR amplification yields. In all probability, this incongruity between bacterial concentration and PCR-yields is caused by insufficiencies in the DNA-extraction protocol. Highly polluted DNA-samples from observation boreholes 4007 and 5001 had a slightly brownish-yellow color, which is caused by humic substances. Humic substances are formed by microbial degradation of organic matter. They are resistant

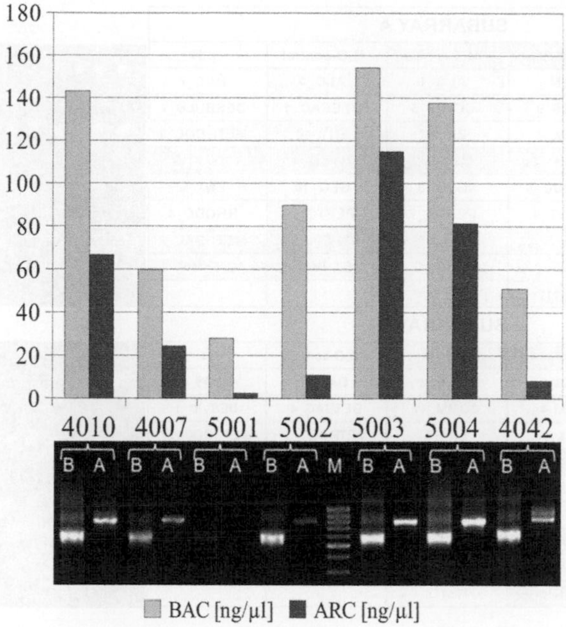

Figure 4.3 PCR-amplification of microbial 16S rRNA genes from groundwater samples.

to further biodegradation and can be divided into three main fractions: humic acids, fulvic acids, and humin. Due to their phenolic and carboxylic acid functional groups, humic substances have the ability to chelate positively charged multivalent ions like Mg^{2+}, Ca^{2+} and FeII, as well as other ions (Ghabbour and Davies, 2001). Since the Mg^{2+}-concentration is a crucial factor for PCR, the chelation of Mg^{2+}-ions by humic substances can inhibit the reaction completely (Tsai and Olson, 1992). The results of this study indicate, that the employed DNA-extraction protocol does not facilitate the complete removal of the ubiquitous humic substances from groundwater samples.

4.3.3 DNA-microarray layout, hybridisation and analysis

Amino modified DNA-probes for the detection of *Bacteria*- and *Archaea*-families, which have been detected in soil and groundwater were designed and covalently bound to epoxy-functionalised glass-slides (Fig. 4.4). Each probe was evaluated by using the probe-check module of the ribosomal Database Project II (Cole *et al.*, 2005). Only probes that yielded more than 95% specific hits in their taxonomic units, i.e. bacterial families, were selected for the microarray design. Hybridisation was carried out under high stringency conditions, i.e. 10°C below the calculated Tm of 52°C at 420 mM Na^+ and 25% formamide (Anderson, 1999). The DNA-microarray developed in this study, facilitated the simultaneous detection of *Bacteria* and *Archaea* and confirmed earlier results obtained with PCR-DGGE (Kerndorff *et al.*, 2008). Hybridisation signals of probe ARC-1115rv and 517r provided the normalisation factor for *Archaea*- and

SUBARRAY A

27f	ARC-333fw	ARC-344fw	H_2O-control	pdT
GAPDH	ß2M	ALC_1	ALC_4	BAC_7
CLOS_2	CLOS_9	COMM_5	DESBAC_1	DESBULB_7
DESVIB_5	DESX_2	ENT_7	ENT_10	METHCOC_2
METHPHIL_1	MICRO_7	MYCO_2	MYCO_4	MYCO_8
PSEUDO_2	PSEUDO_5	GEO_8	GEO_10	ENT_6
NEISS_2	PEPST_4	PEPTO_7	PEPTO_9	RHODO_4
RHODO_9	STREMY_1	STREMY_5	STREMY_8	METHBAC_2
METHBAC_4	METHBAC_6	METHBAC_8	METHMIC_2	METHMIC_4
METSARC_2	ARC-1115rv	517r		

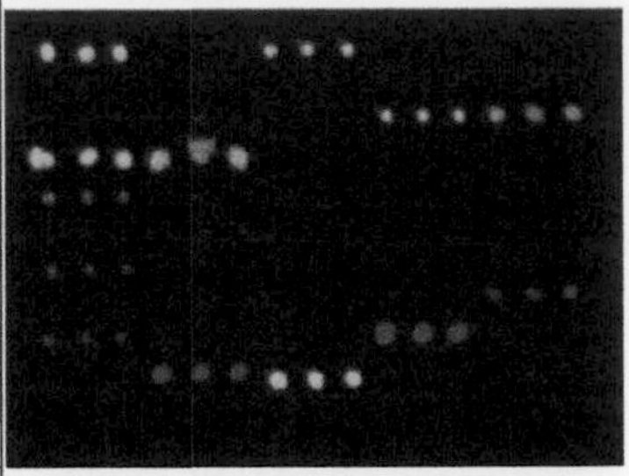

SUBARRAY B

27f	ARC-333fw	ARC-344fw	H_2O-control	pdT
GAPDH	ß2M	ALC_2	BAC_4	CLOS_1
CLOS_7	COMM_4	COMM_11	DESBAC_4	DESVIB_1
DESX_1	DESX_9	ENT_8	METHCOC_1	METHCOC_10
MICRO_2	MYCO_1	MYCO_3	MYCO_7	PSEUDO_1
PSEUDO_4	GEO_1	GEO_9	GEO_12	ENT_9
PEPST_1	PEPTO_2	PEPTO_8	RHODO_2	RHODO_7
RHODO_11	STREMY_3	STREMY_7	METHBAC_1	METHBAC_3
METHBAC_5	METHBAC_7	METHMIC_1	METHMIC_3	METSARC_1
METSARC_3	ARC-1115rv	517r		

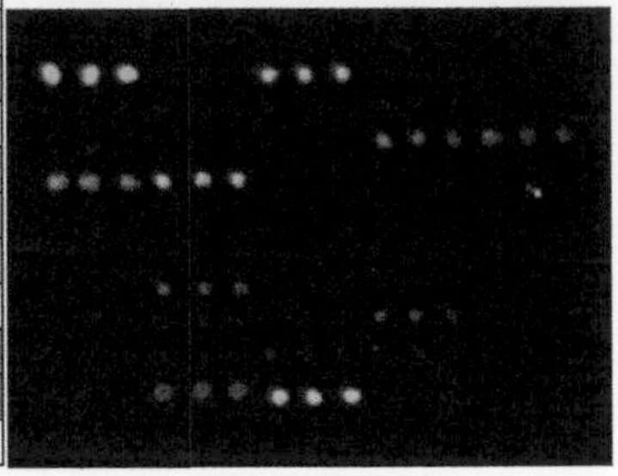

Figure 4.4 Array-layout and hybridisation example.

Bacteria specific hybridisation signals, respectively. Dividing each *Archaea*- and *Bacteria* specific signal by the arithmetic mean of its corresponding normalisation factor resulted in normalised data, representing the relative abundance of *Bacteria*- and *Archaea*-families in the analysed sample. The concentration of microorganisms determined by epifluorescent direct counting (DAPI) served as correction factor for the microarray data and facilitated a semi-quantitative assessment of the microbial communities in the groundwater samples.

The control probes ß2M, GAPDH and pdT did not generate any hybridisation signal above background level, thus proving the stringency of the selected hybridisation conditions. *Mycobacteria* are aerobic microorganisms such as *Bacillaceae*, *Comamonadaceae*, *Pseudomonadaceae*, *Neisseriaceae* and *Streptomycetaceae*. Probes designed for the detection of these *Bacteria*-families did not generate any hybridisation signal above background level. First, these results prove the taxonomic specifity of the selected probes and second, the anoxic conditions in the collected groundwater samples. DNA of sulfate-reducing bacteria from the taxonomic families *Desulfobacteraceae, Desulfobulbaceae, Desulfovibrionaceae* and other sulfate reducing bacteria could be detected in every analysed groundwater sample. The fraction of sulfate-reducing bacteria is significantly increased in contaminated groundwater samples and therefore provides a marker for contamination. In contrast to the ubiquitously prevailing sulfate-reducing bacteria, DNA of methane-producing Archeae from the families *Methanobacteriaceae*, *Methanomicrobiaceae* and *Methanosarcinaceae* can be detected in contaminated groundwater exclusively (Fig. 4.5).

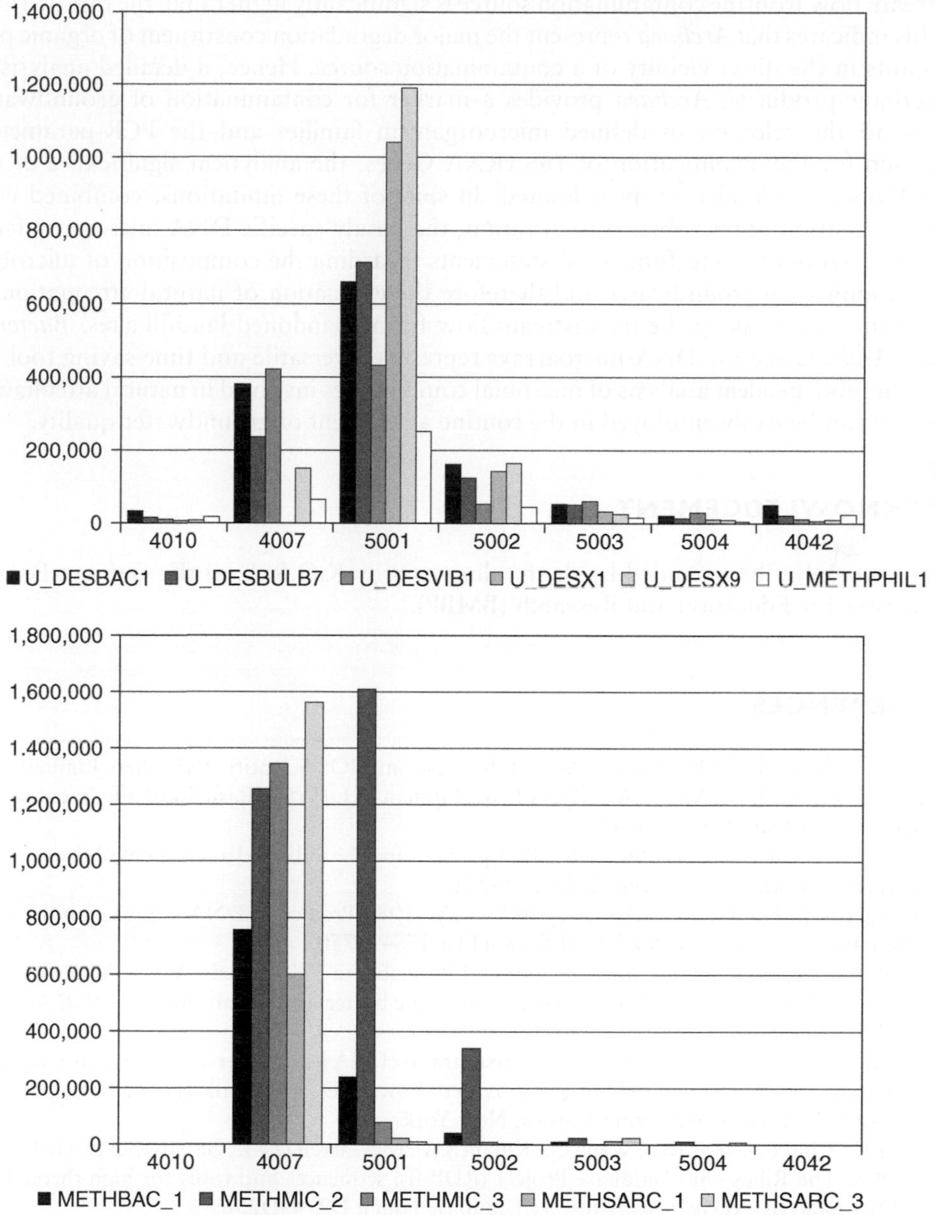

Figure 4.5 Abundance of detectable *Bacteria*- and *Archaea*-families.

4.4 CONCLUSIONS

The decrease of TOC concentration in the analyzed groundwater samples equates to the decrease in concentration of sulfate reducing bacteria down gradient from the contamination source and, therefore presents a suitable marker for bacterial degradation of organic pollutants. The decrease rate of methane-producing *Archaea* along the downstream flow from the contamination source is significantly higher than the rate of TOC. This indicates that *Archaea* represent the major degradation constituent of organic pollutants in the direct vicinity of a contamination source. Hence, a detailed analysis of methane-producing *Archaea* provides a marker for contamination of groundwater. Due to the selection of defined microorganism families and the PCR-parameters chosen for the amplification of 16S rRNA Genes, the analytical significance of the developed DNA-microarray is limited. In spite of these limitations, combined with determination of microbial concentration, the family-specific DNA-microarray facilitates quantitative and functional statements regarding the composition of microbial communities in groundwater and therefore the evaluation of natural attenuation of contaminations along the downstream flow from abandoned landfill sites. *Bacteria*- and *Archaea*-specific DNA-microarrays represent a versatile and time-saving tool for the site-independent analysis of microbial communities involved in natural attenuation which can be easily employed in the routine assessment of groundwater quality.

ACKNOWLEDGEMENT

This work has been funded by the funding priority K.O.R.A. of the German Federal Ministry for Education and Research (BMBF).

REFERENCES

Anderson M.L.M. (1999) Nucleic Acid Hybridization. BIOS Scientific Publishers Limited.

Altschul S.F., Gish W., Miller W., Myers E.W., Lipman D.J. (1990) Basic local alignment search tool. *J Mol Biol* (215): 403–410.

Baker G.C., Cowan D.A. (2004) 16 S rDNA primers and the unbiased assessment of thermophile diversity. *Biochem Soc Trans* 32(2): 218–221.

Breslauer K.J., Frank R., Blöcker H., Marky L.A. (1986)Predicting DNA duplex stability from the base sequence. *Proc Natl Acad Sci* 83(11): 3764–3750.

Brodie E.L., DeSantis T.Z., Moberg Parker J.P., Zubietta I.X., Piceno Y.M., Andersen G.L. (2007) Urban aerosols harbor diverse and dynamic bacterial populations. *Proc Natl Acad Sci* 104(1): 299–304.

Bittner M. (2003) Fluorescent labeling of first-strand cDNA using reverse transcriptase. In *DNA Microaarrays-A molecular cloning manual* (ed. Bowtell D, Sambrook J) Cold Spring Harbor Laboratory Press. Cold Spring Harbor, New York.

Cole J.R., Chai B., Farris R.J., Wang Q., Kulam S.A., McGarrell D.M., Garrity G.M., Tiedje J.M. (2005) The Ribosomal Database Project (RDP-II): sequences and tools for high-throughput rRNA analysis. *Nucleic Acids Res* 33(Database Issue): D294–D296.

Curtis T.P., Sloan W.T., Scannell J.W. (2002) Estimating prokaryotic diversity and its limits. *Proc Natl Acad Sci* 99 (16): 10494–10499.

Eberwine J., Yeh H., Miyashiro Kv, Cao Y., Nair S., Finnell R., Zettel M., Coleman P. (1992) Analysis of gene expression in single live neurons. *Proc Natl Acad Sci* 89: 3010–3014.

Ferguson A.S., Huang W.E., Lawson K.A., Doherty R., Gibert O., Dickson K.W., Whiteley A.S., Kulakov L.A., Thompson I.P., Kalin R.M., Larkin M.J. (2007) Microbial analysis of soil and groundwater from a gasworks site and comparison with a sequenced biological reactive barrier remediation process. *J Appl Microbiol* 102: 1227–1238.

Ghabbour E.A., Davies G. (Editors) (2001) *Humic Substances: Structures, Models and Functions*. Cambridge, U.K.: RSC publishing.

Gliesche C. (1999) Die Mikrobiologie des Grundwasserraumes und der Einfluß anthropogener Veränderungen auf die mikrobiellen Lebensgemeinschaften (Microbiology of the groundwater zone – the influence of athropogenic changes on microbial communities). UBA-Texte 13/99, Forschungsbericht 103 02 898, UBA-FB 98–118, Umweltbundeamt Berlin.

Hendrickx B., Dejonghe W., Boenne W., Brennerova M., Cernik M., Lederer T., Bucheli-Witschel M., Bastiaens L., Verstraete W., Top E.M., Diels L., Springael D. (2005) Dynamics of an oligotrophic bacterial aquifer community during contact with a groundwater plume contaminated with benzene, toluene, ethylbenzene and xylenes: an in situ mesocosm study. *Appl Environ Microbiol* 71(7): 3815–3825.

Kerndorff H., Kühn S., Minden T., Orlikowski D., Struppe, Thomas (2008) Effects of natural attenuation processes on groundwater contamination caused by abandoned waste sites in Berlin. *Environ Geol* 55: 291–301.

Kilb B. (1999) Analyse mikrobieller Lebensgemeinschaften in anaeroben Grundwassersystemen mittels molekulargenetischer Methoden (Analysis of microbial communities in anaerobic groundwater systems via molecular genetic methods). Veröffentlichungen des Instituts für Wasserforschung GmbH in Kooperation mit der Wasserwerke Westfalen GmbH, der Westfälische Wasser- und Umweltanalytik GmbH und der Dortmunder Energie- und Wasserversorgung GmbH, Heft 59.

Koizumi A., Kelly J.J., Nakagawa T., Urakawa H., El-Fantroussi S., Al-Muzaini S., Fukui M., Urushigawa Y., Stahl D.A. (2002) Parallel characterization of anaerobic toluene- and ethylbenzene-degrading microbial consortia by PCR-denaturing gradient gel electrophoresis, RNA-DNA-membrane hybridzation, and DNA-microarray technology. *Appl Environ Microbiol* 68(7): 3215–3225.

Loy A, Lehner A, Lee N, Adamczyk J, Meier H Ernst J, Schleifer KH, Wagner M (2002) Oligonucleotide microarray for 16S rRNA gene-based detection of all recognized lineages of sulfate-reducing prokaryotes in the environment. *Appl Environ Microbiol* 68(10): 5064–5081.

Loy A., Küsel K., Lehner A., Drake H.L., Wagner H.L. (2004) Microarray and functional analyses of sulfate-reducing prokaryotes in low-sulfate, acidic fens reveal coocurrence of recognized genera and novel lineages. (2004) *Appl Environ Microbiol* 70(12): 6998–7009.

Loy A., Schulz C., Lücker S., Schöpfer-Wendels A., Stoecker K., Baranyi C., Lehner A., Wagner M. (2005) 16S rRNA Gene-Based Oligonucleotide Microarray for Environmental Monitoring of the Betaproteobacterial Order "Rhodocyclales". *Appl Environ Microbiol* 71(3): 1373–1386.

Maniatis T., Fritsch E.F., Sambrook J. (1982) *Molecular cloning – A laboratory manual.* Cold Spring Harbor Laboratory Press. Cold Spring Harbor, New York

Murray A.E., Hollibaugh J.T., Orrego C. (1996) Phylogenetic composition of bacterioplankton from two california estuaries compared by denaturing gradient gel electrophoresis of 16S rDNA-fragments. *Appl Environ Microbiol* 62: 2676–26680.

Nozawa-Inoue M., Scow K.M., Rolston D.E. (2005) Reduction of perchlorate and nitrate by microbial communities in vadose soil. *Appl Environ Microbiol* 71(7): 3928–3934.

Peplies J., Glöckner F.O., Amann R. (2003) Optimization strategies for DNA microarray-based detection of bacteria with 16S rRNA-targeting oligonucleotide probes. *Appl Environ Microbiol* 69(3): 1379–1407.

Peplies J., Lau S.C.K., Pernthaler J., Amann R., Glöckner F.O. (2004) Application and validation of DNA microarrays for the 16S rRNA-based analysis of marine bacterioplankton. *Environ Microbiol* 6(6): 638–645.

Peplies J., Lachmund C., Glöckner F.O., Manz W. (2006) A DNA microarray platform based on direct detection of rRNA for characterization of freshwater sediment-related prokaryotic communities. *Appl Environ Microbiol* 72(7): 4829–4838.

Rice P., Longden I., Bleasby A. (2000) EMBOSS: The European Molecular Biology Open Software Suite. *Trends in Genet* 16(6): 276–277.

Röling W.F.M., van Breukelen B.M., Braster M., Bin L., van Verseveld H.W. (2001) Relationships between Microbial Community Structure and Hydrochemistry in a Landfill Leachate-Polluted Aquifer. *Appl Environ Microbiol* 67(10): 4619–4629.

Schena M., Shalon D., Heller R., Chai A., Brown P.O., Davis R.W. (1996) Parallel human genome analysis: Microarray-based expression monitoring of 1000 genes. *Proc Natl Acad Sci* 93: 10614–10619.

Schena M. (2003) *Microarray Analysis.* John Wiley & Sons, Wiley-Liss, USA

Schloss P.D., Handelsman J. (2004) Status of the microbial census. *Microbiol Mol Biol Rev* 68(4): 686–691.

Tas N., van Eekert M.H.A., Schraa G., Zhou J., de Vos W.M., Smidt H. (2009) Tracking functional guilds: "Dehalococcoides" spp. In european river basins contaminated with hexachlorobenzene. *Appl Environ Microbiol* 75(14): 4696–4704.

Weiss J.V., Cozzarelli I.M. (2008) Biodegradation in contaminated aquifers: Incorporating microbial/molecular methods. *Ground Water* 46(2): 305–322.

Wilson K.H., Blitchington R.B., Greene R.C. (1990) Amplification of bacterial 16S ribosomal DNA with polymerase chain reaction. *J Clin Microbiol* 28: 1942–1946.

Tsai Y.L. and Olson B.H. (1991): Rapid method for direct extraction of bacterial DNA from soil and sediments. *Appl Environ Microbiol* 57(4): 1070–1074.

Tsai Y.L., Olson B.H. (1992): Rapid method for separation of bacterial DNA from humic substances in sediments for polymerase chain reaction. *Appl Environ Microbiol* 58(7): 2292–2295.

Chapter 5

Methodology for determining impacts on the chemical status of groundwater bodies: An application in a Mediterranean river basin (Southern Spain)

Damián Sánchez García & Francisco Carrasco Cantos

Centre of Hydrogeology, Department of Geology, Faculty of Science, University of Málaga, Málaga, Spain

ABSTRACT

In this paper a practical interpretation of the criteria established in the European Water Framework Directive to evaluate the impacts on the groundwater chemical status is carried out by means of physicochemical parameters and related threshold values. As a result 67 parameters have been identified to evaluate the chemical status of groundwater, and the threshold values for each of them have been proposed. Groundwater bodies designated as protected areas having a special use have also been taken into account. These include groundwater bodies used for the abstraction of drinking water and groundwater bodies designated as nitrate vulnerable zones. The proposed procedure has been applied in a pilot Mediterranean river basin located in southern Spain. Results show that groundwater bodies in aquifers with intergranular porosity in general have important impacts on their chemical status, whereas those identified in carbonate aquifers have a good chemical status. The main pollutants are: nitrogen and phosphorus compounds (related to agricultural practices), hydrocarbons and metals.

5.1 INTRODUCTION

According to Article 5 of the Water Framework Directive (WFD – European Commission, 2000) all Member States are required to conduct a study determining the impact of human activity on the status of surface water and groundwater. This study consists of two parts: an analysis of the pressures affecting water bodies, and an assessment of the impact produced by these pressures on the status of water bodies. The combined results of these analyses (both pressures and impacts) permits the identification of water bodies at risk of not reaching the environmental objectives set out in Article 4 of the WFD (Barth & Fawell, 2001; Mostert, 2003).

In accordance with the deadlines defined in the WFD, the first analysis of pressures and impacts affecting groundwater bodies was to be completed in December 2004. This first assessment was in many cases carried out with insufficient time or without the necessary data despite the difficulty involved in carrying out the study for the first time which, in many respects, is innovative.

The analysis of pressures and impacts should be reviewed and updated in each European river basin district by December 2013, and thereafter every six years.

The aim of this work is to propose a methodology to evaluate the impact on the chemical status of groundwater bodies, which may be a procedure to be used in the review to be completed in 2013 in all European river basins. This is done by taking into account, in particular, the definition of a good chemical status of groundwaters, the environmental objectives that the directive provides for them, as well as physicochemical data collected through monitoring networks. Furthermore, an example is given that shows the application of this methodology by using groundwater bodies within the Guadalhorce River basin (Southern Spain).

The results of this work can be applied to the second risk assessment due to be carried out by 2013, in the framework of the second cycle of river basin management plans. These results have been derived taking into account recommendations set out in the guidance document No 26 on risk assessment and the use of conceptual models for groundwater (European Commission, 2010b).

Many assessments of impacts are carried out by considering only a few parameters or a single one. Fernández-Ruiz *et al.* (2005) use four variables to determine the impact on groundwater (nitrates, chlorides, sulphates and electrical conductivity). Kay *et al.* (2009) determine the impact derived from agricultural activities by using three parameters (total organic carbon, nutrients and pesticides). Giupponi & Vladimirova (2006), Glavan (2007) and Krause *et al.* (2008) assess impact by using only concentrations of nutrients. Kunkel *et al.* (2007), however, significantly extend the range of hydrochemical parameters considered (40) to evaluate the chemical status of groundwater bodies, whilst Carrasco *et al.* (2008) do so from the water chemistry, nitrogen and phosphorus compounds and various physicochemical parameters.

Andreadakis *et al.* (2007) use mathematical models to identify the impacts instead of data originating from monitoring networks. Comber *et al.* (2008) proposed a method that assesses the presence of an impact on the chemical status of water caused by metals. More recently, Quevauviller (2009) describes legally-binding aspects of the evaluation of the chemical status of groundwaters as well as efforts undertaken at the European level to harmonise analytical methods.

During the two-year period (2005–2006), the European project BRIDGE was carried out to develop a methodology for estimating contaminant threshold values in groundwater bodies. A summary of the methodology proposed can be consulted in Pauwels *et al.* (2007).

Since the publication of the Directive 2006/118/EC on the protection of groundwater (*Groundwater Directive*, European Commission, 2006b), new attention has been paid to terms related to groundwater chemical status, such as background levels (Wendland *et al.*, 2005), threshold values (Kmiecik *et al.*, 2006; Hinsby *et al.*, 2008; Marandi & Karro, 2008; Wendland *et al.*, 2008; Blum *et al.*, 2009; Preziosi *et al.*, 2010), quality standards (Müller, 2008) and pollution trends (Batlle Aguilar *et al.*, 2007).

The Spanish authorities have defined threshold values for only 20 substances (European Commission, 2010a). These are either substances that contribute to the characterisation of at least one groundwater body that is at risk, or substances with high natural background concentrations. In all cases the threshold values were established at the level of the groundwater body, which means that they are only applicable in the bodies of groundwater where they were defined.

5.2 ENVIRONMENTAL OBJECTIVES

5.2.1 Preliminary considerations

As previously stated, Article 5 of the WFD requires that an analysis of pressures and impacts of water bodies is completed. The ultimate goal of this analysis is to identify water bodies at risk of not meeting their environmental objectives. It is necessary to keep this frame of reference in mind when assessing impacts, especially because an impact occurs when a given body of water fails to meet its environmental objectives. This leads to the conclusion that as a preliminary step towards the identification of impacts on the chemical status of groundwater bodies, it is necessary to review the environmental objectives of groundwater bodies, which are defined in Article 4 of the WFD and are further addressed in the Annex V.

5.2.2 Groundwater bodies

The objective that the WFD establishes for groundwater bodies is to achieve a good status. A good groundwater status is achieved when both its chemical and quantitative status can be recognised as being *good*.

The parameters that should be used to evaluate the chemical status of groundwater bodies are electrical conductivity and the concentration of pollutants (Annex V of the WFD). In order to achieve good chemical status, the electrical conductivity must not indicate the existence of salinization or other types of intrusions, and concentrations of pollutants must be below the quality standards established in Directive 2006/118/EC. Furthermore, the chemical and ecological status of surface water bodies and ecosystems that directly depend on the groundwater body should not be impacted (Table 5.1).

5.2.3 Protected areas

Some groundwater bodies constitute protected areas in accordance with Article 6 of the WFD due to a certain use or special protection of their waters. This is the case of groundwater used for drinking water, or areas that have been declared as nitrate vulnerable zones.

Table 5.1 Definition of good groundwater chemical status according to the WFD (Annex V 2.3).

Objective	*Parameters*	*Criteria*
Good chemical status	Conductivity	Not indicative of saline or other intrusion
	Concentration of pollutants	Do not exhibit the effects of saline or other intrusions
		Do not exceed the quality standards applicable under Directive 2006/118/EC
		Do not result in failure to achieve the environmental objectives nor any significant diminution of quality of associated surface waters

The environmental objectives that the directive establishes for these areas are, in addition to those described in the previous section (good chemical and quantitative status), those specified in European Community legislation through which they were designated as protected areas.

5.3 PROPOSED PROCEDURE TO EVALUATE IMPACTS OF GROUNDWATER CHEMICAL STATUS

5.3.1 Background

Both the first impact assessment carried out in 2005, and the next revisions and updates of this assessment, represent initial and general characterisations whose objective is to identify all the chemical substances or physicochemical parameters that can cause a groundwater body to be at risk of failing its environmental objectives. Once the potential risks have been identified, an additional characterisation is to be carried out in order to establish a more precise assessment, as well as to propose solutions by means of Programmes of Measures. Therefore, the analysis of impacts on the groundwater chemical status should be as extensive as possible and, consequently, the list of pollutants and indicator parameters considered to assess impacts should be also as extended as possible.

According to the guidance document on risk assessment (European Commission, 2010b), the second impact analysis due in December 2013 must be built on the work undertaken during the first cycle, especially the clarification on objectives and operational requirements, monitoring data and the first status classifications.

5.3.2 Impact classes proposed

Presumably, the result of an impact assessment on a water body should be one of the following: a) impacts are found, b) no impacts reported or c) no data found. However, there may be situations where these three categories do not sufficiently describe the kind of impact affecting a water body, for example when the concentration of a pollutant is higher than normal values in the water body, although it lies within the permissible limits. In this case the water body meets the environmental quality standard for that parameter, although at the same time there is an impact given that the average concentration has increased significantly.

In order to be able to assess situations such as the example given, the definition of two types of impacts is proposed: an *important* impact and a *slight* impact. An *important* impact will be present when one of the parameters used to assess the chemical status does not meet the quality standard. The *slight* impact is reserved for those cases which do not exceed the quality standard, however the concentration indicates that the natural status of a water body has been altered due to human activity. If neither of these cases applies, the water body will be defined as *no impacts reported*, and where no data are available for evaluation, the classification *no data found* will be assigned.

5.3.3 Impact assessment on the chemical status of groundwater bodies

The requirements for a groundwater body to have a good chemical status can be summarised as:

1 No evidence exists of salinization or seawater intrusion.
2 The concentrations of contaminants do not exceed the quality standards set in the Directive 2006/118/EC (Groundwater Directive).
3 The chemical or ecological status of surface water bodies and terrestrial ecosystems that depend on these groundwater bodies do not deteriorate.

Table 5.2 shows the list of the 67 physicochemical parameters proposed in this work to identify the impacts on the chemical status of groundwater bodies, as well as the threshold values proposed to define the *slight* and *important* impacts.

5.3.3.1 *Salinization or seawater intrusion*

Four physico-chemical parameters are proposed to identify the existence of an impact made by salinization or seawater intrusion (parameters 1 to 4 in Table 5.2): electrical conductivity, chloride, sodium and sulphate concentrations.

The criteria proposed to identify an impact by salinization or seawater intrusion are based on the existence of increasing trends over time with respect to any of these parameters. Given the difficulty of establishing a single quantitative threshold to differentiate between the *important* and *slight* impacts, a distinction based on the following characteristics of increasing trends is proposed: (i) number of control points showing an upward trend, (ii) number of physico-chemical parameters showing an upward trend, (iii) clarity or evidence of trends.

5.3.3.2 *Quality standards established in Directive 2006/118/EC*

Two quality standards are established in Directive 2006/118/EC (Annex I) related to the concentration of nitrates (50 mg/l) and pesticides (0.1 μg/l for a single pesticide and 0.5 μg/l for the sum of pesticides).

The quality standards set out for these two parameters (5 and 6 in Table 5.2) are proposed to identify the existence of an *important* impact given that exceeding this limit would imply that a groundwater body fails to reach good chemical status. In the case of nitrates, that can be found naturally in water, it is proposed to define a *slight* impact when the concentration is between 20 and 50 mg-NO_3/l. With respect to pesticides, which are substances that do not come from natural sources, it was decided to consider its mere existence in water as the evidence of a *slight* impact.

5.3.3.3 *Surface water bodies and associated ecosystems*

One of the requirements of the WFD for a groundwater body to have a good chemical status is that the status of associated surface water bodies and ecosystems is not deteriorated by its action.

Table 5.2 Parameters and criteria for assessing the impacts on the chemical status of groundwater bodies. (All concentrations are expressed in μg/l unless otherwise indicated).

Parameter	Impact: Slight	Impact: Important		Parameter	Impact: Slight	Impact: Important	
1. Electrical conductivity	Upward temporary evolutions			(d) Benzo(g,h,i)-perylene	Presence	MA: $\Sigma > 0.002$	
2. Chloride				(e) Indeno(1,2,3-cd)-pyrene			
3. Sodium				35. Simazine	Presence	MA > 1	>4
4. Sulphate				36. Tributyltin compounds	Presence	MA > 0.0002	>0.0015
5. Nitrates	20–50 mg/l	>50 mg/l		37. Trichloro-benzene	Presence	MA > 0.4	
6. Pesticides	Presence	>0.1 (indiv.)	>0.5 (total)	38. Trichloro-methane	Presence	MA > 2.5	
7. Alachlor	Presence	MA > 0.3	>0.7	39. Trifluralin	Presence	MA > 0.03	
8. Anthracene	Presence	MA > 0.1	>0.4	40. (a) Total DDT	Presence	MA > 0.025	
9. Atrazine	Presence	MA > 0.6	>2.0	(b) P,p-DDT	Presence	MA > 0.01	
10. Benzene	Presence	MA > 10	>50	41. Aldrin	Presence	MA: $\Sigma > 0.01$	
11. Brominated diphenylether	Presence	MA > 0.0005		42. Dieldrin			
12. Cadmium: <40 mg/l $CaCO_3$	–	MA > 0.08	>0.45	43. Endrin			
40–50 mg/l $CaCO_3$	Presence	MA > 0.08	>0.45	44. Isodrin			
50–100 mg/l $CaCO_3$	>0.45	MA > 0.09	>0.60	45. Carbon tetrachloride	Presence	MA > 12	
100–200 mg/l $CaCO_3$	>0.60	MA > 0.15	>0.90	46. Tetrachloro-ethylene	Presence	MA > 10	
200 mg/l $CaCO_3$	>0.90	MA > 0.25	>1.50	47. Trichloro-ethylene	Presence	MA > 10	
13. C10–13 Chloroalkanes	Presence	MA > 0.4	>1.4	48. Chloro-benzene	Presence	MA > 20	
14. Chlorfenvinphos	Presence	MA > 0.1	>0.3	49. Dichloro-benzene	Presence	MA > 20	
15. Chlorpyrifos	Presence	MA > 0.03	>0.10	50. Ethyl-benzene	Presence	MA > 30	
16. 1,2-Dichloroethane	Presence	MA > 10		51. Metolachlor	Presence	MA > 1	
17. Dichloromethane	Presence	MA > 20		52. Terbuthylazine	Presence	MA > 1	
18. Di(2-ethylhexyl)-phthalate	Presence	MA > 1.3		53. Toluene	Presence	MA > 50	

19. Diuron	Presence	MA > 0.2	> 1.8	54. 1,1,1-Trichloro-ethane	Presence	MA > 100
20. Endosulfan	Presence	MA > 0.005	> 0.010	55. Xilene	Presence	MA > 30
21. Fluoranthene	Presence	MA > 0.1	> 1.0	56. Cyanides	Presence	MA > 40
22. Hexachloro-benzene	Presence	MA > 0.01	> 0.05	57. Fluoride	> 1.5 mg/l	MA > 1.7 mg/l
23. Hexachloro-butadiene	Presence	MA > 0.1	> 0.6	58. Arsenic	> 10	MA > 50
24. Hexachloro-cyclohexane	Presence	MA > 0.02	> 0.04	59. Copper: 10 mg/l $CaCO_3$	> 2.5	MA > 5
25. Isoproturon	Presence	MA > 0.3	> 1.0	10–50 mg/l $CaCO_3$	> 11	MA > 22
26. Lead and its compounds	> 10	MA > 7.2		50–100 mg/l $CaCO_3$	> 20	MA > 40
27. Mercury	Presence	MA > 0.05	> 0.07	> 100 mg/l $CaCO_3$	> 60	MA > 120
28. Naphthalene	Presence	MA > 2.4		60. Total chromium	> 50	MA > 50
29. Nickel and its compounds	> 10	MA > 20		61. ChromiumVI	> 1	MA > 5
30. Nonylphenol	Presence	MA > 0.3	> 2.0	62. Selenium	> 1	MA > 1
31. Octylphenol	Presence	MA > 0.1		63. Total zinc: 10 mg/l $CaCO_3$	> 6	MA > 30
32. Pentachloro-benzene	Presence	MA > 0.007		10–50 mg/l $CaCO_3$	> 40	MA > 200
33. Pentachloro-phenol	Presence	MA > 0.4	> 1.0	50–100 mg/l $CaCO_3$	> 60	MA > 300
34. Polyaromatic hydrocarbons:				> 100 mg/l $CaCO_3$	> 100	MA > 500
(a) Benzo(a)pyrene	Presence	MA > 0.05	> 0.10	64. Total phosphorus	12–50	> 50
(b) Benzo(b)fluor-anthene	Presence	MA: $\Sigma > 0.03$		65. Biological oxygen demand	> 2.5 mg/l	> 4.0 mg/l
(c) Benzo(k)fluor-anthene				66. Ammonium	Presence	> 0.5 mg/l
				67. Phosphate	Presence	> 0.5 mg/l

"MA": mean annual concentration; the other values are expressed as maximumallowable concentrations.

Groundwater bodies are aquifers that in many cases discharge through one or more springs or as baseflow to rivers. The springs feed rivers, streams and lakes which, in turn, constitute surface water bodies. There are stretches of surface water bodies (especially at the headwaters of rivers and streams) where springs greatly contributes to the volume of water. Consequently, a deterioration in the status of this groundwater would result in a deterioration of the quality of surface water bodies and associated ecosystems (Castro & Hornberger, 1991; Winter, 1999; Woessner, 2000). For this reason, it was considered necessary to include in the list of parameters used to assess the chemical status of groundwater bodies, those parameters which are necessary to evaluate the chemical status of surface waters. These substances are numbered from 7 to 65 in Table 5.2.

The parameters 7 to 47 in Table 5.2 were obtained from the Directive 2008/105/EC (European Commission, 2008), which defines maximum permissible concentrations allowed in surface water for 33 priority substances (those that pose a significant risk to, or via, the aquatic environment) in addition to 8 other pollutants. In this proposal, exceeding these concentrations –which are expressed as average annual values and as maximum allowable concentrations– is considered as evidence of an *important* impact since they will prevent an associated surface water body to reach a good chemical status.

In regards to *slight* impacts, the 41 substances have been grouped into two types: those that are not naturally found in water (all of them except for cadmium, lead and nickel), for which their mere presence in water has been considered indicative of a *slight* impact, and those that can originate from both natural sources as well as polluting activities. In regards to cadmium, lead and nickel (numbers 12, 26 and 29 respectively in Table 5.2), a *slight* impact is present when concentrations surpass 10 μg/l for lead and nickel, and 0 to 0.9 μg/l (depending on water hardness) for cadmium. In the case of lead, the value corresponds to the maximum concentration recommended by the World Health Organization (2008), whilst values of nickel and cadmium were established within the framework of this work as no previous information was found.

Substances from 48 to 63 in Table 5.2 were obtained from the Royal Decree 995/2000 (Official [Spanish] State Gazette, No. 147, 20.6.2000), which, complying with the request of Directive 2006/11/EC (European Commission, 2006a), defined the quality objectives for certain pollutants in surface water. The maximum concentrations established in the Royal Decree have been interpreted as indicating the existence of an *important* impact. With regards the thresholds that identify the existence of a *slight* impact, its estimate is calculated using the following three criteria:

For chemical substances that do not have a natural origin, their presence in the water reflects the existence of a *slight* impact (cells with the term "Presence" in Table 5.2).

In the case of fluoride, arsenic and total chromium (numbers 57, 58 and 60), the thresholds considered were 1.5 mg/l, 10 μg/l and 50 μg/l respectively, which are the maximum concentrations recommended by the World Health Organization (2008).

Concentrations assigned to the parameters copper, chromium VI, selenium and zinc have been established within the framework of this work.

Finally, the substances 64 and 65 (total phosphorus and biological oxygen demand) were obtained from Annex VIII of the WFD and thresholds considered were taken from the national river ecosystem classifications of Finland and England, respectively. These two classification systems are presented as impact assessment tools in the guidance document on the analysis of pressures and impacts elaborated by the European Commission (2003).

5.3.3.4 *Other pollutants*

Ammonium and phosphate (parameters 66 and 67 in Table 5.2) are not included in the definition of good chemical status for groundwaters, however they can be found in other parts of the WFD such as Annex V 2.4.2 (parameters that must be monitored) and Annex VIII (list of the main pollutants).

Their presence in groundwater has been interpreted as indicative of a *slight* impact since they rarely have a natural origin, whereas concentrations higher than 0.5 mg/l are indicative of an *important* impact (value established within the framework of this work).

5.3.4 Impact assessment in protected areas

In Annex IV of the WFD the types of protected areas to be considered are specified:

1 Areas designated for the abstraction of water intended for human consumption.
2 Areas designated for the protection of economically significant aquatic species.
3 Bodies of water designated as recreational waters such as bathing waters.
4 Nutrient-sensitive areas: (a) areas designated as vulnerable zones under Directive 91/676/EEC and (b) areas designated as sensitive areas under Directive 91/271/EEC.
5 Areas designated for the protection of habitats or species where the water is an important factor in their protection.

Protected areas related to economically significant aquatic species (2), recreational waters (3) and sensitive areas (4b) are addressed in this study because they only concern surface waters. Besides, in the study area there are no protected habitats or species dependent on groundwaters (5).

5.3.4.1 *Water intended for human consumption*

Water intended for human consumption is regulated by the Directive 98/83/EC (European Commission, 1998). Annex I of this Directive establishes the maximum allowable concentrations of a variety of microbiological, chemical and indicator parameters that water must meet in order to be considered as suitable for human consumption. Therefore, the 48 parameters found in this annex are taken into account in assessing the impacts in these protected areas (Table 5.3).

The maximum permissible concentrations set out in the directive are considered as indicative of an *important* impact, except for colour, odour, taste and turbidity (parameters 33, 38, 42 and 46 respectively in Table 5.3). Directive 98/83/EC provides a qualitative threshold for these four parameters ("acceptable for the consumers and without anomalous changes") which makes assessment difficult. For this reason, the use of numerical thresholds was preferred, obtained from the Royal Decree 140/2003 (Official [Spanish] State Gazette, No. 45, 21.2.2003) which incorporated the Directive 98/83/EC into the Spanish legal system.

Table 5.3 Parameters and criteria for assessing impacts in water bodies designated as protected for supplying water for human consumption. (All concentrations are expressed in μg/l unless otherwise indicated).

Parameter	*Impact*		*Parameter*	*Impact*	
	Slight	*Important*		*Slight*	*Important*
Microbiological parameters			25. Selenium	>5	>10
1. Escherichia coli	–	>0 CFU/100 ml	26. Tetrachloroethene	Presence	>10
2. Enterococci	–	>0 CFU/100 ml	27. Trihalomethanes – Total	Presence	>100
Chemical parameters			28. Vinyl chloride	Presence	>0.5
3. Acrylamide	Presence	>0.1	*Indicator parameters*		
4. Antimony	Presence	>5.0	29. Aluminium	>100	>200
5. Arsenic	>5	>10	30. Ammonium	>0.25 mg/l	>0.5 mg/l
6. Benzene	Presence	>1.0	31. Chloride	>125 mg/l	>250 mg/l
7. Benzo(a)pyrene	Presence	>0.01	32. Clostridium perfringen	–	>0 CFU/100 ml
8. Boron	>0.5 mg/l	>1 mg/l	33. Colour	>7.5 mg/l Pt	>15 mg/l Pt
9. Bromate	Presence	>10	34. Conductivity	>1250 μS/cm	>2500 μS/cm
10. Cadmium	>2.5	>5.0	35. pH	<7 or >9	6.5 or 9.5
11. Chromium	>25	>50	36. Iron	>100	>200
12. Copper	>1 mg/l	>2 m g/l	37. Manganese	>25	>50
13. Cyanide	Presence	>50	38. Odour	Dilution index >2 at 25°C	Dilution index >3 at 25°C
14. 1,2-dichloroethane	Presence	>3	39. Oxidisability	>2.5 mg/l O_2	>5.0 mg/l O_2
15. Epichlorohydrin	Presence	>0.1	40. Sulphate	>125 mg/l	>250 mg/l
16. Fluoride	>1 mg/l	>1.5 mg/l	41. Sodium	>100 mg/l	>200 mg/l
17. Lead	>5	>10	42. Taste	Dilution index >2 at 25°C	Dilution index >3 at 25°C
18. Mercury	Presence	>1	43. Colony count 22°	–	No abnormal change
19. Nickel	>10	>20	44. Coliformbacteria	–	>0 CFU/100 ml
20. Nitrate	20–50 mg/l	>50 mg/l	45. Total organic carbon	–	No abnormal change
21. Nitrite	>0.25 mg/l	>0.5 mg/l	46. Turbidity	>2.5 NTU	>5.0 NTU
22. Pesticides	Presence	>0.1	47. Tritium	>50 Bq/l	>100 Bq/l
23. Pesticides – Total	Presence	>0.5	48. Total indicative dose	>0.05 mSv/year	>0.10 mSv/year
24. Polycyclic aromatic hydrocarbons	Presence	>0.1			

Values are expressed as maximum allowable concentrations; "CFU": colony-forming units; "NTU": nefelometric turbidity unit; "Bq": becquerel; "Sv": sievert.

With respect to the *slight* impact, it was considered that it does not apply to parameters where their presence in water indicates an *important* impact (parameters 1, 2, 32 and 44 in Table 5.3) or those that are evaluated in terms of the existence of "anomalous changes" (parameters 43 and 45). In the case of not naturally occurring substances, once again their presence in water indicates the existence of a *slight* impact. Finally, for the rest of substances a less demanding threshold was selected, usually corresponding to 50% of the threshold associated with *important* impacts.

5.3.4.2 *Nitrate Vulnerable Zones*

According to Directive 91/676/EEC (European Commission, 1991), Nitrate Vulnerable Zones are areas where runoff flows into surface waters that have, or may have, a nitrate concentration higher than 50 mg-NO_3/l, or into groundwaters that contain or could contain more than 50 mg-NO_3/l of nitrate. Consequently, the only parameter to be considered in this type of protected areas is the nitrates concentration, and exceeding the concentration limit will indicate an *important* impact. The existence of a *slight* impact is established at a concentration between 20 and 50 mg-NO_3/l of nitrates.

5.4 RESULTS OF THE APPLICATION OF THE METHODOLOGY IN A CASE STUDY AREA

5.4.1 Description of the case study area and data used

The pilot area where the proposed methodology to identify impacts has been applied is the Guadalhorce River basin. It is located in the Málaga province (southern Spain) (Fig. 5.1). Covering an area of almost 3200 km^2, it is one of the main Mediterranean basins in southern Spain.

The climate of the region is characterised by rainy winters and dry and hot summers. The mean annual values of precipitation range from less than 400 mm to 1100 mm in mountainous areas.

Rocks that outcrop in the Guadalhorce River basin (Fig. 5.1) belong to the Betic Cordillera. The southern sector comprises rocks of the Internal Zone of the Cordillera, divided into the Alpujarride, Malaguide and Dorsal complexes. The Flysch unit lies between the Internal and External Zones of the Cordillera, and within the External Zone, the Subbetic domain and the Triassic rocks are the most representative elements.

Groundwater in the study area occurs in carbonate, arenaceous and evaporitic rocks. The groundwater bodies were defined in a previous study by Sánchez *et al.* (2009), in which groundwater bodies were delineated in the Guadalhorce River basin, 24 groundwater bodies were delineated (Fig. 5.2). The nature of the materials where these water bodies are located is variable: 14 carbonate, 9 detrital and 1 evaporitic materials.

Identifying impacts on the chemical status of groundwater bodies in the pilot river basin was carried out from the analysis of physico-chemical data from three control networks: a self-monitored network established as part of this work, and the monitoring networks of the Spanish Geological Survey and the Andalusian Water Agency. In total 301 points were controlled, from which 1541 samples were taken within the period from 1974 to 2007 (Table 5.4).

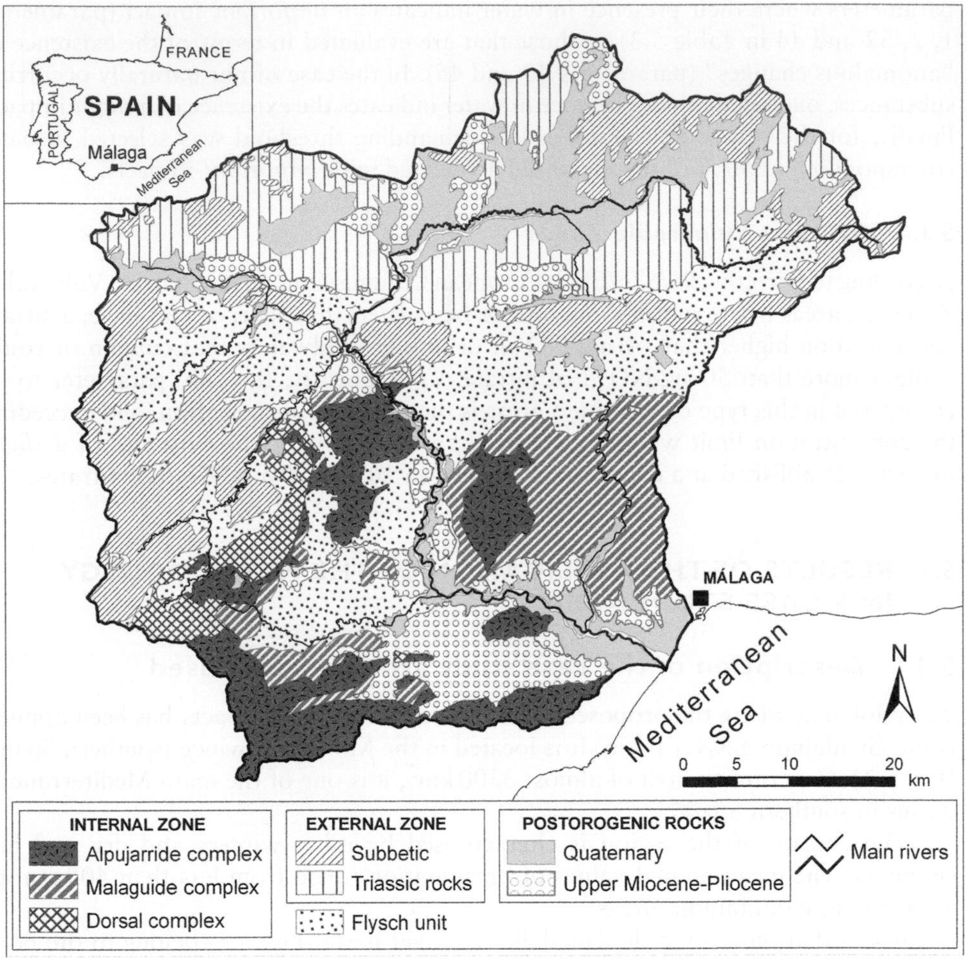

Figure 5.1 Location and geologic map of the Guadalhorce River basin.

A total of 105 physico-chemical parameters were analysed in the three control networks including electrical conductivity, temperature, pH, dissolved oxygen, redox potential, total organic carbon, major ions, nitrogen and phosphorus compounds, metals, hydrocarbons, organophosphate and organochlorine pesticides, volatile organic compounds, trihalomethanes and triazines.

5.4.2 Impacts on groundwater chemical status

To assess the impact on the chemical status of groundwater bodies, data from the monitoring networks and the list of physico-chemical parameters and threshold values given in Table 5.2 were used.

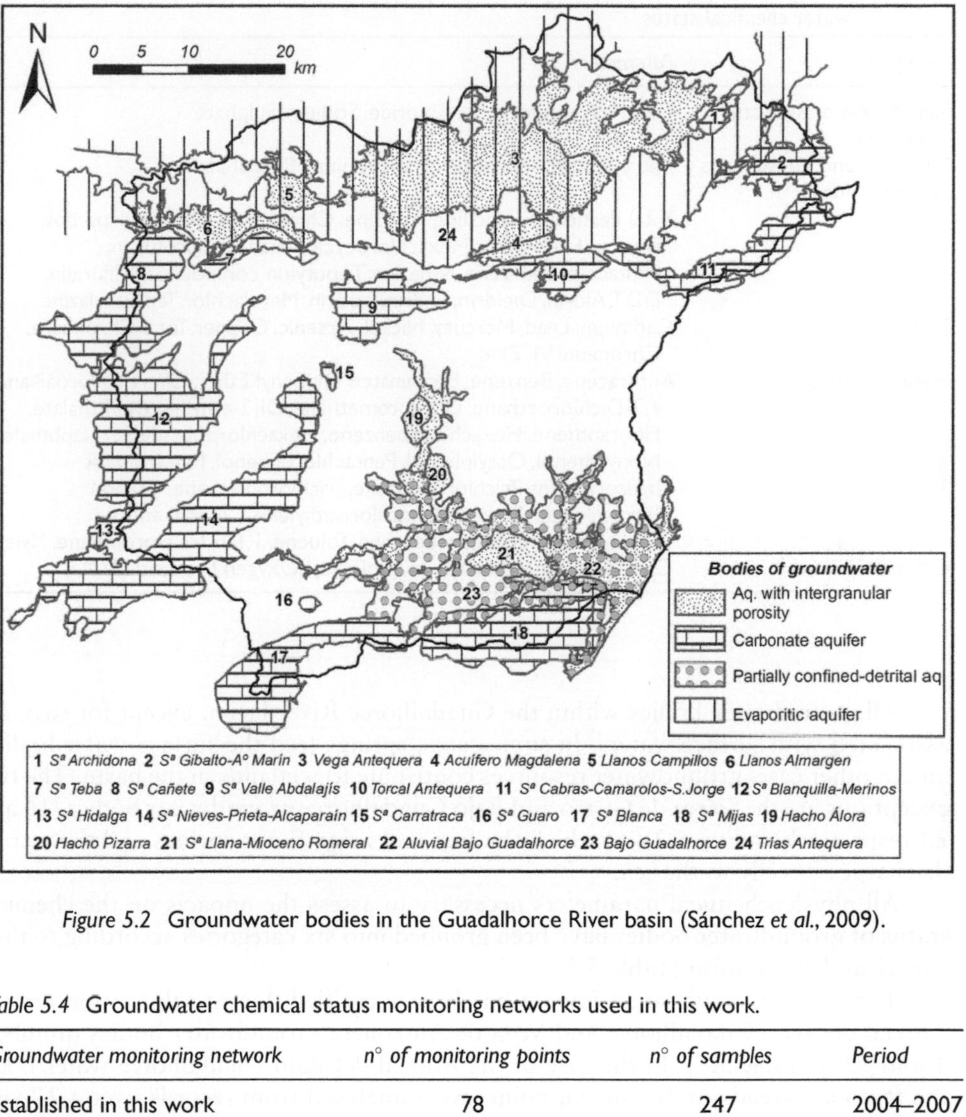

Figure 5.2 Groundwater bodies in the Guadalhorce River basin (Sánchez *et al.*, 2009).

Table 5.4 Groundwater chemical status monitoring networks used in this work.

Groundwater monitoring network	*n° of monitoring points*	*n° of samples*	*Period*
Established in this work	78	247	2004–2007
Spanish Geological Survey	202	1215	1974–2003
Andalusian Water Agency	21	79	2002–2004
Total	301	1541	1974–2007

Before commencing the assessment, it is necessary to identify groundwater bodies that feed a surface water body or an ecosystem. In this case, it would be necessary to analyse also the physico-chemical parameters which assess the chemical status of surface water bodies (parameters 7 to 65 in Table 5.2). Groundwater bodies that are not associated with any surface water body or ecosystem do not need to be analysed for these parameters.

Table 5.5 Categories of physicochemical parameters proposed to assess the impacts on the groundwater chemical status.

Category	*Parameters*
Salinization or seawater intrusion	Electrical conductivity, Chloride, Sodium, Sulphate
Nitrogen and phosphorus compounds	Nitrate, Total Phosphorus, Ammonium, Phosphate
Pesticides	Total Pesticides, Alachlor, Atrazine, Chlorfenvinphos, Chlorpyrifos, Diuron, Endosulfan, Hexachlorocyclohexane, Isoproturon, Pentachlorobenzene, Simazine, Tributyltin compounds, Trifluralin, DDT, Aldrin, Dieldrin, Endrin, Isodrin, Metolachlor, Terbuthylazine
Metals	Cadmium, Lead, Mercury, Nickel, Arsenic, Copper, Total Chromium, Chromium VI, Zinc
Hydrocarbons	Anthracene, Benzene, Brominated Diphenyl Ether, C_{10-13} Chloroalkanes, 1,2-Dichloroethane, Dichloromethane, Di(2-ethylhexyl) Phthalate, Fluoranthene, Hexachlorobenzene, Hexachlorobutadiene, Naphthalene, Nonylphenol, Octylphenol, Pentachlorophenol, Polyaromatic hydrocarbons, Trichlorobenzene, Trichloromethane, Carbon Tetrachloride, Tri- and Tetrachloroethylene, Chloro- and Dichlorobenzene, Ethylbenzene, Toluene, 1,1,1-Trichloroethane, Xylene
Others	Cyanide, Fluoride, Selenium, Biological Oxygen Demand (BOD)

All groundwater bodies within the Guadalhorce River basin, except for two, are associated with surface waters. In some cases, springs feed the surface water bodies, and in other cases groundwater resources contribute to wetlands in the basin. The two exceptions are the Sierra de Guaro and Bajo Guadalhorce groundwater bodies (16 and 23 respectively in Fig. 5.2), of which the first lacks significant springs and the second discharges directly to the sea.

All physicochemical parameters necessary to assess the impacts on the chemical status of groundwater bodies have been grouped into six categories according to their nature and application (Table 5.5).

The only cases where an impact has been identified due to salinization are the Aluvial del Bajo Guadalhorce and Vega de Antequera groundwater bodies (numbers 3 and 22 in Table 5.6). In the case of the Aluvial del Bajo Guadalhorce water body, the temporal trends of 16 control points were analysed from records over a 27 year period. In six control points, an increasing trend over time for electrical conductivity, chloride and sodium was identified (Fig. 5.3a, b and c). In view of these results, this groundwater body was classified as an *important* impact (Table 5.6).

In the Vega de Antequera water body, 13 control network points have been analysed with a 24 year data series, out of these only one point shows an increasing temporal trend of chloride and sodium concentrations (Fig. 5.3d). Given the localised nature of the salinization, this water body was assigned a *slight* impact (Table 5.6).

Data analysis on the chemical status related to nitrogen and phosphorus compounds has led to the conclusion that an *important* impact exists in four groundwater bodies (all of them of a detrital nature), mainly due to concentrations of nitrates but also because of ammonium and phosphate contents (Table 5.6). Moreover, in four other

Table 5.6 The impact on the chemical status of groundwater bodies in the Guadalhorce River basin. (**II**: important impact; SI: slight impact; NI: no impacts reported; empty cell: no data found; "–": not applicable).

	Chemical status						*Drinking water*						
Body of groundwater	Salinization	N-P compounds	Pesticides	Metalsv	Hydrocarbons	Other	Salinization	Bacteria	N and pesticides	Metals	Hydrocarbons	Other	*Vulnerable zones*
1. Sierra de Archidona		SI	SI	NI	**II**	NI	NI		SI	NI	**II**	NI	–
2. Sierra de Gibalto-Arroyo Marín	NI	SI			NI	NI	NI		SI			NI	–
3. Vega de Antequera	SI	**II**	**II**	**II**	**II**	**II**	–	–	–	–	–	–	**II**
4. Acuífero de la Magdalena		NI				NI	–	–	–	–	–	–	–
5. Llanos de Campillos		**II**				NI	–	–	–	–	–	–	–
6. Llanos de Almargen		**II**				NI	–	–	–	–	–	–	–
7. Sierra de Teba	NI	SI	NI	NI	**II**	NI	SI		SI	NI	**II**	NI	–
8. Sierra de Cañete	NI	SI			NI	NI	NI		SI			NI	–
9. Sierra del Valle de Abdalajís	NI	NI		**II**	NI	NI	SI		NI	SI		NI	–
10. Torcal de Antequera	NI	NI		NI		NI	NI		NI	NI		NI	–
11. Sierras Cabras-Camarolos-San Jorge	NI	NI		NI	NI	NI	NI		NI	NI		NI	–
12. Sierras Blanquilla-Merinos	NI	NI		NI		NI	NI		NI	NI		NI	–
13. Sierra Hidalga	NI	NI				NI	–	–	–	–	–	–	–
14. Sª Nieves-Prieta-Alcaparaín	NI	NI		NI	NI	NI	NI		NI	NI		NI	–
15. Serrezuela de Carratraca		NI			NI	NI	–	–	–	–	–	–	–
16. Sierra de Guaro		NI		–	–	–	NI		NI			NI	–
17. Sierra Blanca	NI	NI		NI		NI	NI		NI	NI		NI	–
18. Sierra de Mijas	NI	NI			NI	NI	NI		NI			NI	–
19. Hacho de Álora		NI				NI	–	–	–	–	–	–	–
20. Hacho de Pizarra		NI				NI	NI		NI			NI	–
21. Sierra Llana-Mioceno de El Romeral		NI				NI	NI		NI			NI	–
22. Aluvial del Bajo Guadalhorce	**II**	**II**	**II**	**II**	**II**	**II**	**II**		**II**	**II**	**II**	**II**	**II**
23. Bajo Guadalhorce		NI		–	–	–	NI		NI	NI		NI	NI
24. Trías de Antequera	NI	NI		NI		NI	–	–	–	–	–	–	–

groundwater bodies a *slight* impact has been identified due to nitrates concentrations (Table 5.6).

Only four groundwater bodies have pesticide data (numbers 1, 3, 7 and 22 in Fig. 5.2 and Table 5.6). Out of these, two groundwater bodies were identified with an *important* impact and one groundwater body with a *slight* impact, due to water concentrations exceeding the threshold value with respect to total pesticides, atrazine, endosulfan and pentachloro-benzene.

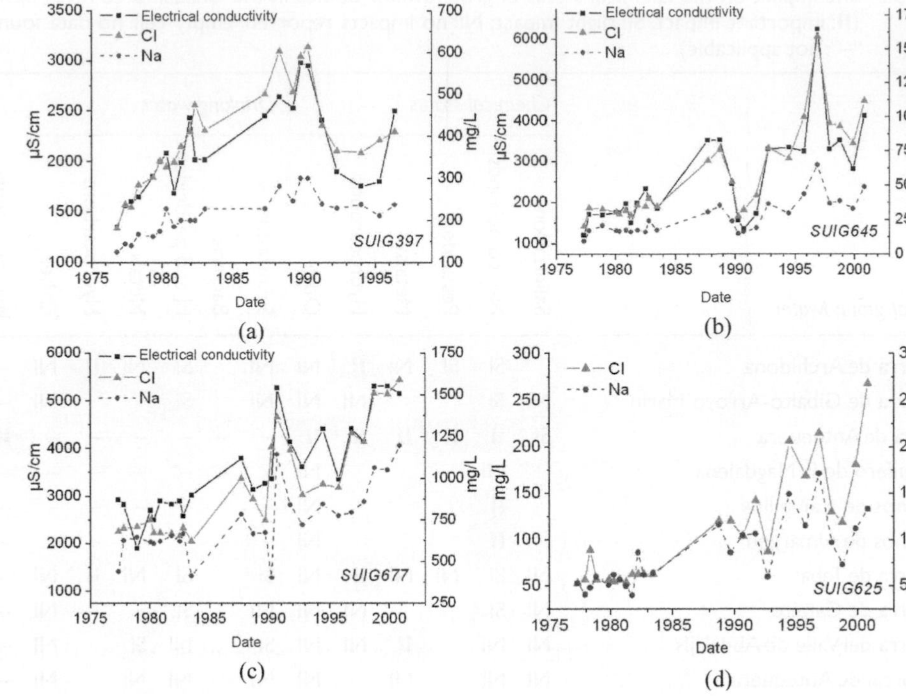

Figure 5.3 Temporal evolution of electrical conductivity and concentrations of sodium and chloride in three control points of the Aluvial del Bajo Guadalhorce water body (a, b and c) and one control point in the Vega de Antequera water body (d).

Three groundwater bodies were identified with *important* impacts associated with metal concentrations (Table 5.6), in particular lead, copper and zinc.

Out of the 26 hydrocarbons evaluated, 16 of them have been identified at concentrations above the threshold values, which has led to the definition of an *important* impact in the case of four groundwater bodies (Table 5.6). In the Aluvial del Bajo Guadalhorce groundwater body (number 22), monitoring points that detected hydrocarbons are located near industrial areas and the Malaga airport. The main hydrocarbons responsible of these impacts are dichloromethane, hexachlorobutadiene, trichloro-benzene, trichloro-methane, toluene and xilene.

With regards to the 'Other components' category (Table 5.5), two groundwater bodies were defined as bearing an *important* impact due to the concentrations of selenium (Table 5.6).

5.4.3 Impacts in protected areas

5.4.3.1 *Groundwater used for drinking water*

Out of the 24 groundwater bodies in the Guadalhorce River basin, 16 are used for water intended for human consumption (numbers 1, 2, 7 to 12, 14, 16 to 18 and 20 to

Table 5.7 Categories of physicochemical parameters proposed to assess the impacts on drinking water protected areas.

Category	*Parameters*
Salinization	Electrical conductivity, Chloride, Sodium, Sulphate
Bacteria	Escherichia coli, Enterococci, Clostridium perfringens, Colony count 22°, Coliform bacteria
Nitrogen compounds and pesticides	Nitrate, Nitrite, Ammonium, Pesticides, Total Pesticides
Metals	Antimony, Arsenic, Boron, Cadmium, Chromium, Copper, Lead, Mercury, Nickel, Aluminium, Iron, Manganese
Hydrocarbons	Acrylamide, Benzene, Benzo(a)pyrene, 1,2-Dichloroethane, Epichlorohydrin, Polycyclic aromatic hydrocarbons, Tetrachloroethylene, Total Trihalomethanes, Vinyl Chloride
Others	Bromate, Cyanide, Fluoride, Selenium, Colour, pH, Odour, Oxidisability, Taste, Total Organic Carbon, Turbidity, Tritium, Total indicative dose

23 in Fig. 5.2). The impacts on the status of these water bodies were evaluated based on the parameters and criteria set out in Table 5.3, which have been grouped into six different categories (Table 5.7).

An *important* impact associated to salinization has been identified in the Aluvial del Bajo Guadalhorce groundwater body (*Drinking water* columns in Table 5.6; body of groundwater number 22), because the concentrations of chloride, sulphate and sodium as well as the values of conductivity exceed the established thresholds.

In the Sierra de Teba and Sierra del Valle de Abdalajís groundwater bodies (7 and 9 respectively in Table 5.6) a *slight* impact has been identified also due to salinization processes. In the Sierra de Teba water body this impact may be due to ingress of water from Venta River into the aquifer in the area known as Tajo del Molino, where the river runs directly above permeable rocks (Carrasco *et al.*, 2007) (Fig. 5.4). Water from Venta River in this stretch has a conductivity higher than 2000 μS/cm and chloride, sulphate and sodium contents between 250 and 500 mg/l. With respect to the Sierra del Valle de Abdalajís groundwater body, a monitoring point was detected with chloride, sulphate and electrical conductivity values above the threshold. This salinization seems to be related to water infiltrating from Guadalhorce reservoir or the Villaverde lake, which contain water with a high degree of mineralisation (Fig. 5.5).

In the absence of bacterial data, it was not possible to assess the impacts arising from the presence of bacteria in groundwater bodies (Table 5.6).

With regard to nitrogen compounds and pesticides category, the Aluvial del Bajo Guadalhorce is the only groundwater body where an *important* impact associated with all parameters (nitrates, nitrites, ammonium and pesticides) was observed. In another four groundwater bodies a *slight* impact was identified, mainly due to nitrates concentrations (Table 5.6).

Metals are responsible for an *important* impact and a *slight* impact due to manganese concentrations. With regards to hydrocarbons, an *important* impact was detected in three groundwater bodies due to the concentrations of benzene, 1,2-dichloroethane and trihalomethanes. Finally, an *important* impact was detected in the Aluvial del Bajo Guadalhorce groundwater body (number 22 in Table 5.6) due to the selenium content.

5.4.3.2 *Vulnerable zones*

Three groundwater bodies are located in Nitrate Vulnerable Zones within the Guadalhorce River basin (numbers 3, 22 and 23 in Fig. 5.2 and Table 5.6).

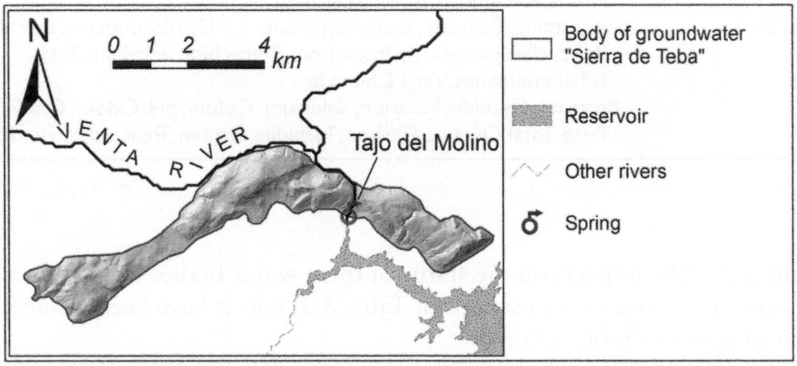

Figure 5.4 The area of Tajo del Molino where the Venta River runs through limestones in the Sierra de Teba groundwater body.

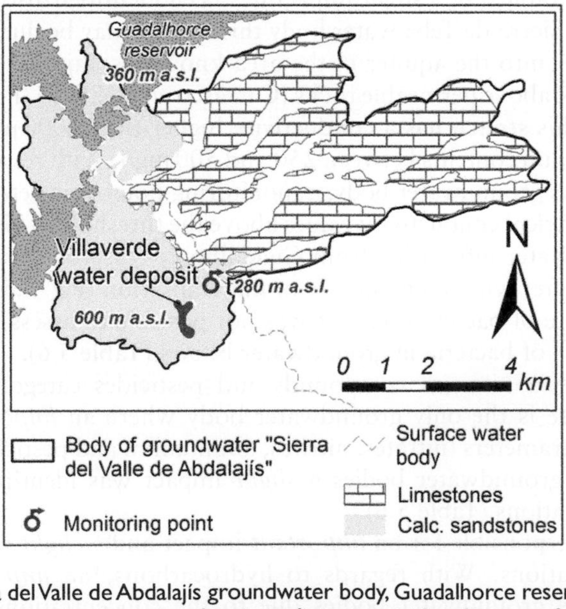

Figure 5.5 The Sierra del Valle de Abdalajís groundwater body, Guadalhorce reservoir, Villaverde water deposit and monitoring point showing salinization, with indication of topographic heights.

In Nitrate Vulnerable Zones, only the concentration of nitrate must be analysed in groundwater bodies to verify whether or not an impact exists. Since threshold values used in Nitrate Vulnerable Zones are the same as those used to assess impacts on the chemical status of groundwater bodies, the results are the same, which is: an *important* impact on two of these groundwater bodies (Vega de Antequera and Aluvial del Bajo Guadalhorce, numbers 3 and 22) and *no impact reported* in the case of the Bajo Guadalhorce groundwater body (number 23).

5.5 DISCUSSION OF RESULTS

According to the results, in the river basin 7 groundwater bodies (29%) have at least one *important* impact, in 2 groundwater bodies (8%) at least one *slight* impact has been identified, and in the remaining 15 water bodies (63%) the available data do not indicate the presence of any impact. However, in 4 of these 15 groundwater bodies, data scarcity led to them being classified as *no data found*, instead of *no impacts reported* (Table 5.8 and Fig. 5.6).

Out of the 7 water bodies where an *important* impact was observed, the Aluvial del Bajo Guadalhorce groundwater body (number 22 in Fig. 5.6) has the greatest range of parameters rated with an *important* impact: 12 of 13 that were evaluated. This is due to its proximity to the town of Malaga, which has many industrial, commercial and transport facilities (highways, airports).

Groundwater bodies subject to as many impacts as the Aluvial del Bajo Guadalhorce water body may require exemptions to the compliance with the environmental objectives of the WFD, such as less stringent objectives or extension of deadline beyond 2015 or 2021 (depending on the application of the first or second

Table 5.8 Distribution of groundwater bodies in the Guadalhorce River basin depending on the type of impact identified in its chemical status, with indication of the nature of each groundwater body (C: carbonate; D: detrital; E: evaporitic).

Important impact	*Slight impact*	*No impact*	*No data*
Sierra de Archidona (C)	Sierra de Gibalto-Arroyo Marín (C)	Torcal de Antequera (C)	Acuífero de la Magdalena (D)
Vega de Antequera (D)	Sierra de Cañete (C)	Sª Cabras-Camarolos-San Jorge (C)	Hacho de Álora (D)
Llanos de Campillos (D)		Sierras Blanquilla-Merinos (C)	Hacho de Pizarra (D)
Llanos de Almargen (D)		Sierra Hidalga (C)	Sª Llana-Mioceno de El Romeral (D)
Sierra de Teba (C)		Sª Nieves-Prieta-Alcaparaín (C)	
Sª del Valle de Abdalajís (C)		Serrezuela de Carratraca (C)	
Aluvial Bajo Guadalhorce (D)		Sierra de Guaro (C)	
		Sierra Blanca (C)	
		Sierra de Mijas (C)	
		Bajo Guadalhorce (D)	
		Trías de Antequera (E)	

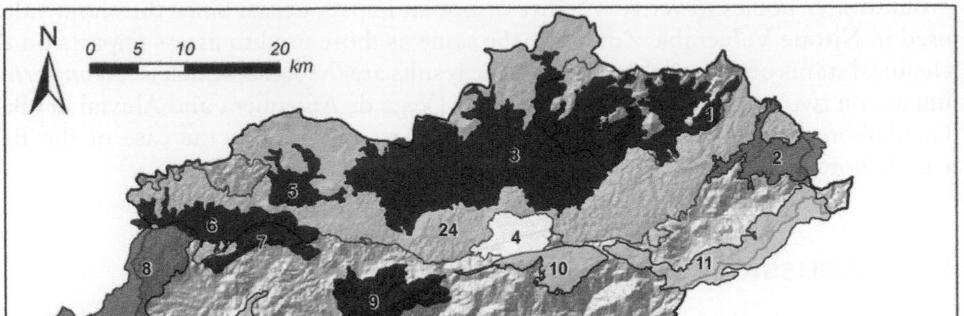

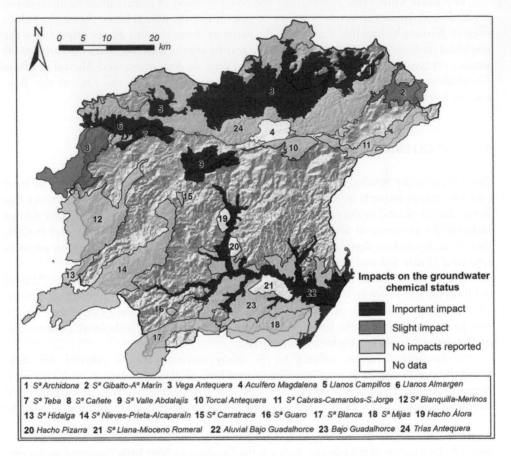

Figure 5.6 Application of the proposed methodology to evaluate impacts on the groundwater chemical status in the case study area.

river basin management plan, respectively). However, for all cases where an exemption is applied, all measures that can be taken should still be taken to reach the best status possible (European Commission, 2009).

In the Llanos de Campillos and Llanos de Almargen groundwater bodies (5 and 6 in Fig. 5.6), an *important* impact exists (due to nitrogen and phosphorus) that may be related to local intensive livestock production.

The high concentrations of nitrogen and phosphorus compounds in water are mainly responsible of the impacts on the chemical status of groundwater bodies: 8 of them have an impact due to this reason. These compounds, especially nitrate concentrations, cause the main impacts on the chemical status of other groundwater bodies in Europe (Mohaupt *et al.*, 2005; Andreadakis *et al.*, 2007; Pintar *et al.*, 2007; Carrasco *et al.*, 2008, among others).

Groundwater bodies defined as *no impacts reported* correspond to areas barely affected by human activities. With the exception of two groundwater bodies, they are all found in carbonate aquifers. In these areas, the abrupt relief, difficult access, depth to the water table and lack of arable land are factors that reduce human activities when compared to detrital aquifers. The characteristics of detrital aquifers favour the development of human activities. This explains why out of the five groundwater bodies in detrital aquifers and with sufficient data to assess impacts, four were defined with an *important* impact.

With regards to the 16 groundwater bodies used for the abstraction of drinking water, in six of them a *slight* or *important* impact was identified mainly due to nitrate concentrations.

5.6 CONCLUSIONS

This work presents a methodology for assessing the impacts on the chemical status of groundwater bodies. The proposed procedure was developed on the basis of the environmental objectives of the WFD for groundwater bodies. Following this, criteria based on a series of physicochemical parameters and threshold values were established, arising from their respective environmental objectives, from which the existence of an impact was determined. In the case of groundwater bodies, 67 parameters have been proposed, 48 for water intended for human consumption and one (nitrate) for vulnerable zones.

Risk assessments, and in particular impact analysis, is necessary as the WFD will continue to require forward predictions of complex environmental conditions and processes. Unlike the first risk assessment carried out in 2004, which had to make predictions based on relatively sparse data and often with only a broad knowledge of the operational requirements of the WFD Article 4 objectives, the second and subsequent risk assessments can be built on the work of the previous river basin planning cycle.

Impact analysis on the chemical status represents a comprehensive tool aimed at identifying any pollutant that might put a groundwater body at risk of failing to meet its environmental objectives. For this reason, impact analysis should not only focus on substances for which Member States have defined threshold values, but for a wider range of pollutants.

The result of the application of this methodology in the Guadalhorce River basin was the identification of seven water bodies with an *important* impact, two with a *slight* impact, eleven with *no impact reported* and four with *no data found*. In regards to the factors that cause these impacts, the nitrogen and phosphorus compounds (mainly nitrate) are the most remarkable. Their origin is most likely related to the agriculture and livestock activities carried out in these areas. Groundwater bodies of a detrital nature have the most significant impacts. All groundwater bodies except for two with *no impacts reported* are of a carbonate nature, this is due to the scare development of human activities in these areas.

The results of this study show the existence of *important* impacts on the chemical status of groundwater bodies used for the abstraction of drinking water, which could make them not comply with the environmental objectives established under Article 4 of the WFD. According to the concept of prioritisation from guidance document No 20 on exemptions to the environmental objectives, measures should be based on relevant criteria such as consequences or costs of non action, for example for the protection of drinking water supplies. Therefore, the programme of measures that Member States must ensure compliance with the objectives of the WFD and should focus on these bodies of groundwater, since in many cases they constitute the only supply of drinking water, especially in areas where surface water resources are scarce or non-existent.

ACKNOWLEDGEMENTS

This work forms part of the projects REN2003-01580 and CGL2008-04938 of DGICYT, the Research Group RNM-308 of the Andalusian Government and the associated unit IGME-GHUMA "Unidad de Estudios Hidrogeológicos Avanzados". We thank Dr. G. Malina and an anonymous reviewer for their constructive comments and suggestions, which helped us to significantly improve the manuscript.

REFERENCES

Andreadakis A., Gavalakis E., Kaliakatsos L., Noutsopoulos C., Tzimas A. (2007) The implementation of the Water Framework Directive (WFD) at the river basin of Anthemountas with emphasis on the pressures and impacts analysis. *Desalination* 210:1–15.

Barth F., Fawell J. (2001) The Water Framework Directive and European Water Policy. *Ecotox Environ Safe* 50:103–105.

Batlle Aguilar J., Orban P., Dassargues A., Brouyère S. (2007) Identification of groundwater quality trends in a chalk aquifer threatened by intensive agriculture in Belgium. *Hydrogeol J* 15(8):1615–1627.

Blum A., Legrand H., Grath J., Scheidleder A., Broers H.-p., Tomlin C., Ward R. (2009) Threshold Values and the Role of Monitoring in Assessing Chemical Status Compliance. In: Quevauviller P., Fouillac A.-M., Grath J., Ward R. (eds) *Groundwater Monitoring.* John Wiley & Sons Ltd, Chichester, UK.

Carrasco F., Sánchez D., Vadillo I. (2007) Atlas hidrogeológico de la provincia de Málaga [Hydrogeological Atlas of the Málaga province], volume 2, chapter Sierra de Teba-Almargen-Campillos. Instituto Geológico y Minero de España y Diputación Provincial de Málaga, 95–100.

Carrasco F., Sánchez D., Vadillo I., Andreo B., Martínez C., Fernández L. (2008) Application of the European water framework directive in a Western Mediterranean basin (Málaga, Spain). *Environ Geol* 54:575–585.

Castro N.M., Hornberger G.M. (1991) Surface-subsurface water interactions in an alluvial mountain stream channel. *Water Resour Res* 27:1613–1621.

Comber S.D.W., Merrington G., Sturdy L., Delbeke K., van Assche F. (2008) Copper and zinc water quality standards under the EU Water Framework Directive: The use of a tiered approach to estimate the levels of failure. *Sci Total Environ* 403:12–22.

European Commission (1991) Council Directive 91/676/EEC concerning the protection of waters against pollution caused by nitrates from agricultural sources. *Official Journal*, L 375, 31.12.1991, 1–8.

European Commission (1998) Council Directive 98/83/EC on the quality of water intended for human consumption. *Official Journal of the European Communities*, L 330, 5.12.1998. 32–54.

European Commission (2000) Directive 2000/60/EC of the European Parliament and of the Council Establishing a Framework for Community Action in the Field of Water Policy. *Official Journal of the European Communities*, L 327, 22.12.2000, 1–72.

European Commission (2003) Analysis of Pressures and Impacts. Guidance document n°3. Office for Official Publications of the European Communities. Produced by Working Group 2.1–IMPRESS. Common Implementation Strategy for the Water Framework Directive (2000/60/EC), 150 pp.
European Commission (2006a) Directive 2006/11/EC of the European Parliament and of the Council on pollution caused by certain dangerous substances discharged into the aquatic environment of the Community. *Official Journal of the European Union*, L 64, 4.3.2006, 52–59.
European Commission (2006b) Directive 2006/118/EC of the European Parliament and of the Council on the protection of groundwater against pollution and deterioration. *Official Journal of the European Union*, L 372, 27.12.2006, 19–31.
European Commission (2008) Directive 2008/105/EC of the European Parliament and of the Council on environmental quality standards in the field of water policy, amending and subsequently repealing Council Directives 82/176/EEC, 83/513/EEC, 84/156/EEC, 84/491/EEC, 86/280/EEC and amending Directive 2000/60/EC of the European Parliament and of the Council. *Official Journal of the European Union*, L 348, 24.12.2008, 84–97.
European Commission (2009) Guidance Document on Exemptions to the Environmental Objectives. Guidance document n°20. Office for Official Publications of the European Communities. Technical Report 2009 – 027. Common Implementation Strategy for the Water Framework Directive (2000/60/EC), 42 pp.
European Commission (2010a) Annex 3 to the Commission Staff Working Document accompanying the Report from the Commission in accordance with Article 3.7 of the Groundwater Directive 2006/118/EC on the establishment of groundwater threshold values. Information on the Groundwater Threshold Values of the Member States, 76 pp.
European Commission (2010b) Guidance on Risk Assessment and the Use of Conceptual Models for Groundwater. Guidance document n°26. Office for Official Publications of the European Communities. Technical Report 2010 – 042. Common Implementation Strategy for the Water Framework Directive (2000/60/EC), 67 pp.
Fernández-Ruiz L., Danés-Castro C., Ocaña-Robles L. (2005) Metodología de evaluación preliminar de presiones e impactos en las masas de agua subterránea [Methodology for a preliminary analysis of pressures and impacts on bodies of groundwater]. In: López-Geta J.A., Rubio-Campos J.C., Martín-Machuca M. (eds) *VI Simposio del Agua en Andalucía, Sevilla*, Spain. Volume 14, 1197–1208.
Giupponi C., Vladimirova I. (2006) Ag-PIE: A GIS-based screening model for assessing agricultural pressures and impacts on water quality on a European scale. *Sci Total Environ* 359:57–75.
Glavan M. (2007) Investigation of the impact of land use management scenarios on diffuse source nutrients in the River Axe catchment, PhD Thesis, Cranfield University, UK, 206 pp.
Hinsby K., Condesso de Melo M.T., Dahl M. (2008) European case studies supporting the derivation of natural background levels and groundwater threshold values for the protection of dependent ecosystems and human health. *Sci Total Environ* 401(1–3):1–20.
Kay P., Edwards A.C., Foulger M. (2009) A review of the efficacy of contemporary agricultural stewardship measures for ameliorating water pollution problems of key concern to the UK water industry. *Agr Syst* 99(2–3):67–75.
Kmiecik E., Stach-Kalarus M., Szczepanska J., Twardowska I., Stefaniak S., Janta-Koszuta K. (2006) Assessment of groundwater chemical status based on aggregated data from a monitoring network exemplified in a river drainage basin. In: *Progress in Biomedical Optics and Imaging – Proceedings of SPIE*. Volume 6377. Article number 63770M.
Krause S., Jacobs J., Voss A., Bronstert A., Zehe E. (2008) Assessing the impact of changes in landuse and management practices on the diffuse pollution and retention of nitrate in a riparian floodplain. *Sci Total Environ* 389:149–164.

Kunkel R., Wendland F., Hannappel S., Voigt H.J., Wolter R. (2007) The influence of diffuse pollution on groundwater content patterns for the groundwater bodies of Germany. *Water Sci Technol* 55(3):97–105.

Marandi A., Karro E. (2008) Natural background levels and threshold values of monitored parameters in the Cambrian-Vendian groundwater body, Estonia. *Environ Geol* 54(6):1217–1225.

Mohaupt V., Richter S., Rohrmoser W. (2005) Ergebnisse der Bestandsaufnahme zur Wasserrahmenrichtlinie – Der Zustand der Gewässer in der Bundesrepublik Deutschland [Results of the impact analysis according to the water framework directive – The status of the water bodies in Germany]. GWF Wasser – Abwasser 146(10):718–722.

Mostert E. (2003) The European Water Framework Directive and water management research. *Phys Chem Earth* 28:523–527.

Müller D. (2008) Establishing Environmental Groundwater Quality Standards. In: Quevauviller P., Borchers U., Thompson C., Simonart T. (eds) *The Water Framework Directive: Ecological and Chemical Status Monitoring*. John Wiley & Sons Ltd, Chichester, UK.

Pauwels H., Muller D., Griffioen J., Hinsby K., Melo T., Brower R. (2007) BRIDGE – Background criteria for the identification of groundwater thresholds. Publishable final activity report. http://www.wfd-bridge.net.

Pintar M., Globevnik L., Bremec U. (2007) Harmonisation of water management and agricultural policies in Slovenia. *Journal of Water and Land Development* 11:31–44.

Preziosi E., Giuliano G., Vivona R. (2010) Natural background levels and threshold values derivation for naturally As, V and F rich groundwater bodies: A methodological case study in Central Italy. *Environmental Earth Sciences* 61(5):885–897.

Quevauviller P. (2009) Evaluation de l'état chimique des eaux de surface et souterraines au titre de la directive cadre sur l'eau – normes de qualité et surveillance [Evaluation of the chemical status of surface and ground waters under the Water Framework Directive – Quality standards and monitoring]. *Houille Blanche* 4:72–76.

Sánchez D., Carrasco F., Andreo B. (2009) Proposed methodology to delineate bodies of groundwater according to the European water framework directive. Application in a pilot Mediterranean river basin (Málaga, Spain). *J Environ Manage* 90(3):1523–1533.

Wendland F., Hannappel S., Kunkel R., Schenk R., Voigt H.J., Wolter R. (2005) A procedure to define natural groundwater conditions of groundwater bodies in Germany. *Water Sci Technol* 51(3–4):249–257.

Wendland F., Berthold G., Blum A., Elsass P., Fritsche J.-G., Kunkel R., Wolter R. (2008) Derivation of natural background levels and threshold values for groundwater bodies in the Upper Rhine Valley (France, Switzerland and Germany). *Desalination* 226(1–3):160–168.

Winter T.C. (1999) Relation of streams, lakes and wetlands to groundwater flow systems. *Hydrogeol J* 7:28–45.

Woessner W.W. (2000) Stream and fluvial plain ground-water interactions: rescaling hydrogeologic thought. *Groundwater* 38(3):423–429.

World Health Organization (2008) *Guidelines for Drinking-water Quality*. Third edition. Incorporating the first and second addenda. Volume 1, Recommendations. Geneva, 515 pp.

Chapter 6

Heavy metals removal from contaminated groundwater using permeable reactive barriers with immobilised membranes

Iwona Zawierucha[1], *Cezary Kozlowski*[1] *& Grzegorz Malina*[2]

[1] *Institute of Chemistry, Environmental Protection and Biotechnology, Jan Dlugosz University of Czestochowa, Czestochowa, Poland*

[2] *Department of Hydrogeology and Engineering Geology, AGH University of Science and Technology, Cracow, Poland*

ABSTRACT

The performance of polymer inclusion membrane (PIM) within a permeable reactive barrier was evaluated for the removal of cadmium and zinc from the synthetic groundwater at their initial concentrations of 25 mg/l. The facilitated transport of Zn(II) and Cd(II) through PIM containing 7-dinonylnaphthalene-1-sulfonic acid (DNNS) as an ion carrier, and o-nitrophenyl pentyl ether (ONPPE) as a plasticiser allowed for reducing the Cd(II) and Zn(II) concentrations in groundwater below permissible limits for drinking water. The recovery factor values of Zn(II) and Cd(II) ions for the transport through PIM with 0.50 M of DNNS, were 98% and 91%, respectively.

6.1 INTRODUCTION

Many toxic heavy metals have been discharged into the environment as industrial wastes causing serious soil and water pollution (Inglezakis *et al.*, 2003). They are also common groundwater contaminants at industrial and military installations (Erdem *et al.*, 2004). More strict environmental regulations on the discharge of heavy metals require various technologies for their removal (Kocaoba, 2007). Groundwater contaminated with heavy metals is typically treated by "pump and treat" that is neither a cost-effective nor sustainable approach. Permeable reactive barriers (PRBs) (Fig. 6.1) seem to provide an effective and sustainable alternative for the *in situ* treatment of groundwater contaminated with heavy metals (Diels *et al.*, 2002; Suponik, 2009). Such barriers can be horizontal or vertical (USEPA, 1997), therefore, they can be installed both perpendicularly to the groundwater flow direction (Fronczyk, 2006) and infiltration through the unsaturated zone (Malina, 2011). Moreover PRBs can be a simple trench filled with reactive material, or may be of the "funnel and gate" type where groundwater flow in the aquifer is channeled by impermeable sidewalls to a reactive zone or a reaction vessel (USEPA, 1998; Barton *et al.*, 2004). The commonly used techniques for toxic ions removal or reduction of their concentrations in the aqueous phase involve precipitation, ion-exchange, membrane filtration and sorption (Pehlivan & Altun, 2007). The materials used in reactive barriers should not

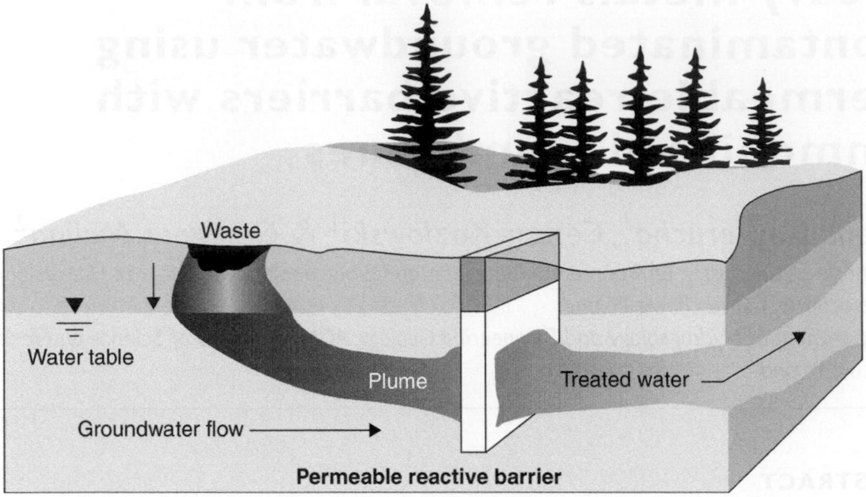

Figure 6.1 The concept of a permeable reactive barrier (PRB) (Diels *et al.*, 2002).

cause counter-productive reactions, and toxic intermediary products, or products that react with contaminants to detrimental effect. Likewise, the material should not interrupt the flow of groundwater, be of low cost, and have a long service life (Navarro *et al.*, 2006).

Liquid membranes (LMs) and, recently polymer inclusion membranes (PIMs), can be an attractive alternative for the selective removal of metal ions, such as: Zn(II), Cd(II), Cr(VI), Cr(III), Ni(II) and Fe(II) from aqueous solutions (Pospiech & Walkowiak, 2010). Transport of these metal species through PIM can be described as the simultaneous extraction and re-extraction operations proceeding in a single stage (Walkowiak *et al.*, 2000). PIMs are formed by casting a solution containing an ion carrier, a plasticiser and a base polymer, such as cellulose triacetate (CTA) or polyvinyl chloride (PVC), to form a thin, flexible and stable film. The resulting membrane is used to separate source and receiving phases (Kozlowski & Walkowiak, 2002; Kozlowski & Walkowiak, 2004). PIMs retain most of the advantages of LMs, while exhibiting excellent stability and versatility. Moreover, mechanical properties of PIMs are quite similar to those of filtration membranes, thus enable PIM-based systems to exhibit many advantages, such as: ease of operation, minimum use of hazardous chemicals and flexibility in membrane composition, to achieve the desired selectivity and separation efficiency (Nghiem *et al.*, 2006). The application of membranes for removal of toxic metals from groundwater have been subjected to numerous studies (Chiarizia *et al.*, 1992; Choi & Kim, 2003; Galan *et al.*, 2008; Alaguraj *et al.*, 2009). The PIMs technology (the tubular modules formed from immobilised membranes) can be used to pre-treatment of landfill leachates in order to minimise the risk of groundwater contamination and/or to decrease the concentration of heavy metals in subsoil. The PIMs facilitates the *in situ* removal of metal ions from aqueous solution by ion exchange process (Zawierucha & Malina, 2011). Compared to ion-exchange resins, ion exchange membranes are very promising alternative materials for metals removal because they are not compressible, and can eliminate the internal diffusion limitations caused by resins used in flow-by mode (Nasef & Yahaya, 2009).

In this chapter, the performance of PIMs technology as a separation method within a PRB for Cd(II) and Zn(II) removal from groundwater was investigated. The influence of process parameters, such as: pH of source phase, volume ratio of source/receiving phase, molar concentration of DNNS in PIMs on the counter-ion transport was also studied.

6.2 MATERIALS AND METHODS

6.2.1 Materials

Dichloromethane, o-nitrophenyl pentyl ether (ONPPE) as a plasticiser, and cellulose triacetate (CTA) as a support were purchased from Fluka and used without further purification. The ion carrier, i.e. 3,7-dinonylnaphthalene-1-sulfonic acid (DNNS) was obtained from the Aldrich Company. The synthetic contaminated groundwater was prepared by dilution of chloride and nitrate salts of test metals ($ZnCl_2$ and $Cd(NO_3)_2$ from POCh Gliwice) with deionised water, leading to their initial concentrations of 25 mg/l.

6.2.2 Methods

6.2.2.1 *Polymer inclusion membrane preparation*

A solution of the support CTA, the ion carrier (DNNS), and the plasticiser (ONPPE) in dichloromethane was prepared. A portion of this solution was poured into a membrane mold comprised of a 6.0 cm glass ring attached to a glass plate with CTA dichloromethane glue. The organic solvent was allowed to evaporate overnight, and the resulting membrane was separated from the glass plate by immersion in cold water. The membrane was soaked in the aqueous solution of HCl for 12 hours and stored in deionised water. The thickness of membrane containing 4.0 cm^3 of ONPPE/1.0 g CTA was equal to 28 μm as measured by a digital ultrameter (A2002M type from Inco-Veritas) with 0.1 μm standard deviation over four readings. The resulting membrane contained (wt): 71% CTA, 20% DNNS, and 9% ONPPE. The effective surface area of membrane was of 6.0 cm^2.

6.2.2.2 *Membrane transport experiments*

The transport experiments through PIM were carried out in a permeation cell, in which the membrane film was tightly clamped between two cell compartments (Fig. 6.2). Both, the source and receiving aqueous phases (50 cm^3 each) were stirred at 600 rpm with synchronous motors. As the source and receiving phases: 0.001 M Zn(II) and Cd(II), and 1.0 mol dm^{-3} HCl aqueous solutions were used, respectively. At the pre-concentration procedure the volumes of source, and receiving aqueous phases were equal to 500 and 50 cm^3, respectively. This membrane system was applied in experiments described in section 6.3.1.1. The apparatus applied for pre-concentration of

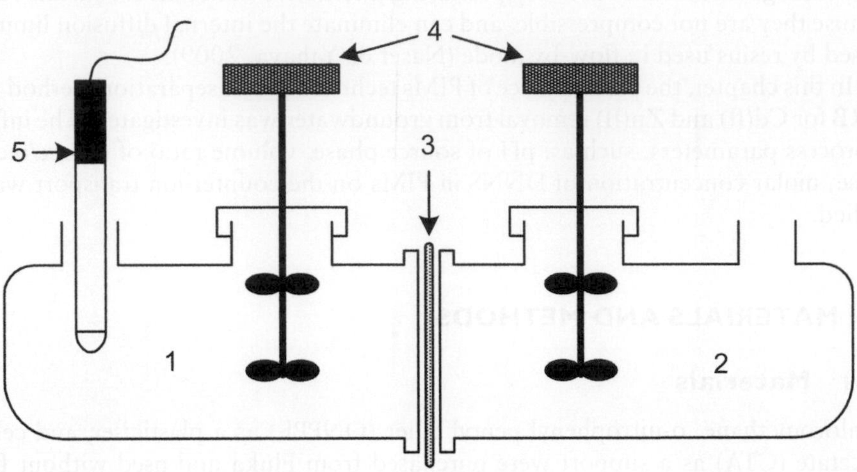

Figure 6.2 Diagram of the transport cell: 1 – source phase, 2 – receiving phase, 3 – membrane, 4 – mechanical stirrers, 5 – pH electrode.

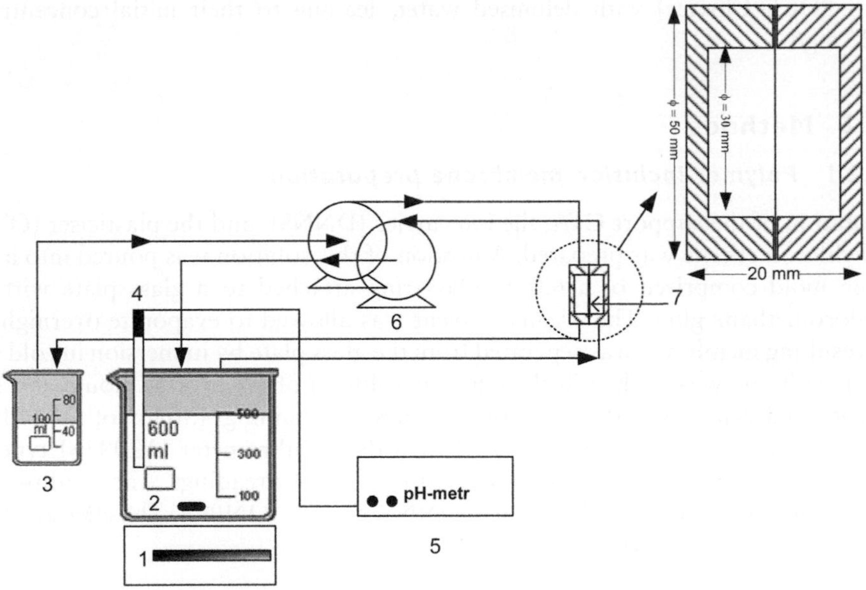

Figure 6.3 Diagram of the transport experiments across PIM: 1 – magnetic stirrer, 2 – source phase, 3 – receiving phase, 4 – pH electrode, 5 – pH-meter, 6 – peristaltic pump, 7 – membrane module.

investigated metal ions is shown in Fig. 6.3. In this case, a membrane module was used, to which source and receiving aqueous phases were pumped from appropriate tanks with peristaltic pumps (PP1B-05A type, Zalimp, Poland) at the rate of 100 cm^3/min. Samples of the aqueous source phase were removed periodically via a sampling port with a syringe, and analyzed by atomic absorption spectrophotometer (Unicam Solaar

939) with graphite tube, to determine the metal ion concentrations. All experiments were carried out at temperature of $20 \pm 0.5°C$.

6.2.3 Parameters of metal ions transport

The kinetics of the transport across PIMs is described as a first-order reaction for metal-ion concentration (Kozlowski, 2006):

$$\ln \frac{c}{c_i} = -kt \tag{6.1}$$

where c is the molar metal ion concentration at given time in the source phase (M), c_i is the initial molar concentration of metal ion in the source phase (M), k is the rate constant (s^{-1}), and t is the time of transport (s).

To calculate k values, the plots of $\ln(c/c_i)$ vs. time were prepared. The relation of $\ln(c/c_i)$ vs. time was linear, as confirmed by high values of determination coefficients (r^2), *i.e.* ≥ 0.98.

The permeability coefficient (P) was calculated as:

$$P = -\frac{V}{A} \cdot k \tag{6.2}$$

where V is the volume of the aqueous source phase, and A is the area of the effective membrane.

The initial flux (J_i) was determined as:

$$J_i = P \cdot c_i. \tag{6.3}$$

To describe the efficiency of metal ion removal from the source phase, the recovery factor (RF) was calculated as:

$$RF = \frac{c_i - c}{c_i} \cdot 100\% \tag{6.4}$$

The selectivity coefficient S, was defined as the ratio of initial fluxes for M1 and M2 metal ions, respectively:

$$S = \frac{J_{i,M1}}{J_{i,M2}} \tag{6.5}$$

The reported kinetic parameter values correspond to the average of three replicates; with the standard deviation within 2%.

6.3 RESULTS AND DISCUSSION

6.3.1 Transport by polymer inclusion membranes

The composition of PIM, *i.e.* type of membrane support, type and concentration of ionic carrier and plasticiser, plays the main role in the metal ions transport. At first, the influence of concentration plasticiser (*i.e.* o-nitrophenyl pentyl ether – ONPPE) immobilized into in cellulose triacetate support with DNNS as the ionic carrier on Zn(II) and Cd(II) ions transport was studied. Blank experiments, in the absence of a carrier, yielded no significant flux across PIM composed only with a support and a plasticiser. As it was proved by Bartsch & Way (1998), there is a linear relationship between the number of carbon atoms in the alkyl group of the series of alkyl o-nitrophenyl ethers and the alkali metal cations flux of transport across the CTA membranes. The pentyl o-nitrophenyl ether was found as the best plasticiser among studied series of alkyl o-nitrophenyl ethers.

6.3.1.1 *Competitive transport of Zn(II) and Cd(II) with DNNS*

The most significant factor determining metal ions transport through a liquid membrane is the property of an ion carrier, and the most important property of a carrier is in this context its acid strength (Cox, 2004). Alkylsulfonic acids, generally have lower pK_a values than the corresponding alkylcarboxylic acids, thereby prefer extraction of cations from considerably more acidic media (<0.5 M H^+). DNNS used for the investigated metal ions transport is a strong organic acid (Otu & Westland, 1991; Hogfeldt *et al.*, 1981).

Fig. 6.4 presents the RF values of investigated metal ions from the source phase, determined after 24 hours transport through PIM with DNNS at the concentration range 0.05–0.50 M. The Zn(II) and Cd(II) ions were extracted more efficiently, since their RF values for the transport through PIM with 0.50 M of DNNS, were of 98% and 91%, respectively. The most remarkable differences between RF values were noted when the membrane contained 0.10–0.30 M of DNNS. Under these conditions, the transport through the membrane shows a relatively good selectivity of Zn(II) over Cd(II) ions, and for 0.2 M DNNS the $S_{zn/Cd}$ was equal to 5.

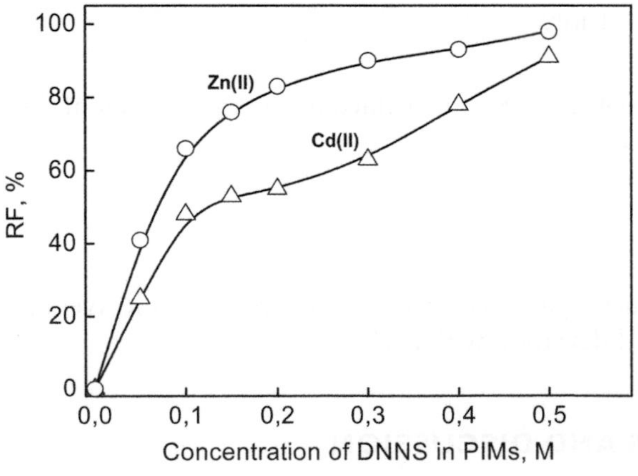

Figure 6.4 Recovery factors obtained for Zn(II) and Cd(II) after 24 h of transport across PIM with DNNS at different concentrations. The source phase: 50 cm^3 of model aqueous solution, the receiving phase: 50 cm^3 of 1.0 M HCl solution, PIM: 4.0 cm^3 ONPOE/1.0 g CTA at different carrier concentrations.

Table 6.1 The initial fluxes, selectivity orders, and selectivity coefficients for Zn(II) and Cd(II) transport across PIMs with DNNS.

pH of source phase	*Metal ions*	*Initial fluxes, J_i, $\mu mol/m^2s$*	*Selectivity order: Zn(II) > Cd(II)*
1	Cd(II)	2.6	Selectivity coefficients
	Zn(II)	4.3	Zn(II)/Cd(II) = 1.6
1.5	Cd(II)	2.9	2.0
	Zn(II)	5.7	
2.0	Cd(II)	3.1	2.3
	Zn(II)	7.0	
3.0	Cd(II)	4.0	2.3
	Zn(II)	9.0	
4.0	Cd(II)	7.0	1.7
	Zn(II)	12.0	
5.0	Cd(II)	9.3	1.5
	Zn(II)	14.0	

The zinc(II) ions were transported with a highest rate, and the selectivity order of this process was as follows: Zn(II) > Cd(II). The process efficiency increased with the ionic carrier saturation of liquid phase membrane. The accumulation of metal ions into immobilized membrane with DNNS is not significant due to extraction and back-extraction processes occurred at interface of membrane and aqueous phases. The maximal concentration of metal ions determined based on mass balance was less than 0.1% of initial concentration.

The aqueous source phase acidity is constantly increased by protons and metal ions exchange during counter-transport in the membrane system, and pH of aqueous source phase is a significant factor determining the transport selectivity. The pH of the source phase during transport through PIM containing DNNS became stable on the level of 2.5 after 6 hours.

The acidity of source phase was stable and kept on constant range of pH = 1–5 using pH-stat. During transport across the immobilised membrane, the cations such as Zn(II) and Cd(II) were preferentially removed through PIM with DNNS. The decrease of pH values of the source aqueous phase resulted in the increase of the Zn(II) rate transport. The metal ions were transported due to the protonation and deprotonation reactions of the carrier (DNNS), which occurred because of the difference between pH of the source phase and the receiving phase. This behaviour can be explained by low protonation constant of the acidic carrier. On the basis of linear relations of $\ln(c/c_i)$ versus the process time, the values of rate constant of transport through PIM with DNNS were determined. The kinetic parameters and the process selectivity were calculated, and the results listed in Table 6.1.

The process selectivity is defined by a sequence of metal ions transport fluxes, which decrease in the following order: Zn(II) > Cd(II). The maximum value of the initial flux for Zn(II) ions at pH = 5 was equal to 14 $\mu mol/m^2s$. Cd(II) ions were transported at a lower rate, and the maximum J_i values for pH = 5 and 4 were of 9.3 and 7.0 $\mu mol/m^2s$, respectively.

The values of initial fluxes obtained by Kozlowska et al. (2007) for the transport of Zn(II), Cd(II) and Pb(II) ions across the PIM with di(2,4,4-trimethylpentyl)

Table 6.2 Zn(II) and Cd (II) removal from synthetic groundwater using PIM with DNNS.

Number of run	Initial Zn(II)/Cd(II) concentration in the source phase (mg/l)	Volume of source phase (cm^3)	Volume ratio of source/ receiving phase	Time of process (hr)	Residual concentration in the source phase (mg/L)	
					Zn(II)	Cd(II)
1	25.0/25.0	50	1:1	6	5.0	1.0
2	5.0/1.0	500	10:1	12	0.1	0.005

dithiophosphinic acid confirm low selectivity removal of zinc and cadmium ions. In their experiments the values of initial fluxes of Zn(II), Cd(II) and Pb(II) ions were equal to 0.94, 1.49 and 2.59 $\mu mol/m^2s$, respectively.

6.3.1.2 *Pre-concentration by polymer inclusion membranes*

Generally, the rate of transferred mass of metal ions through PIM depends on equilibrium of metal complexes with DNNS in the aqueous phase/membrane boundary layer, and also on the source/receiving phase volume ratio. For run no. 1, at the source/receiving phase volume ratio equal to 1.0, it was possible to reduce the Zn(II) and Cd(II) concentrations from 25.0 to 5.0 mg/l and 1.0 mg/l in 6 hours, respectively (Table 6.2). However, the concentration of Cd(II) in the aqueous phase still exceeded drinking water standards in Poland – the permissible limit for cadmium is equal to 0.005 mg/l (Dz.U.2007.61.417).

Results of run no. 2, when the source/receiving phase volume ratio was of 10: 1, show that after 12 hours it was possible to reduce the Zn(II) concentration from 5.0 to 0.1 mg/l and the Cd(II) concentration from 1.0 to 0.005 mg/l. After 12 hours the Cd(II) concentration in source aqueous phase was reduced below the permissible limit for drinking water.

Kozlowski and Walkowiak (2007) in the competitive transport of Co-60, Sr-90, and Cs-137 from the aqueous sodium nitrate solution through PIM containing CTA as the support, ONPPE as the plasticizer, and DNNS as the ion carrier found the selectivity order: Co(II) > Cs(I) > Sr(II). After 72 hours, the RFs for Co(II), Cs(II), and Sr(II) were equal to 98%, 91%, and 51%, respectively. The results indicate that the transport through PIM is a suitable method for lowering metal ion concentrations in wastewaters and contaminated drinking water.

6.4 CONCLUSIONS

This study demonstrated that zinc(II) and cadmium(II) can be effectively removed from groundwater by transport through a barrier with the PIM with DNNS as an ion carrier and ONPPE as a plasticiser. The recovery factors after 24 hours of process for Zn(II) and Cd(II) were of 98% and 91%, respectively. Transport through PIM allowed for reducing Zn(II) and Cd(II) concentrations in the source aqueous phase to 0.1 and 0.005 mg/L, respectively, i.e. below the permissible limits for drinking water in Poland.

Thus, the application of PIMs can be considered as an effective alternative for heavy metals removal from contaminated groundwater, and the immobilisation of specific ion carriers on the reactive material within PRB – as a novel approach in groundwater remediation at contaminated sites.

REFERENCES

Alaguraj M., Pakanivelu K., Velan M. (2009) Removal of Cu(II) using liquid emulsion membrane. *Int J ChemTech Research*, 1(3):722–726.

Barton C.S., Stewart D.I., Morris K., Bryant D. (2004) Performance of three resin-based materials for treating uranium-contaminated groundwater within a PRB. *J Hazard Material*, B116:191–204.

Bartsch R.A., Way J. (1996) Chemical Separation with Liquid Membranes. *ACS Symposium Series 642 Amer Chem Soc*, Washington DC.

Chiarizia R., Horwitz E.P., Hodgson K.M. (1992) Removal of inorganic contaminants from groundwater. Use of Supported Liquid Membranes. In: *Environmental Remediation* (Eds) Vandegrift G.F., Reed D.T., Tasker I.R., ACS Symposium Series, 509:22–33.

Choi D.W., Kim Y.H. (2003) Cadmium removal using hollow fiber membrane with organic extractant. *Korean J Chem Eng*, 20(4):768–771.

Cox M. (2004) Solvent Extraction in Hydrometallurgy. In: *Solvent Extraction Principles and Practice* (Eds) Rydberg J., Cox M., Musikas C., Choppin G.C., Marcel Dekker, 455–505.

Diels L., Van der Lelie N., Bastiaens L. (2002) New developments in treatment of heavy metal contaminated soils. *Reviews in Environmental Science and Bio/Technology* 1:75–82.

Dz.U. 2007.61.417 Rozporzadzenie Ministra Zdrowia z dnia 29 marca 2007 r w sprawie jakosci wody przeznaczonej do spozycia przez ludzi.

Erdem E., Karapinar N., Donat R. (2004) The removal of heavy metal cations by natural zeolites. *J Colloid Interf Sci* 280:309–314.

Fronczyk J. (2006) Permeable sorption barriers: parameters of experimental materials. Przeglad Naukowy Inzynieria i Ksztaltowanie Srodowiska. Wyd. SGGW Warszawa Rocznik XV Zeszyt 1(33): 85–94.

Galan B., Castaneda D., Ortiz I. (2008) Integration of ion exchange and non-dispersive solvent extraction processes for the separation and concentration of Cr(VI) from ground waters, *J Hazard Mater* 152:795–804.

Hogfeldt E., Chiarizia R., Danesi P.R., Soldatov V.S. (1981) Structure and ion-exchange properties of dinonylnaphthalenesulfonic acid and its salts. *Chem Scr* 18:13–18.

Inglezakis V.J., Loizidou M.D., Grigoropoulou H.P. (2003) Ion exchange of Pb^{2+}, Cu^{2+}, Fe^{3+} and Cr^{3+} on natural clinoptilolite: selectivity determination and influence of acidity on metal uptake. *J Colloid Interf. Sci* 261:49–54.

Kocaoba S. (2007) Comparison of Amberlite IR 120 and dolomite's performances for removal of heavy metals. *J Hazard Mater* 147(1–2):488–496.

Kozlowska J., Kozlowski C., Koziol J.J. (2007), Transport of Zn(II), Cd(II), and Pb(II) across CTA plasticized membranes containing organophosphorous acids as an ion carriers. *Separation and Purification Technology* 57:26–30.

Kozlowski C. (2006) Facilitated transport of metal ions through composite and polymer inclusion membranes. *Desalination* 198:140–148.

Kozlowski C., Walkowiak W. (2002) Removal of chromium(VI) from aqueous solutions by polymer inclusion membranes. *Wat Res* 36:4870–4876.

Kozlowski C., Walkowiak W. (2004) Transport of Cr(VI), Zn(II) and Cd(II) ions across polymer inclusion membranes with tridecyl(pyridine) oxide and tri-n-octylamine. *Separation Science and Technology* 39:3127–3141.

Kozlowski C., Walkowiak W. (2007) Competitive transport of cobalt-60, strontium-90, and cesium-137 radioisotopes across polymer inclusion membranes with DNNS. *J Membrane Science* 297:181–189.

Malina G. (2011) Risk reduction of soil and groundwater at contaminated areas. Zaklad Poligraficzny Mos-Łuczak sp. (ISBN 83-89696-94-0978), PZiTS o/wielkopolski, Poznan.

Nasef M.M., Yahaya A.H. (2009) Adsorption of some heavy metals ions from aqueous solutions on Nafion 117 membrane. *Desalination* 249:677–681.

Navarro A., Chimenos J.M., Muntaner D., Fernandez I. (2006) Permeable reactive barriers for the removal of heavy metals: Lab – scale experiments with low-grade magnesium oxide. *Groundwater Monitoring and Remediation* 26(4):142–152.

Nghiem L.D., Mornane P., Porter J.D., Perera J.M., Cattral R.W., Kolev S.D. (2006) Extraction and transport of metal ions and small organic compounds using polymer inclusion membranes (PIMs). *J Membrane Science* 281:7–41.

Otu E., Westland A.D. (1991) Solvent extraction with sulphonic acids. *Solvent Ext Ion Exch* 9:875–883.

Pehlivan E., Altun T. (2007) Ion-exchange of Pb^{2+}, Cu^{2+}, Zn^{2+}, Cd^{2+}, and Ni^{2+} ions from aqueous solution by Lewatit CNP 80. *J Hazard Mater* 140:299–307.

Pospiech B., Walkowiak W. (2010) Studies of iron(III) removal from chloride aqueous solutions by solvent extraction and transport through polymer inclusion membranes with D2EHPA. *Physicochemical Problems of Mineral Processing* 44:195–204.

Suponik T. (2009) The use of zero-valent iron in PRB technology. In: *Reclamation and revitalization of demoted areas*, G. Malina (ed) Zaklad Poligraficzny Mos-Luczak sp. J., Poznan:161–172.

USEPA (1997) Permeable Reactive Barriers Action Team. EPA 542-F-97-0120.

USEPA (1998) Permeable Reactive Barrier Technologies for Contaminant Remediation. EPA/600/R-98/125.

Walkowiak W., Bartsch R.A., Kozlowski C., Gega J., Charewicz W.A., Amiri-Eliasi B (2000) Separation and removal of metal ionic species by polymeric inclusion membranes. *J Radioanalytical and Nuclear Chemistry* 246:643–650.

Zawierucha I., Malina G. (2011) The use of PRB with ion exchange resins for cleaning up heavy metals contaminated groundwater. In: *Reclamation and revitalization of demoted areas*, G. Malina (ed) Zaklad Poligraficzny Mos-Luczak sp. J., Poznan:175–184.

Chapter 7

Event based monitoring and early warning system for groundwater resources in alpine karst aquifers

Hermann Stadler[1], Albrecht Leis[1], Markus Plieschnegger[1], Paul Skritek[2] & Andreas H. Farnleitner[3]

[1]*Joanneum Research, Institute of Water, Energy and Sustainability, Department for Water Resources Management, Graz, Austria*

[2]*University of Applied Sciences – Technikum Wien, Institute of Telecommunication and Internet Technologies, Vienna, Austria*

[3]*Vienna University of Technology, Institute of Chemical Engineering, Department of Applied Biochemistry and Gene Technology, Vienna, Austria*

ABSTRACT

Spring waters from alpine karst aquifers are important drinking water resources. As their water quality can change very rapidly during event situations, water abstraction management has to be performed in near real-time. Four summer event sets (2005–2008) at alpine karst springs were investigated in detail in order to evaluate the spectral absorption coefficient at 254 nm (SAC254) as a real-time early warning proxy for fecal pollution. For the investigation Low-Earth-Orbit (LEO) Satellite-based data communication between portable hydro meteorological measuring stations and an automated microbiological sampling device were used. This allowed a dissemination of the data via the internet and registered users could also download the data. It was demonstrated that it is possible to use SAC254 as a real-time proxy parameter for estimating the extent of fecal pollution after establishing specific spring and event-type calibrations that take into consideration the variability of the occurrence and the transferability of fecal material. To enhance the possibilities of such early warning systems, on-site analyses of the isotopic composition of the spring water were carried out isochronous with other hydrological parameters and with high time resolution. The new developed wavelength scanned Cavity Ring Down Spectroscope (CRDS) was adapted for on-site use. During snowmelt 2010, the system showed the whole dynamic of environmental isotopes, physicochemical and microbial parameters at the karst spring LKAS2.

7.1 BACKGROUND AND AIM OF RESEARCH

Water resources from alpine and other mountainous karst aquifers have an important role in water supply in many European countries. As regulated by the WFD (Water Framework Directive; EC 2000), karstic catchments require sustainable protection. The increasing impact in such regions and the different utilisation in the watersheds of karst springs are important reasons to establish early warning systems and quality assurance networks for water supplies. These systems rely heavily on in-situ measurements providing online and near real-time data. With a satellite based network

Figure 7.1 Location of the investigation area.

of measuring and sampling stations it is possible to carry out precipitation triggered event monitoring campaigns at different karst springs (Stadler *et al.*, 2010) combining on-line measurements of hydrological parameters with field-laboratory based analyses of microbial fecal indicators (Stadler *et al.*, 2008).

The targets of the study were (1) to investigate the dynamic of chemical parameters, environmental isotopes and microbial fecal pollution indicators at a high resolution time scale during hydrological events, (2) to evaluate the previously investigated parameter SAC254 as an appropriate real-time pollution proxy for optimized spring water abstraction management within an early warning system and (3) to implement also automated sampling of event-causing precipitation in the catchment area to carry out isotopic analyses.

7.2 HYDROGEOLOGICAL SETTINGS OF THE STUDY AREA

The alpine and mountainous karst system that was studied is located in the Northern Calcareous Alps (Figure 7.1) in Austria reaching altitudes up to approximately 2300 m above sea level (masl.). The spring LKAS2 is situated at an altitude of approximately 650 masl.

As LKAS2 drains a mainly Triassic limestone aquifer and it complies with a typical limestone spring type according to D'Amore (1993), having well developed karst conduits (Stadler & Strobl, 1997). The mean discharge 1995–2009 was 5.240 l/s including a high range of variations resulting in a discharge ratio (Q_{max}/Q_{min}) of 1:40. The mean water residence time was estimated between 0.8 to 1.5 years, using an exponential model based on investigations of environmental isotopes. According to this, the discharge response after precipitation is very rapid (some hours). The estimated alpine catchment covers an area of about 70 km^2, the mean altitude is 1780 masl. (Stadler & Strobl, 1997). Vegetation comprises summer pastures, natural calcareous alpine swards with open krummholz and forests (Dirnböck *et al.*, 1999).

7.3 METHODS

7.3.1 LEO-Satellite system

Based on extensive technical and cost comparisons and validation measurements, e.g. (Stadler & Skritek, 2003), the ORBCOMM LEO Satellite system was chosen. ORBCOMM is a "Little-LEO" system, with 30 servicing satellites in 6 orbit planes of 800 km altitude. It provides bi-directional "short message" data-transfer at 2.4/4.8 kbps, with data blocks preferably less than 100 bytes. ORBCOMM operates at frequencies about 140 MHz, providing large satellite footprints, and requires only low-cost/low-power equipment, allowing, e.g., simple whip-antennas as well as small solar-panels for power supply and transmission even from within forests. The ORBCOMM modem transmits data to the satellite, from where down-link transmission is performed either directly to one of the Gateway Earth Stations (GES) or as "globalgrams" (data stored in the satellite and forwarded to earth when the satellite passes the desired GES). The GES emails the data to the receiver via the internet or re-transmits it to any "nomadic" ORBCOMM modem again via satellite (ORBCOMM, 2011).

7.3.2 Assembling, cross-linked stations and data streams

The precipitation station (PS) is located in the catchment area of the spring, where the event sampling will be carried out. It is equipped with a tipping bucket, a data logger and a LEO-Satellite modem (Figure 7.2). It can be supplemented with additional meteorological sensors and sampling devices. The monitoring and sampling site at the spring (spring sampling station, SSS) is equipped with an additional data logger, a pressure probe to register the changes in discharge, two automatic sampling units (one for the reference sample and one for the periodic samples) and a LEO-Satellite modem for real-time control and data transmission. It can be supplemented with additional hydrological or meteorological sensors. Central node of this network is the Central Monitoring Station (CMS). Here all data and information are collected, information to technical service team is disseminated and the data base is updated.

Precipitation Station (PS). It records rainfall and other meteorological data. From the intensity and the recorded amount of precipitation a specific trigger criterion is derived. If this trigger-level is exceeded, the PS activates one or more SSS via satellite (Data Stream 1, Figure 7.1) to take the reference sample. This happens before the event affects the discharge of the spring. The CMS is also informed via satellite by receiving periodic data sets from the PS to observe the further trend of precipitation (Data Stream 2, Figure 7.2).

Spring Sampling Station (SSS). As soon as the activation data-set is received, the automatic sampling unit takes the reference sample. The status is sent to the CMS

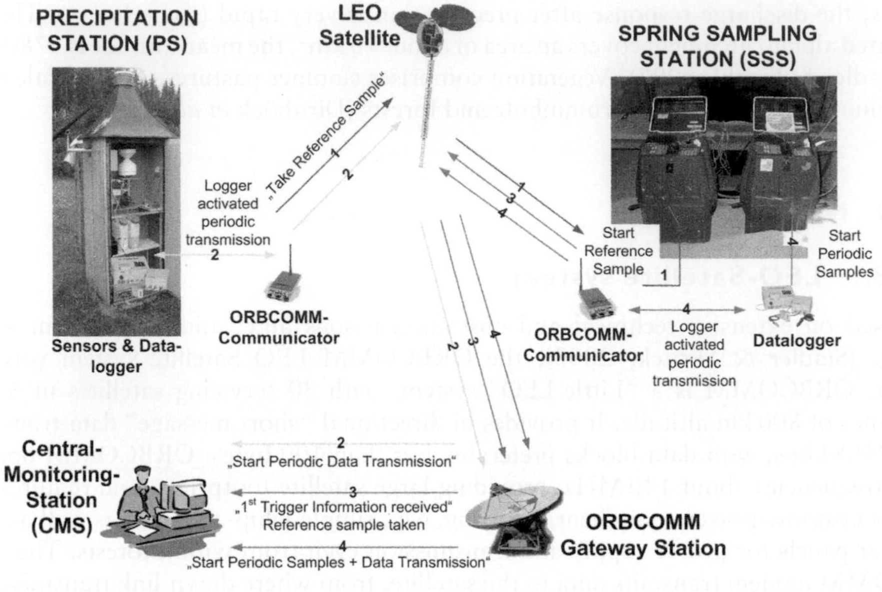

Figure 7.2 Block-Diagram of Assembling: stream of data and information of an event-triggered LEO Satellite Hydrology Network.

(Data Stream 3, Figure 7.2). This procedure can be repeated several times, depending on the number of sampling bottles in the automatic sampling device. This is necessary because at this moment it is not clear, whether the upcoming event at the spring is worth sampling.

The SSS waits during a specified period of time for the increase of the discharge, which is the second trigger event. The trigger level is derived from the increase of the gauge height within a period of time and is chosen according to the characteristics of the spring. This trigger criterion is activated by the data logger. If the predefined trigger level is exceeded, the periodic sampling within the event sampling starts automatically and the status information and measured values are continuously sent via satellite to the CMS (Data Stream 4, Figure 7.2).

Central Monitoring Station (CMS) and Web-Interface. The information from all stations is collected by the CMS. Additionally the local service team (LST) is informed from the CMS automatically of important events such as the onset of a rainfall event (1st trigger) and starting of the sampling procedure at the SSS (2nd trigger) via GSM (Global System for Mobil Communication). Depending on the sampling time increment and the number of bottles in the automatic samplers the LST can plan the next visit at the SSS to maintain the station. The CMS provides an online Internet-Portal (Figure 7.3) for access to these environmental data. It is built around the server-based operating system Debian, a reliable freeware, providing perfect interaction and performance with the server. Among others, the server comprises a RAID-system (Redundant Array of Independent Disks) for fault-tolerant operation.

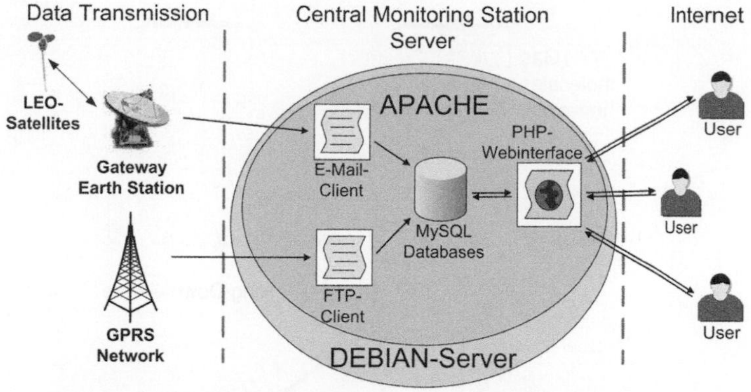

Figure 7.3 Data dissemination at the CMS via Web-interface.

To provide on-line communication with access to the stored measurement data via the Internet, an ApacheWeb server was implemented on the Debian-Server. The dynamically generated online website can be viewed under http://wrms007.joanneum.at. Using password-protection, several access levels to the data and visualization are feasible for different user groups, e.g., general public access to environmental information or individual access for specific in-depth data for research-project co-workers (Heiner, 2005).

7.3.3 Microbial monitoring

The event at the karst spring LKAS2 was caused by a thunderstorm with 40.2 mm precipitation measured in the watershed at 1520 masl. The samples at the spring (n = 157) were taken with automatic sample devices from August 21 to August 31 in 2009 stored at ambient spring water conditions and treated for the different analyses not later than 24 hours after sampling. The rain water was stored after automatic sampling in an air-tight container for 16 hours before treating. *E. coli* was analysed by the Colilert system (IDEXX) directly at a field laboratory (Stadler *et al.*, 2008). Hydrological in situ measured on-line parameters were collected with an increment of 15 minutes. To study microbial fecal pollution *E. coli* was chosen as an indicator. In contrast to other standard fecal indicators, various previous investigations highlighted its excellent applicability as a general fecal pollution indicator in alpine karstic environment (i.e. high prevalence and abundance in human, livestock and wildlife excreta, low or no prevalence in alpine soils, half life time of *E. coli* in spring water in the range of the average event period length, Farnleitner *et al.*, 2010).

7.3.4 Isotopic investigations

For on-site investigation of the isotopic composition of the spring water during an event the newly developed high sensitive gas analyser, the Cavity Ring Down Spectroscopy (CRDS) system (Berden *et al.*, 2000) was adapted and used during the snowmelt event

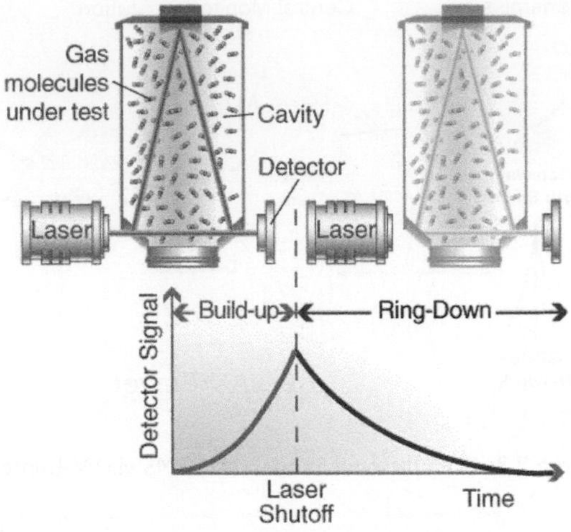

Figure 7.4 Schematic of Picarro CRDS analyzer showing how a ring down measurement is carried out.

in 2010 at the karst spring LKAS2. The system leads to a new class of on-site capable measuring devices, the wavelength-scanned (WS)-CRDS (Gupta *et al.*, 2009).

Nearly every small gas-phase molecule (e.g., CO_2, H_2O, H_2S, NH_3) has a unique near-infrared absorption spectrum. At sub-atmospheric pressure, this consists of a series of narrow, well-resolved, sharp lines, each at a characteristic wavelength. Because these lines are well-spaced and their wavelength is well-known, the concentration of any species can be determined by measuring the strength of this absorption, i.e. the height of a specific absorption peak. But in conventional infrared spectrometers, trace gases provide far too little absorption to measure, typically limiting sensitivity to the parts per million at best. CRDS avoids this sensitivity limitation by using an effective path length of many kilometers. It enables gases to be monitored in seconds or less at the parts per billion level, and some gases at the parts per trillion level.

In CRDS, the beam from a single-frequency laser diode enters a cavity defined by two or more high reflectivity mirrors. Picarro analysers use a three-mirror cavity as shown in Figure 7.4 to support a continuous travelling light wave. This provides a superior signal to noise ratio compared to a two-mirror cavity that supports a standing wave. When the laser is on, the cavity quickly fills with circulating laser light. A fast photo-detector senses the small amount of light leaking through one of the mirrors to produce a signal that is directly proportional to the intensity in the cavity.

Furthermore, the final concentration data are particularly robust because they are derived from the difference between ring down times and are independent of laser intensity fluctuations or the absolute laser power.

If a gas species that absorbs the laser light is introduced into the cavity, a second loss mechanism within the cavity (absorption) is introduced. This accelerates the ring down time compared to a cavity without any additional absorption due to a targeted gas species (Figure 7.5). Picarro instruments automatically and continuously calculate

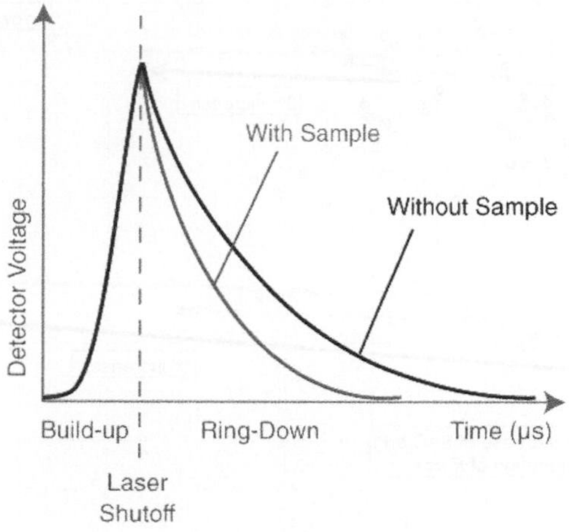

Figure 7.5 Ring down characteristic.

and compare the ring down time of the cavity with and without absorption due to the target gas species. This produces precise, quantitative measurements that take into account any intra-cavity losses that may be changing over time.

Light intensity as a function of time in a CRDS system with and without a sample having resonant absorbance. This demonstrates how optical loss (or absorption by the gas) is rendered into a time measurement (PICARRO, 2011).

7.4 RESULTS AND DISCUSSION

The integration of on-line measured data, laboratory and field-laboratory analyses, all of them recovered with high time resolution, allows a deep insight to these sensitive aquatic systems. Especially the combination with environmental isotopes (Figure 7.7 and Figure 7.9) generates new knowledge of the dynamics, mass transport conditions with different transfer behaviour of the particular substances being of fundamental importance for the sensible use of early warning systems. As an example the correlation between SAC254 and *E. coli* during the course of the event is shown in Figure 7.6. The possibilities to use (on-line) SAC254 as a proxy parameter of (laboratory analysed) *E. coli* are described in detail in Stadler *et al.* (2010).

Very important for the use of SAC254 as an early warning proxy is the lead time of SAC254 to *E. coli*, where an increase of the SAC254 happens, but no important raise of the amount of *E. coli*. This enables enlarged reactions times for water abstraction management.

The comparison of the Oxygen-18 as a proxy of the aquifer-dynamic and turbidity or the SAC254 as indicators of mass transport of substances with different transfer behaviour show the very different reaction of these parameters (Figure 7.7).

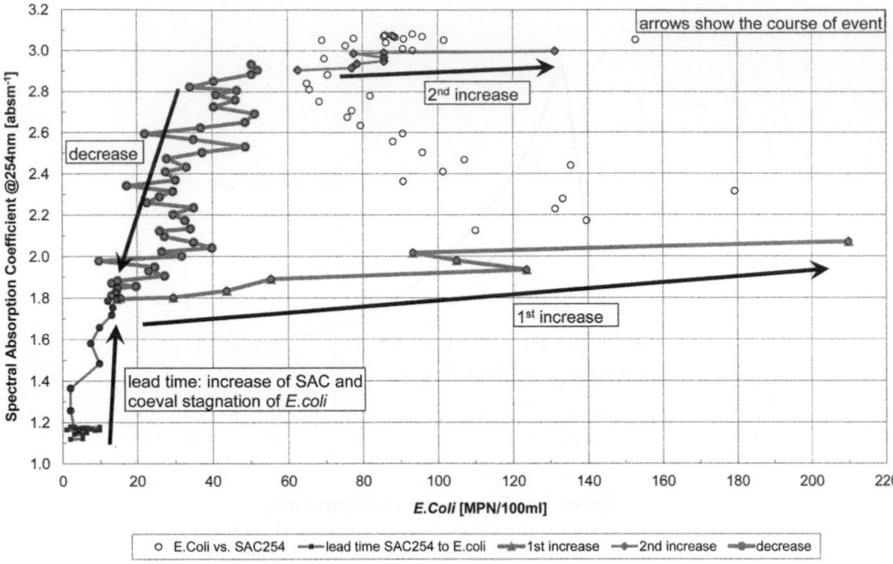

Figure 7.6 Correlation of SAC and *E. coli* during the observed event. The arrows show the course over the period of the event.

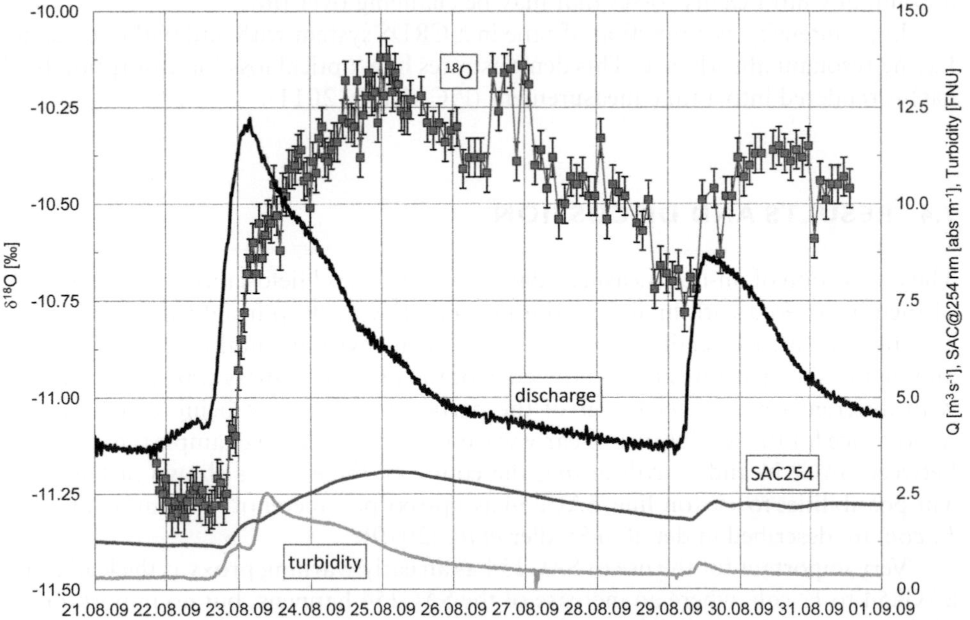

Figure 7.7 Course of parameters during the investigated event. Combination of laboratory analyzes (^{18}O) and in-situ measured parameters (discharge, turbidity and SAC254).

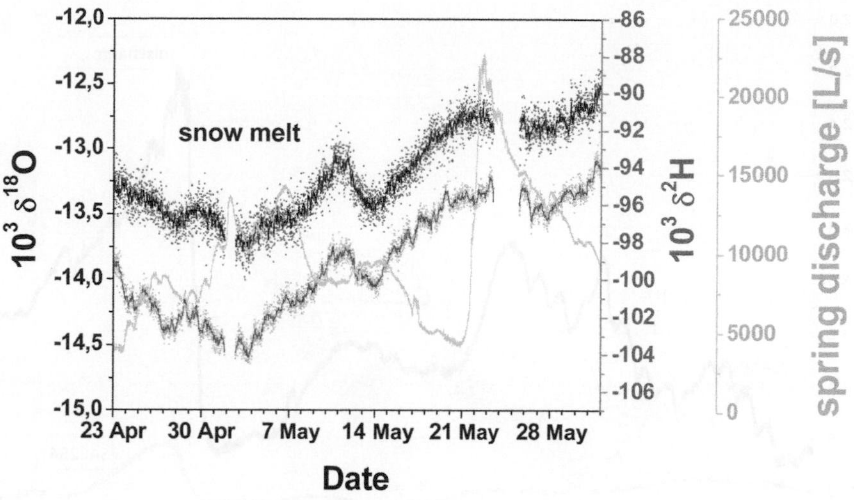

Figure 7.8 Course of ^{18}O, D and discharge during snowmelt (on-line measurements).

The course of the parameters during the event shows that hydraulic reaction occurs before changings of mass transport parameters are visible at the karst spring LKAS2 (Figure 7.7). The changing fingerprint of the water, indicated by its isotopic composition, can give insight to transport phenomena as well as mixing and dispersion processes in the karst aquifer.

The different behaviour of turbidity and SAC254 is mainly caused by the different provenance of the components of the particular parameter. SAC254 causing components are mainly surface related, whereas turbidity components arise mainly from deposits in karst conduits. These components are activated in the karst system by increasing pressure, flow velocity and shear forces. For this reason the reaction of turbidity reaches the spring earlier than changes of SAC254.

In Figure 7.8 the course of Deuterium, Oxygen18 and discharge during spring 2010 is shown. All parameters were measured on-site with a time increment of 15 minutes. The increase of discharge during the first days of May is caused by snowmelt, the rapid increase of discharge after May 21 was activated by a rainfall event, which brought about 100 mm precipitation during one day. The different behaviour of the isotopes after the rainfall event in contrast to behaviour during the snowmelt is observable. The maximum of the depleted isotope content of water from winter precipitation occurs on 3 May. The damped reaction of D and ^{18}O after the rainfall event on 21 May cannot be elucidated, due to missing analyses of the event causing precipitation.

Figure 7.9 show the reaction of the karst spring during the same time period as in Figure 7.8, but now combined with the early-warning proxy SAC254. The different behaviour of SAC254 during snowmelt and after a rainfall event is as apparent as the different reaction of the ^{18}O variation. The possibility of on-site measurements of the isotopic composition of spring water with high time resolution offers detailed insights to the behaviour of the karst system.

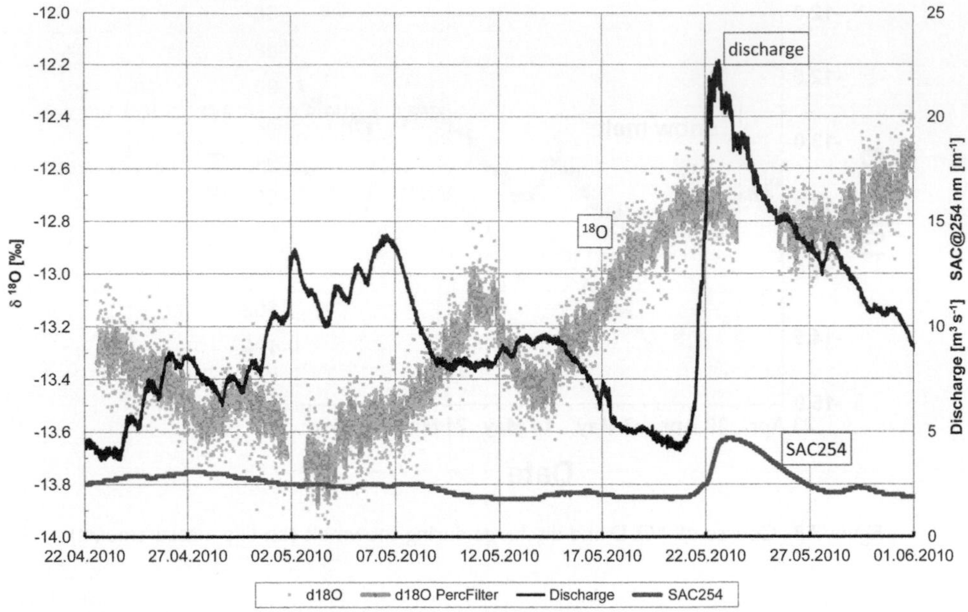

Figure 7.9 Course of the in-situ measured parameters Oxygen-18, SAC254 and Discharge.

7.5 RESUME AND OUTLOOK

It was shown that both, quality parameters of karst spring water and isotopic composition can be measured in-situ (direct in the outflow of the spring, using appropriate probes) respectively on-site (at the location of the spring in a bypass of the spring outflow) with high time resolution. The combination of such measurements with laboratory analyses or field-laboratory analyses can bring further information of the aquifer system with special regards to early warning systems. The availability of such different parameters in near real time offers new possibilities in designing customised early warning systems. The accuracy, reliability and robustness of such systems can be increased significantly.

The efficient distribution of data in an M2M compatible system offers a new time horizon to supervise water quality within a responsive early warning system for water abstraction management.

ACKNOWLEDGEMENTS

The work was carried out in close cooperation with the Vienna Waterworks (Dr. Wolfgang Zerobin) and supported by the area management of the Vienna Waterworks in Wildalpen (Ing. Christoph Rigler). Our thanks go also to Mr. Hermann Kain for technical support. Parts of the project were financially supported by the Austrian Federal Ministry for Transport, Innovation and Technology.

REFERENCES

Berden G., Peeters R., Meijer G. (2000): Cavity ring-down spectroscopy: Experimental schemes and applications [Review], *International Reviews in Physical Chemistry* 19(4): 565–607.

D'Amore F., Scandiffio G., Panichi C. (1983): Some observations on the chemical classification of ground water. *Geothermics* 12: 141–148.

Dirnböck T., Dullinger S., Gottfried M., Grabherr G. (1999): Die Vegetation des Hochschwab – Alpine und Subalpine Stufe. Mitt Naturwissen Verein Stmk 129: 111–251.

EC–Official Journal (2000): Directive 2000/60/EC of the European Parliament and of the council of 23 October 2000 establishing a framework for Community action in the field of water policy http://circa.europa.eu/Public/irc/env/wfd/library?l=/framework_directive&vm=detailed&sb=Title.

Farnleitner A.H., Ryzinska-Paier G.H., Reischer M.M., Burtscher S., Knetsch A.K.T., Kirschner T., Dirnböck G., Kuschnig R.L., Mach R., Sommer R. (2010): Escherichia coli and enterococci are sensitive and reliable indicators for human, livestock and wildlife faecal pollution in alpine mountainous water resources., *J Appl Microbiol*: 109/5: pp. 1599–1608.

Gupta P., Noone D., Galewsky J., Sweeney C., Vaughn B.H. (2009): Demonstration of high-precision continuous measurements of water vapor isotopologues in laboratory and remote field deployments using wavelength-scanned cavity ring-down spectroscopy (WS-CRDS) technology. *Rapid Communications in Mass Spectrometry* 23(16): pp. 2534–2542.

Heiner W. (2005): Online-system for hydro-meteorological data transmission via LEO-satellites (in German). Diploma thesis, p. 126, University of Applied Sciences Technikum-Wien, Vienna

Stadler H., Strobl E. (1997): Karstwasserdynamik und Karstwasserschutz Zeller Staritzen. Graz, Institute of Hydrogeology and Geothermics, *Joanneum Research*, p. 171.

Stadler H., Skritek P. (2003): Remote Water Quality Monitoring "on-line" using LEO Satellites. *Wat Sci Tech* 47(2): 197–204.

Stadler H., Skritek P., Sommer R., Mach R., Zerobin W., Farnleitner A.H. (2008): Microbiological monitoring and automated event sampling at karst springs using LEO-satellites. *Wat Sci Tech* 58(4): 899–909.

Stadler H., Klock E., Skritek P., Mach R., Zerobin W., Farnleitner A.H. (2010): The spectral absorption coefficient at 254 nm as a near real time early warning proxy for detecting faecal pollution events at alpine karst water resources. *Wat Sci Tech* 62(8): 1898–1906.

PICARRO (2011): URL: http://www.picarro.com; last visited 2011/01/25.

ORBCOMM (2011): URL: http://www.orbcomm.com; last visited 2011/01/20.

Chapter 8

Cryopegs in the Yakutian diamond-bearing province (Russia)

Sergey V. Alexeev, Ludmila P. Alexeeva & Alexander M. Kononov
Institute of the Earth's Crust SB RAS, Irkutsk, Russia

ABSTRACT

The cryopegs (negative temperature saline water and brines) occur in the Yakutian diamond-bearing province in Russia. Three groups of cryopegs are distinguished by their geochemical and isotopic properties. The cryopegs may occur: (i) as a result of evaporation and concentration of paleoseawater (group A), (ii) as a result of leaching of halite strata (group B), or (iii) in the process of geochemical evolution under mixing, leaching and water-rock interactions (group C).

8.1 INTRODUCTION

Cryopegs (intra-permafrost and sub-permafrost negative-temperature chloride saline waters and brines) are a major component of cryolithosphere circulation in the cooled sedimentary rock and kimberlite pipes, forming aquifers that are regionally distributed. The interaction between the frozen rock and cryopegs, the ability of negative-temperature waters to move through permafrost diluting ice inclusions in rocks, as well as the considerable decrease in temperature during the cryopegs' migration has caused deep cooling in the geological section (Pinneker *et al.*, 1989). The origin of cryopegs is one of great scientific interest and now is widely discussed among scientists. This study helps in understanding cryopeg formation and, may be useful in diamond exploration and brine drainage disposal.

8.2 GEOLOGY AND GEOCRYOLOGY

The Yakutian diamond-bearing province has a total area of 840 000 km^2 and is located in the northern part of the Siberian platform (Fig. 8.1). Within the boundaries, the geological section consists of sedimentary rocks underlain by an Archaean crystalline basement. The thickness of the sedimentary cover varies from 2 to 3 km. Although the sedimentary rocks in the Yakutian province range in age from Proterozoic (Vendian) to Jurassic, they are dominated by Cambrian sediments. In the central part of the province the Vendian rocks are mainly dolomites interlayered with marls and sandstones. The Cambrian sediments consist of dolomites and limestones interlayered by argillites, clayey limestones and gritstones.

In the southern part of the province the basement is covered by Vendian-Early Cambrian sediments that are made up primarily of dolomites interlayered with

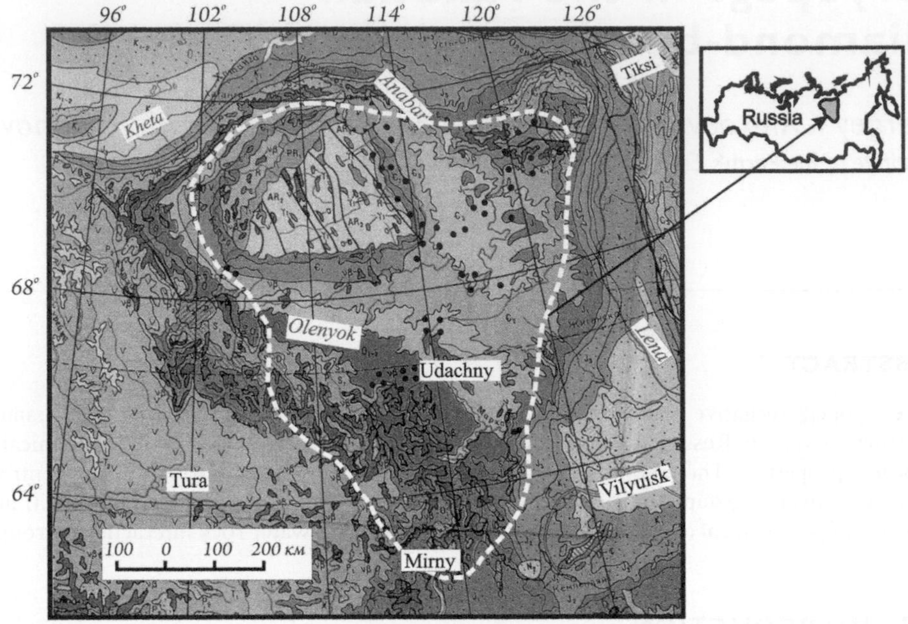

Figure 8.1 Location of the Yakutian diamond-bearing province. Black circles show the position of kimberlite pipes.

argillites, anhydrites, limestones and gritstones. The Cambrian sediments consist of dolomites and limestones interbedded with marls, argillites, anhydrites and halite. Thin layers of Jurassic sandstones and clays are exposed at the surface. In the province, there are numerous Middle Paleozoic kimberlite pipes and Late Paleozoic-Early Mesozoic intrusions that are confined to tectonic fault zones in the area.

The Yakutian diamond-bearing province is unique because of the extreme cooling and is characterised by continuous permafrost, low mean annual rock temperatures (from −2.9 to −8.8°C in the north and from −1.2 to −4.0°C in the south), high rock thermal conductivity (2.2 to 5.2 W/(m °K) and low (0.008 to 0.027 W/m^2) intensity of heat flow (Balobaev, 1991; Duchkov & Balobaev, 2001). These characteristics have caused the formation of the thermal anomaly field and low thermal gradients. In the central part of the province the position of zero isotherm varies from 720 up to 1450 m depth. In the south, it is at depths of 340 to 820 m. The cryolithozone represents the interlayering of ice-rich permafrost, dry permafrost and cooled rocks. The cooled rocks are saturated with the cryopegs (Fig. 8.2).

8.3 HYDROGEOLOGY AND HYDROGEOCHEMISTRY

The water samples (more than 500) collected from exploration boreholes were analysed for chemical composition by Inductively Coupled Plasma-Mass Spectrometry/Inductively Coupled Plasma-Atomic Emission Spectrometry (ICP-MS/ICP-AES) and Ion

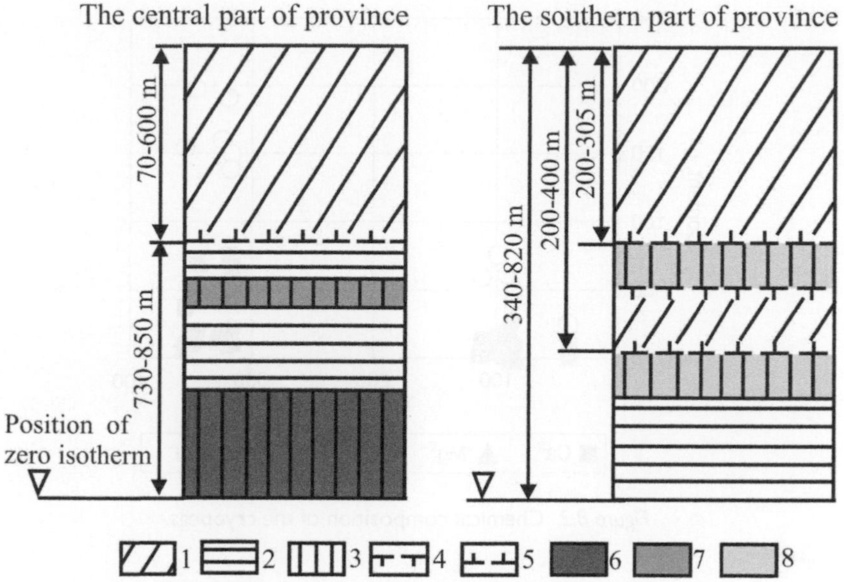

Figure 8.2 The generalized permafrost structure and cryopegs position in the Yakutian diamond-bearing province. 1 – ice-rich permafrost; 2 – "dry" permafrost; 3 – basal cryopegs; 4 – permafrost table; 5 – permafrost base; 6 – Cl Mg-Ca cryopegs (group A samples); 7 – Cl Mg-Ca or Ca-Mg or Ca-Na cryopegs (group C samples); 8 – Cl Na cryopegs (group B samples).

Chromatography (IC) for cations and anions, respectively. Samples of high concentration were diluted 20 times before they were analyzed. The detection limits for Ca, Na, K, Mg, Li and Sr of the ICP-MS/ICP-AES method are: 100, 200, 10, 100, 0.1 and 0.1 μg/l, respectively. The detection limits for: Br, Cl and SO_4 of the IC method are: 100, 200 and 100 μg/l, respectively.

The Yakutian diamond-bearing province is characterised by different types of cryopegs. Two hydrochemical zones can be distinguished in the vertical section of the central part of the province. Groundwater of the Upper Cambrian aquifer within the sedimentary strata and the Middle Palaeozoic kimberlite aquifer are contained in the first zone. The zone is represented by saline waters and diluted brines. The chemical composition of groundwater is principally chloride. The cations balance is: Ca > Mg > Na > K or Mg > Ca > Na > K or Na > Ca > Mg > K. The total dissolved substances (TDS) values vary from 31 to 252 g/l (Fig. 8.3). Groundwater samples were collected from depths of 110 to 650 m, while the thickness of the first zone is limited up to 20 m. Temperatures of the cryopegs vary from −4.0 to −3.0°C. The second zone contains concentrated brines of the Middle Cambrian aquifer within sedimentary strata and the Middle Paleozoic kimberlite aquifer. Groundwater is pumped from depths 600 to 1450 m. TDS values range from 224 to 404 g/l and increase depending on the groundwater occurrence. 98% of the anion balance is represented by chloride. The mean concentrations of cations (in %) are: Ca – 50–70, Na – 15–30, Mg – 15–25, K – 3–5. The temperatures of cryopegs vary from −2.6 to −1.0°C.

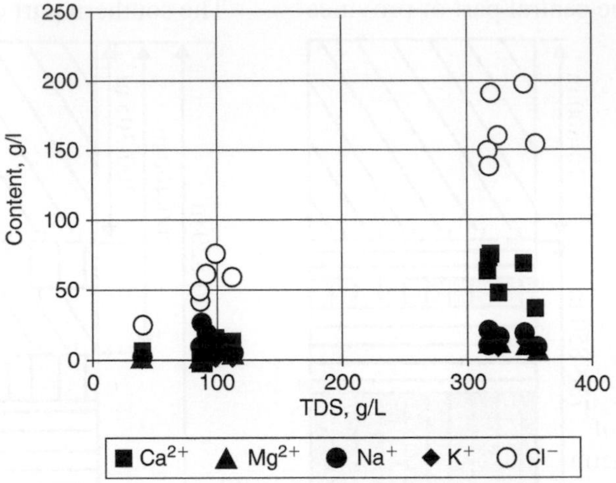

Figure 8.3 Chemical composition of the cryopegs.

On the south of the Yakutian diamond-bearing province cryopegs occur at the depths of 200–400 m. The groundwater is represented by saline waters and brines. Their occurrence corresponds to the Early-Middle Cambrian salt-bearing aquifer and the Middle Paleozoic kimberlite aquifer. The chemical composition of groundwater is also chloride and predominantly sodium. The cations balance is: $Na \gg Ca > Mg > K$. TDS values range between 28 and 165 g/l. The groundwater has a high H_2S content, up to 90 to 120 mg/l. The cryopeg temperatures vary from −2.5 to −0.5 °C.

8.4 STABLE ISOTOPE SIGNATURES

The ^{18}O, ^{2}H, ^{37}Cl and ^{81}Br stable isotopes were analyzed by Isotope Ratio Mass Spectrometry (IRMS) in the University of Waterloo (Canada). The analytical precisions for ^{18}O, ^{2}H, ^{37}Cl and ^{81}Br isotopes are: 0.2‰, 1.0‰, 0.1‰ and 0.1‰, respectively.

The δ^2H and $\delta^{18}O$ results range from −171 to −61.7‰ and from −21.4 to −2‰, respectively. The pattern obtained from the δ^2H and $\delta^{18}O$ values is similar to that previously reported by Pinneker *et al.* (1987). The $\delta^{37}Cl$ values range between −0.40 and +1.3‰. This range is within the known variation for Cl stable isotopes of formation waters (Kaufmann *et al.*, 1993; Eastoe *et al.*, 2001; Frape *et al.*, 2004). The $\delta^{81}Br$ values have a wide variation and range between −0.80 and +2.31‰. This variation is larger than the previously reported range (0.00 to +1.80‰) for Br stable isotopes for natural samples (Eggenkamp & Coleman, 2000; Shouakar-Stash *et al.*, 2005). The different cryopeg groups are not distinguished from each other solely on their chemical composition; they are also distinguishable according to isotopic characteristics.

Group A: the δ^2H and $\delta^{18}O$ values of the cryopeg samples range between from −70.5 to −61.7‰ and from −5.52 to −2.0‰, respectively (Fig. 8.4). The $\delta^{37}Cl$ (Fig. 8.5)

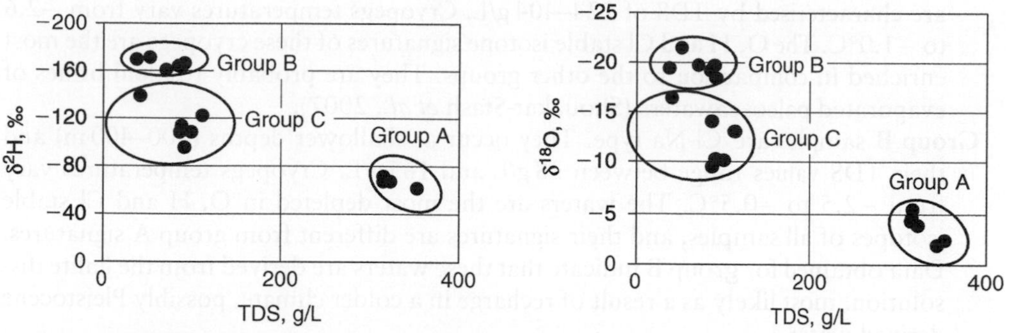

Figure 8.4 Stable isotopic composition (δ^2H and $\delta^{18}O$) of cryopegs of the Yakutian diamond-bearing province. The cryopegs are distinguished on three groups: Group A (Cl Mg-Ca), Group B (Cl Na), Group C (Cl Ca-Mg or Mg-Ca or Ca-Na).

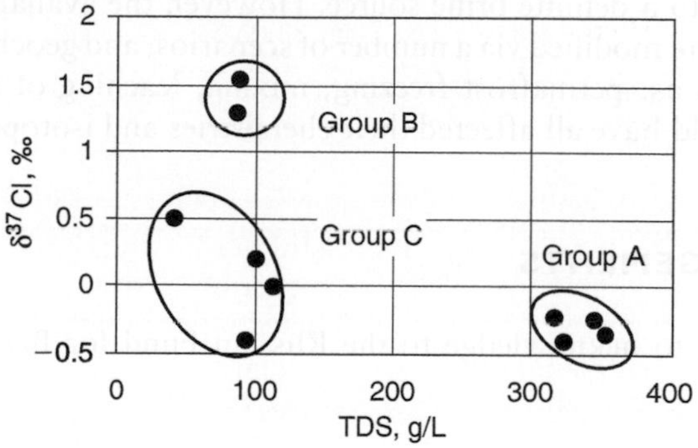

Figure 8.5 Plot of the $\delta^{37}Cl$ versus TDS for the cryopegs.

and $\delta^{81}Br$ values range between −0.4 and −0.2‰, and between −0.13 and +0.24‰, respectively.

Group B: the δ^2H and $\delta^{18}O$ values of the cryopeg samples range from −171 to −113.2‰ and from −21.4 to −9.57‰, respectively. The $\delta^{37}Cl$ value of one sample is +1.54.

Group C: the δ^2H and $\delta^{18}O$ values of this cryopeg samples range from −139.6 to −95.9‰ and from −16.45 to −10.2‰, respectively. The $\delta^{37}Cl$ and $\delta^{81}Br$ values range between −0.4 and +1.3‰, and between −0.8 and +2.31‰, respectively.

8.5 CONCLUSIONS

The cryopegs of the Yakutian diamond-bearing province are classified into three different groups based on their chemical composition and isotopic features (H, O, Cl and Br stable isotopes).

Group A includes Cl/Mg-Ca brines. They are pumped from depths 600 to 1450 m and are characterised by TDS of 224–404 g/L. Cryopegs temperatures vary from −2.6 to −1.0°C. The O, H and Cl stable isotope signatures of these cryopegs are the most enriched in comparison to the other groups. They are probably residual brines of evaporated paleoseawaters (Shouakar-Stash *et al.*, 2007).

Group B samples are Cl-Na type. They occur at shallower depths (200–400 m) and their TDS values range between 28 g/L and 165 g/L. Cryopegs temperatures vary from −2.5 to −0.5°C. The waters are the most depleted in O, H and Cl stable isotopes of all samples, and their signatures are different from group **A** signatures. Data obtained for group **B** indicate that these waters are derived from the halite dissolution, most likely as a result of recharge in a colder climate, possibly Pleistocene derived water.

Group C consists of Cl/Ca-Mg or Mg-Ca or Ca-Na type waters. Groundwater samples were collected from the depths of 110 to 650 m and their salinity varies from 31 to 252 g/l. Cryopeg temperatures vary from −4.0 to −3.0°C. The isotopic data are distributed between the isotopic values of group **A** and group **B**. Data interpretation could not lead to a definite brine source. However, the available data suggest that these waters were modified via a number of scenarios, and geochemical evolutionary processes, such as: permafrost freezing, mixing, leaching of salt and water–rock interaction could have all affected their chemistries and isotopes.

ACKNOWLEDGEMENTS

The authors wish to acknowledge to the Russian Fund for Basic Research (Project 08-05-00086).

REFERENCES

Balobaev V.T. (1991) Geothermy of permafrost of the Northern Asia. Nauka, Novosibirsk (in Russian).

Duchkov A.D., Balobaev V.T. (2001) The evolution of the thermal and phase condition in the Siberian cryilithozone // Environment global changes–2001. SB RAS publ.house, Novosibirsk (in Russian).

Eastoe C.J., Long A., Land L.S., Kyle J.R. (2001) Stable chlorine isotopes in halite and brine from the Gulf Coast Basin: Brine genesis and evolution. *Chem Geol* 176: 343–360.

Eggenkamp H.G.M., Coleman M.L. (2000) Rediscovery of classical methods and their application to the measurement of stable bromine isotopes in natural samples. *Chem Geol* 167: 393–402.

Frape S.K., Blyth A., Blomqvist R., McNutt R.H., Gascoyne M. (2004) Deep Fluids in the continents; II. Crystalline Rocks. In: Drever J.I. (Ed.), Surface and Ground Water, Weathering and Soils, In: *Treatise on Geochemistry*, vol. 5. Elsevier: 541–580.

Kaufmann R.S., Frape S.K., McNutt R., Eastoe C. (1993) Chlorine stable isotope distribution of Michigan Basin formation waters. *Appl. Geochem.* 8: 403–407.

Pinneker E.V., Borisov N.V., Kustov U.I., Brandt S.B., Dneprovskaya L.V. (1987) New data about isotope composition of oxygen and hydrogen of brines of the Siberian Platform. *Water Resour.* 3:105–115 (in Russian).

Pinneker E.V., Alexeev S.V., Borisov V.N. (1989) The interaction of brines and permafrost. *WRI-6 International Symposium. Proceedings*. Balkema, Rotterdam, 557–560.
Shouakar-Stash O., Alexeev, S.V. Frape, S.K. Alexeeva, L.P. Drimmie, R.J. (2007) Geochemistry and stable isotopic signatures, including chlorine and bromine isotopes, of the deep groundwaters of the Siberian Platform, Russia. *Appl Geochem* 22: 589–605.
Shouakar-Stash O., Frape S.K., Drimmie R.J. (2005) Determination of bromine stable isotopes using continuous-flow isotope ratio mass spectrometry. *Anal Chem* 77: 4027–4033.

Chapter 9

Hydrogeological studies in diapiric-layering salt formation: The eastern part of the Catalonian Potassic Basin

Fidel Ribera[1], Helena Dorca[1], Neus Otero[2], Jordi Palau[2] & Albert Soler[2]

[1] *Fundación Centro Internacional de Hidrología Subterránea (FCIHS), C/Provença, Barcelona, Spain*
[2] *Grup de Mineralogia Aplicada i Medi Ambient, Facultat de Geologia-Universitat de Barcelona, Departament de Cristallografia, Mineralogia i Dipòsits Minerals, Barcelona, Spain*

ABSTRACT

Nowadays, the hydrogeological studies in the Catalan Potassic Basin (CPB), located 70 km NW of the city of Barcelona, focus on the characterisation of the Tertiary and Quaternary alluvial aquifers in order to define a conceptual model of their behaviour. The hydrogeochemical and isotopic data revealed a regional fresh water input in the shallow Tertiary and Quaternary alluvial aquifers coexisting with two main salinization sources: 1) a saline natural source related to the Eocene marine salt layers which produce a salinity increase with depth, mainly in the Tertiary formation, and 2) an anthropogenic saline point contamination related basically with mine dumps (usually associated with organic compounds). Furthermore, the conceptual model allows understanding the role of groundwater flow and geologic faults in some salt mine subsidence problems in the CPB.

9.1 INTRODUCTION AND GEOGRAPHICAL SETTING

The Catalan Potassic Basin (CPB), located 70 km NW of the City of Barcelona, is usually described as a part of the Ebro basin, a Tertiary regional thick regressive-sedimentary basin that includes marine, evaporitic-transitional and continental facies, overlain by unconsolidated Quaternary alluvial sediments and affected by Alpine-related tectonic structures (faults and folding). The main surface water courses in the CPB are the Llobregat and Cardener Rivers (Fig. 9.1). The current hydrogeological knowledge of this area is poor and is basically limited to the shallower formations (less than 50 m deep).

The natural salinity of the Quaternary alluvial aquifer is low (with CE range between 1000 to 2000 μS/cm) and mainly related to the river and the lateral groundwater inflow from the Tertiary aquifers. In the Tertiary aquifers the salinity, basically related to Cl^-, SO_4^{2-}, Na^+, K^+ and Mg^{2+} ions, increases with depth (Ribera *et al.*, 2010). The piezometric relation between these two aquifers controls the hydrochemistry in the basin. Both aquifers, especially the alluvial one, are used for urban supply or local irrigation, and the Llobregat River is the main recharge source for the lower

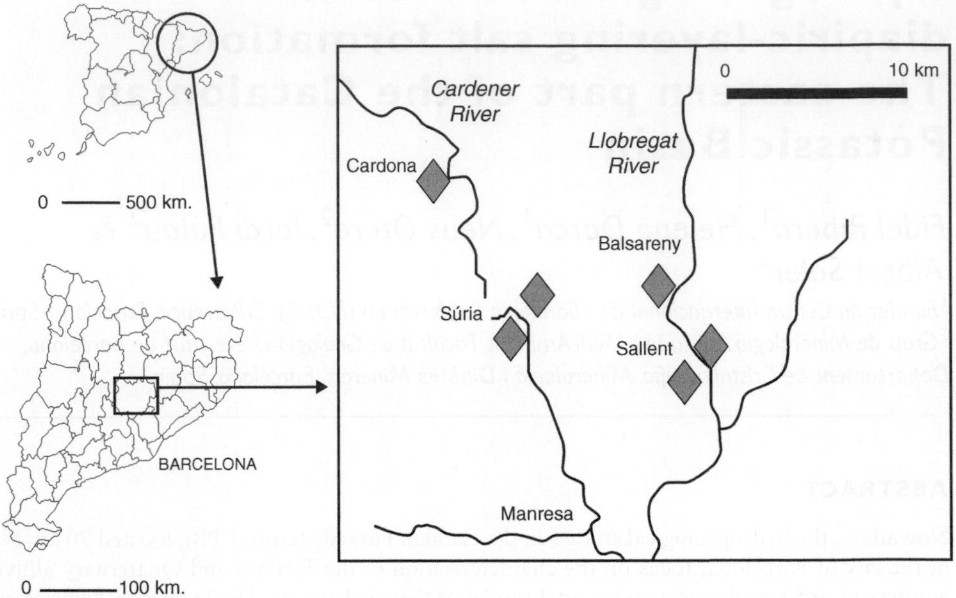

Figure 9.1 Geographical map of the area showing the rivers and the location of the actual main salts mining works.

valley and deltaic aquifers, located 60 km SE, which are a strategic fresh water reservoir for Barcelona Metropolitan area.

The intense underground mining activity in the CPB for K-salts mining (Sylvite and Carnallite) in Sallent, Cardona and Súria since the last century, has generated abandoned mine galleries and cavities, and saline springs and/or the salinization of fresh springs (Fig. 9.2a). Groundwater contamination has also been caused mainly from dumps (Fig. 9.2b).

In the area, the existence of subsidence and dissolution sinkholes areas around the towns of Sallent and Cardona (Fig. 9.2c) are probably modifying the local hydraulic gradient (Ribera *et al.*, 2010).

The objective of this chapter is to reflect the main conclusions from the latest hydrogeological investigations in the CPB carried out during 2008–2010. That work underpins the future establishment of integrated water management for the area, which includes the coexistence of the water resource protection, and the minimisation of subsidence-collapsing processes, with the continuation of the actual modern mining activities in a sustainable context.

9.2 HYDROGEOLOGICAL BACKGROUND

Nowadays, the hydrogeological work in the CPB area is focused on the characterisation of the Tertiary and Quaternary alluvial aquifers in order to define a conceptual

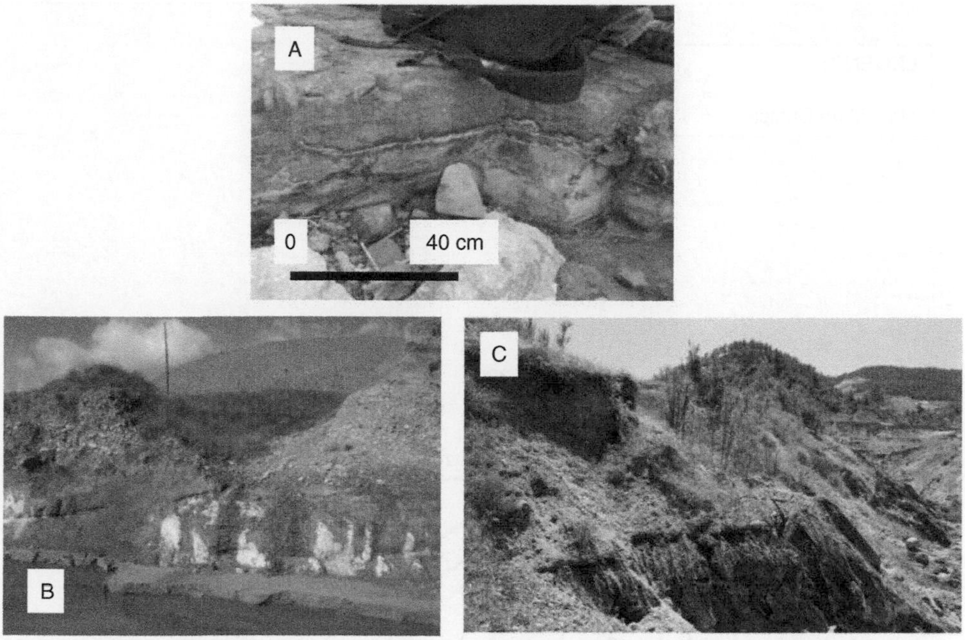

Figure 9.2 A) High salinity groundwater flowing across the carbonate Tertiary fracture system and related white halite (close to the location of Sallent (see Fig. 9.1)). B) Salt dumps mainly composed of halite, with minor concentrations of K-chloride minerals, and mud flotation tails (Dorca *et al.*, 2009). Analysis of mud flotation samples revealed the presence of volatile and semi-volatile organic compounds C) Locations were underground potash mining provoked subsidence and collapse problems. In these areas, salt dissolution by groundwater flow plays a critical role, e.g. part of the Cardener alluvial meander in Cardona, collapsed in 2002 due to failure of a salt cavern beneath it.

groundwater flow model. Several investigation wells were drilled and piezometers were installed at different depths in both aquifers. Geophysical data, groundwater levels, hydraulic parameters, thermal, hydrochemical and isotopic data were obtained. The regional piezometric surface of the Tertiary aquifer units described a flow direction from N-S to NW-SE (Fig. 9.3). However, several irregularities occur: (1) the existence of preferential flow across the main fracture zones locally modifies the general distribution of the hydraulic gradient (Fig. 9.3). In some of these areas (e.g. Llobregat River, in Sallent) the piezometric level could be periodically higher than the river bed surface level; (2) permanent main rivers (Llobregat and Cardener) and their second or third order ephemeral tributaries are the principal drain systems of the alluvial and Tertiary aquifers; and (3) the bigger saline dumps locally modify the piezometric surface and act as a recharge dome.

The distribution of the hydraulic parameters showed lower transmissivity (T) values in Tertiary formations (from 0.1 to 15 m^2/d) and higher values in alluvial sediments (from 1 to 400 m^2/d). Otherwise, the transmissivity in fractured Tertiary areas could be 1 or 2 orders of magnitude higher than the regional average.

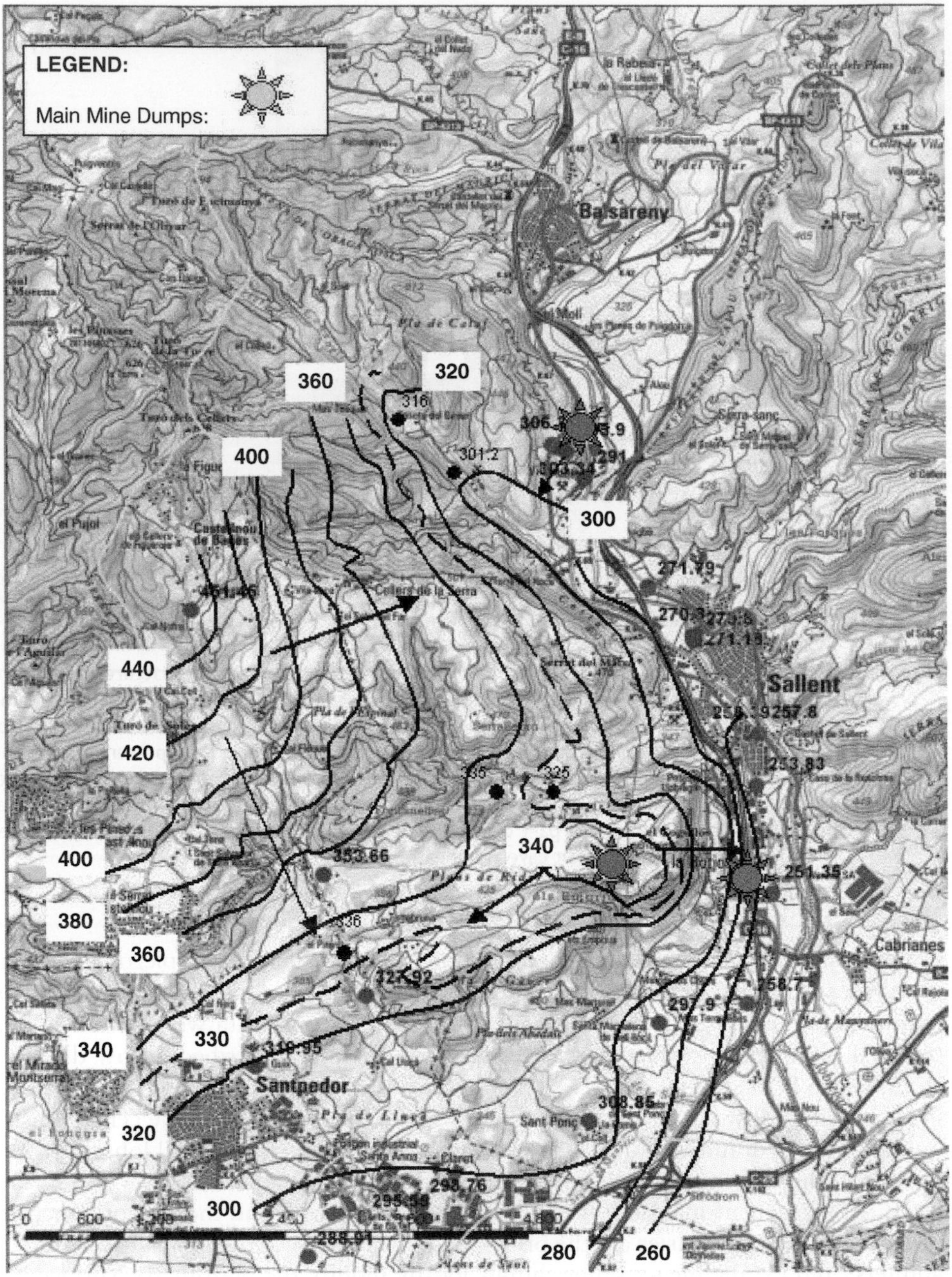

Figure 9.3 Regional piezometric map of the Tertiary aquifer in Llobregat-CPB area (data obtained Partially from Escorcia, *et al.*, 2009).

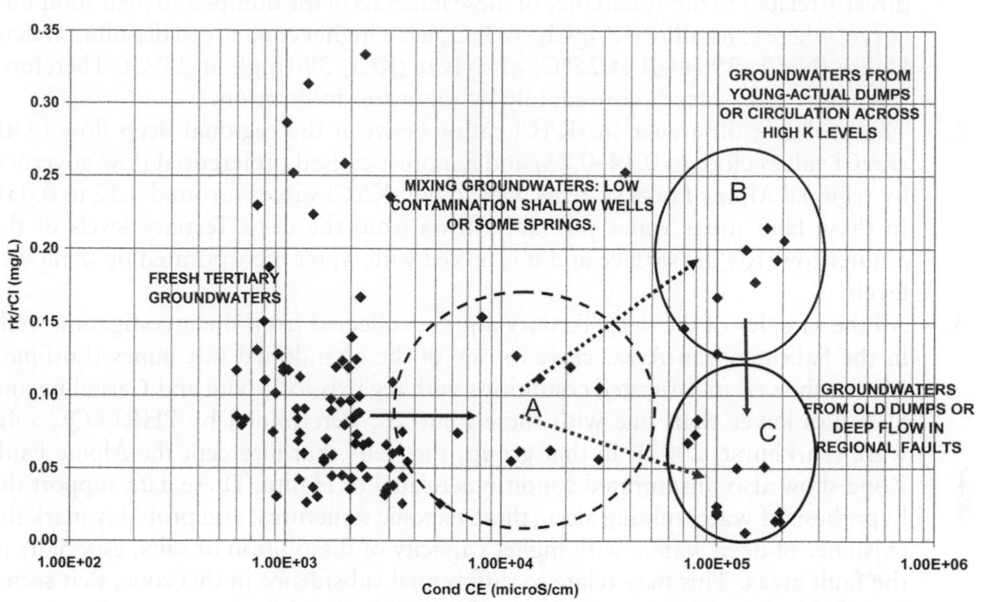

Figure 9.4 Ion ratio (rK/rCl) versus electrical conductivity. Group A: Fresh regional waters partially affected by salinization. Group B: saline groundwater not affected by selective removal of K (water recharged from actual dumps or Tertiary regional deep water circulating along K-rich formations), Group C: saline groundwater affected by selective removal of K-rich minerals (water recharged from old dumps or preferential flux along regional faults).

9.3 HYDROGEOCHEMISTRY AND ISOTOPIC STUDIES

The hydrogeochemical and isotopic data reveal a regional fresh water meteoric input in the shallow Tertiary and Quaternary alluvial aquifers, coexisting along with two main saline sources (Fig. 9.4):

1. A natural saline source related to the Eocene marine salt layers that showed a salinity increase with depth (up to 200 000 mg/l of Cl^- and 34 000 mg/l of K^+ in the deeper levels) across the Tertiary formation.
2. Anthropogenic saline point sources mainly related with K-mine dumps and other mine facilities (usually associated with the presence of volatile and semi-volatile organic compounds). In this context, the impact of the saline plumes on the aquifers is found in the shallower parts.

The ratio between K^+ and Cl^- (rK/rCl) has been applied to define other hydrogeological controls. The rK/rCl values, ranging from 0.01 to 0.35, reveal the following main effects:

1. The existence of a rK/rCl specific signature related to the age of the dumps, where old dumps have lower rK/rCl ratios than the new ones (Fig. 9.4). This ratio is

directly related to the abundance of these minerals in the dumps and their solubility curve, where Carnallite ($KMgCl_3 \cdot 6H_2O$) has a higher capacity of dissolution than Halite (NaCl, 390 mg/l at 25°C) or Sylvite (KCl, 380 mg/l at 25°C). Therefore, Carnallite is dissolved more rapidly by rainwater infiltration.

2. The probable difference in rK/rCl ratios between the regional deep flow (with rK/rCl values close to 0.14–0.25) and a more localised preferential flow governed by regional Alpine faults systems with lower rK/rCl values (around 0.12 to 0.05). In these fault areas water probably flows from the deep Tertiary levels of the aquifer towards the surface and it is mixed with water incorporated by shallower levels.
3. All the samples of the deep Tertiary aquifer collected from the investigation wells in the Sallent urban Area, close to one of the abandoned salt mines (Enrique's Mine), showed unsaturated conditions with regards to Sylvite and Carnallite and SI values lower than one with these minerals, determined by PHREEQC software (Parkhurst, 2002). In that group, the wells that intercept the Alpine Fault Zone show also unsaturated conditions related to Halite. These data support the hypothesis of water mixing along these tectonic structures, and probably mark the existence of deep waters with higher capacity of dissolution of salts, especially in the fault areas. This may relate to differential subsidence in this zone, that seems to be related to the Enrique's mine galleries.
4. A modification of the rK/rCl ratio influenced by the input of NPK (nitrogen-phosphorous-potassium) fertilisers. High nitrate concentrations, related to high dispersion of rK/rCl ratio, were detected in the shallower freshwater aquifers, coexisting in some samples with values up to 150 mg/L of NO_3, and even the deeper wells showed measurable nitrate concentrations, up to 5 mg/l, in wells of more than 100 m deep (FCIHS, 2010b).

The isotopic composition of dissolved sulfate ($\delta^{34}S$ and $\delta^{18}O$) has been applied as a tool to distinguish the origin of salinity in the CPB (Otero & Soler, 2002; 2003). The shallow Tertiary wells (no. 1, 3, 13 and 111 in Fig. 9.5) show a $\delta^{34}S$ ranging from +4 to +10‰ and a $\delta^{18}O$ from 13.8 to +15.2‰ indicating a mix between natural evaporitic sulphate (Gypsum Upper Tertiary) and modern fertilizer sulphate (Fig. 9.5). The influence of K^+-mine tailings in shallow wells (no. 18 or 20 in Fig. 9.5) is also apparent. With regards to the deep Tertiary wells, well S-3 showed isotopic values in agreement with natural evaporite rock interaction (Gypsum Upper Tertiary) with a minor contribution of sulfate from fertilizers (around 10%). Wells S-1 and S-2 fall in the mixing area between the three sources: a) natural evaporite rock interaction (Gypsum Upper Tertiary), b) fertiliser sulfate, and c) mine tailings and/or Halite Deep Level. In S-1 and S-2 wells is not possible to distinguish between the latter two sources, mine tailings and/or Halite-Sylvite Deep Level, since they show a similar isotopic signature.

9.4 CONCEPTUAL MODEL

Direct rainfall recharge is the main input of water to the CPB Tertiary aquifers, complemented by lateral groundwater inputs mainly from the north. This unit behaves as

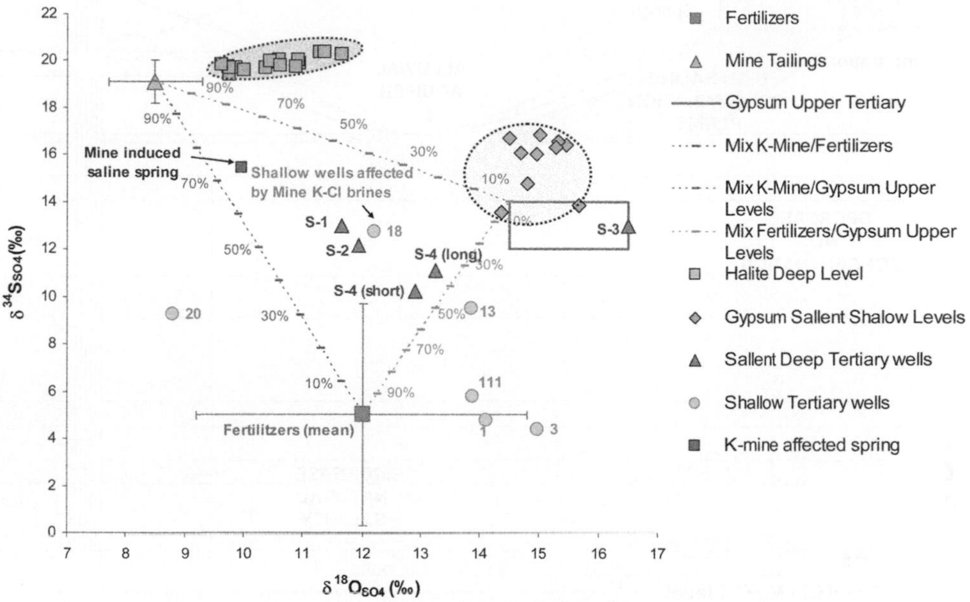

Figure 9.5 δ^{34}S vs. δ^{18}O of dissolved sulphate. The main natural and anthropogenic sources are depicted, together with the studied wells. Fertilizers data are from Vitòria *et al.* (2004), mine tailing values are from Otero and Soler (2002, 2003), data of evaporite bedrock (Gypsum from Sallent shallower levels and halite from deep levels) are from the present study. Mixing lines are represented between the main end-members.

a typical multilayered aquifer. Porosity is essentially secondary, where water flows by strata discontinuities or major faults. The regional hydraulic parameters of the Tertiary aquifer are low, with T values from around 0.01 to 1 m^2/d, and the existence of Alpine faults which promote a local increment in hydraulic conductivity. In some hydraulic assays it could be two orders of magnitude higher than regional values. In these fault areas, an increase of the vertical thermal gradient of water in wells, compared with the regional ones was also detected.

Rivers are the main natural drainage system of the Tertiary aquifer (Fig. 9.6). The main part of the groundwater is fresh water, without important Cl^- or Na^+ concentrations and moderate amounts of SO_4^{2-} (between 144 to 1440 mg/L of SO_4). However, point and diffuse saline springs related to contamination from mine dumps and/or the natural piezometric regime occur. These saline springs mainly seem to appear in the Llobregat River, close to the town of Sallent (Fig. 9.1), and in the upper part of Conangles River (2 km north of Sallent). They are related in both situations to fault systems. The deeper parts of the Tertiary aquifer have higher Cl^-, Na^+ or K^+ concentrations and lower NO_3^- concentrations relative to the upper parts. This vertical distribution is locally modified by contamination related to the mining activities, with high salinity plumes moving across the upper Tertiary formations. If the geological and piezometric conditions are favourable, the saline contamination plumes follow the same preferential pathways as the regional flow, producing in some cases the salinisation of fresh springs.

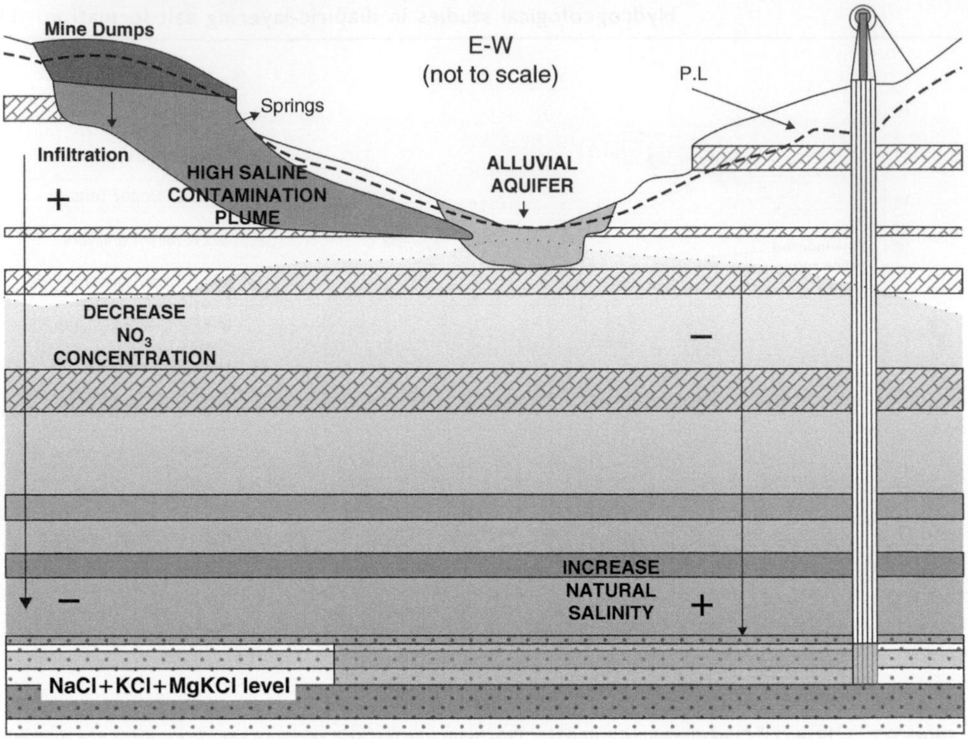

Figure 9.6 Cross-conceptual section across the main aquifers in the CPB zones and mine dumps. P.L: Piezometric Level.

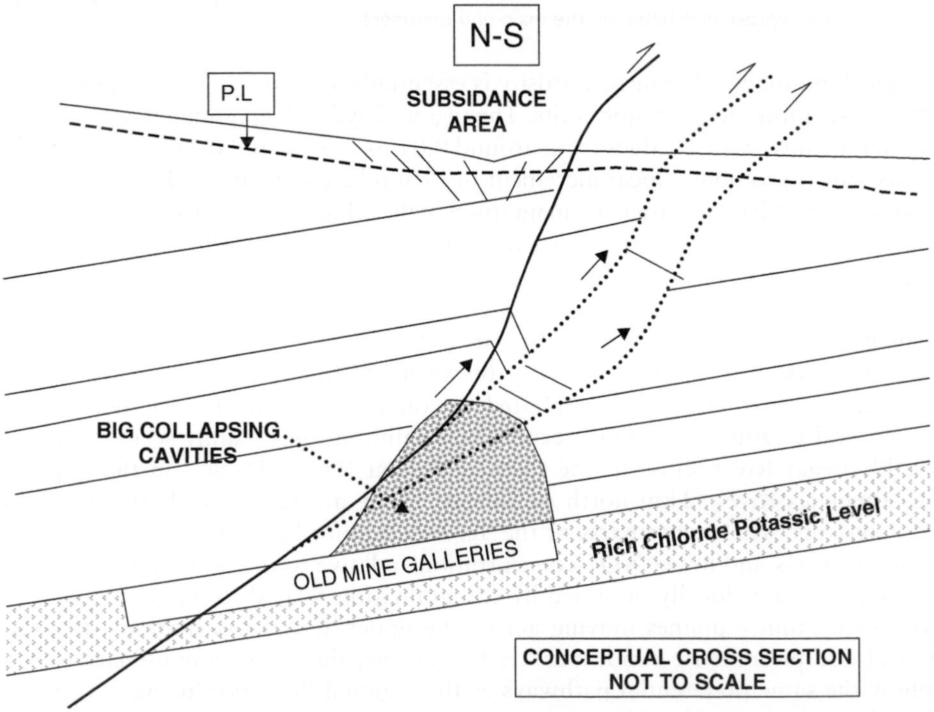

Figure 9.7 Conceptual relation between subsidence and groundwater in the Llobregat Sector of the CPB. P.L: Piezometric Level.

The large faults and/or other types of discontinuity, such as the lithological contact between the roof of the salt levels and the rest of the Tertiary series, are preferential flow paths where the unsaturated water containing potassium chloride minerals occurs. This type of water has a high salt dissolution potential and may induce or accelerate problems related to instability and collapse in some old potash mines in the CPB (Fig. 9.7).

To solve these problems, in some cases, engineering and socio-economic solutions have been proposed and executed, including the restoration of some tailings dumps. In the case of subsidence, preventive actions have been taken.

ACKNOWLEDGEMENT

This work was funded by de contract FBG305318 with Servei Geologic de Catalunya and partially by the projects CICYT-CGL2008-06373-C03-01 of the Spanish Government and the project 2009SGR00103 from the Catalan Government. We would like to thank the Serveis Científics Tècnics of the Universitat de Barcelona.

REFERENCES

Dorca H., Ribera F., Escuder R., Palau J., Otero N., Puig R., Soler A. (2009) Caracterización hidrogeológica de la zona de influencia de la antigua escombrera de Vilafruns (Balsareny, Provincia de Barcelona, España). *Contaminación y protección de los recursos hídricos*. AIH-GA. IV Seminario hispanoamericano sobre temas actuales de la hidrología subterránea. 73–85 ISBN-978-987-1082-40-7. 1 ed. Buenos Aires. Argentina.

Escorcia J., Ferreira do Rosario F., Marsily G. (2009) Estudio hidrogeológico de los acuíferos terciarios y cuaternarios del sector de Sallent, Balsareny, Santpedor, Súria y Callús. 240 pp., 43 CIHS, FCIHS. Barcelona.

Otero N., Soler A. (2002) Sulphur isotopes as tracers of the influence of potash mining in groundwater salinization in the Llobregat Basin (NE Spain). *Water Research* 36: 3989–4000.

Otero N., Soler A. (2003) Stable isotopes of dissolved sulphate as tracers of the origin of groundwater salinization in the Llobregat River (NE Spain). *Extended Synopsis of the International Symposium on Isotope Hydrology and Integrated Water Resources Management, IAEA*, Vienna, 2003, 190–191.

Parkhurst D. (2002) PHREEQC Version. 2. A computer program for Speciation, Batch-reaction, One-Dimensional Transport and Inverse Geochemical Calculation. *USGS*. Fact Sheet FS-031-02.

Ribera F., Dorca H., Martínez P., Piña J., Otero N., Palau J., Soler N. (2010) Estudio hidrogeológico de la Cuenca Potásica Catalana en el entorno de Sallent y la Antigua Mina Enrique. *Boletín Geológico y Minero*. 120, no. 4: 607–616

Vitòria L., Otero N., Soler A., and Canals A. (2004) Fertilizer Characterisation: Isotopic Data (N, S, O, C, and Sr). *Environmental Science and Technology* 38: 3254–3262.

Chapter 10

Hypogenic karst development in a hydrogeological context, Buda Thermal Karst, Budapest, Hungary

Anita Erőss[1]*, Judit Mádl-Szőnyi*[1] *& Anita É. Csoma*[2]

[1]*Department of Physical and Applied Geology, Institute of Geography and Earth Sciences, Eötvös Loránd University, Budapest, Hungary*

[2]*ConocoPhillips, Subsurface Technology, Basin and Sedimentary Systems, Houston, TX, USA*

ABSTRACT

According to the recent developments in the speleogenetic theories, hypogenic karsts and caves are viewed in the regional flow system concept and can be considered as manifestations of flowing groundwater. Therefore a comprehensive hydrogeological study was carried out for the characterisation of processes acting today at the discharge zone of the Buda Thermal Karst. Methods included hydrogeochemical and mineralogical investigations. Among the results of the study, several processes were identified which can be responsible for cave development and formation of minerals. Furthermore, the role of the adjacent sedimentary basin was re-evaluated. These results bring a new insight into the processes acting at a regional discharge zone, which could be responsible for hypogenic cave development. The Buda Thermal Karst system can be considered as the type area and in the same time the modern analogue for hypogenic karsts.

10.1 INTRODUCTION

As groundwater plays a crucial role in karst development, karst phenomena, caves and precipitates can be considered as geological manifestations of flowing groundwater (Tóth, 1999; Klimchouk, 2007). Two principal categories of continental karst systems – the epigenic and hypogenic karsts – are distinguished based on their position within the groundwater flow system (Klimchouk, 2007). The "classical" epigenic karst systems are predominantly local systems, and/or parts of recharge segments of intermediate and regional systems. Hypogenic karst areas are associated with discharge regimes of regional or intermediate flow systems.

According to Tóth's (2009) "virtual spring" concept, and his idea of groundwater being a geological agent (Tóth, 1999), several processes can be present and active simultaneously at a regional discharge zone where flow systems of different orders, conveying waters with different temperature, chemical composition and redox-state, meet. However, the predominance of one or other of the individual processes may determine the main character of the observed discharge phenomena (e.g. caves). The investigation of the effects and manifestations of flow systems plays a crucial role in the understanding of the flow system itself (Tóth, 1999; Tóth, 2009).

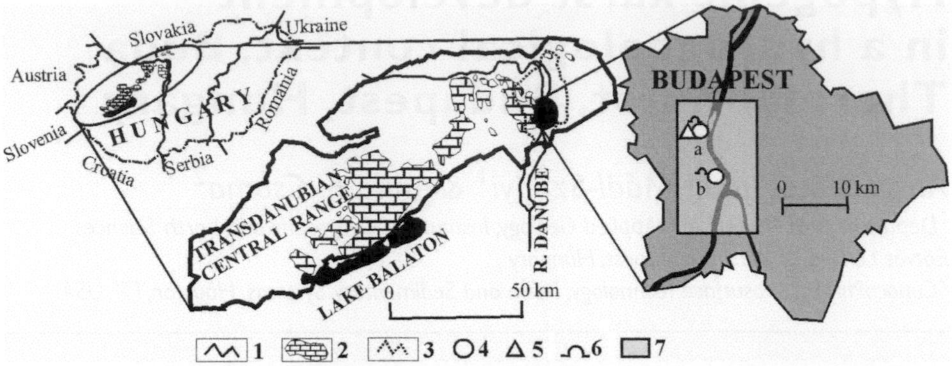

Figure 10.1 Location of the Buda Thermal Karst in the Transdanubian Central Range and the study area in Budapest. 1: Subsurface boundary of Mesozoic carbonates, 2: Uncovered Mesozoic carbonates, 3: Buda Thermal Karst, 4: Warm and hot springs (30–60°C), 5: Lukewarm springs (20–30°C), 6: Hypogenic caves, 7: Study area, a: Rózsadomb discharge area, b: Gellért Hill discharge area.

The recently developed "virtual spring" concept of Tóth (2009) was applied as a basic approach of the study presented in this chapter. Being a marginal area at the boundary of uplifted carbonates and a sedimentary basin, the Buda Thermal Karst serves as a discharge zone of the regional fluid flow. This implies that it may receive fluid components from several sources resulting in a wide range of discharge features including springs, caves, and mineral precipitates. The identification and understanding of recently active processes and their manifestations in this hypogenic karst area will help to identify and understand palaeo phenomena both in the Buda Thermal Karst and in other hypogenic karst areas with similar settings. To achieve this, a comprehensive hydrogeological study was carried out for the characterisation of processes acting today at the discharge zone of the Buda Thermal Karst. Two areas of the discharge zone were chosen as study areas: the Rózsadomb and Gellért Hill (Fig. 10.1). In these areas fluids with various temperatures and chemical compositions discharge, and palaeo and recently developing hypogenic caves occur. In addition, deep wells on the Pest side were previously used for the characterisation of the upwelling hydrothermal fluids.

10.2 STUDY AREA

The Buda Thermal Karst forms the NE extremity of the Transdanubian Central Range (TCR) (Fig. 10.1). The lithological continuity of Triassic carbonate series of the TCR facilitates the hydraulic continuity of the groundwater body of the TCR. The aquifer system of the TCR is characterised mainly by gravity driven groundwater flow (Alföldi, 1982). One of the regional discharge zones of the system is located in Budapest, and it is separated by a step-faulted boundary from the subsided basin to the east (Pest) and the uplifted hilly range in the west (Buda). The course of the Danube River follows this boundary and represents the base level of erosion (Fig. 10.1).

Within the regional discharge zone this study focuses on the Rózsadomb and Gellért Hill areas (a and b on Fig. 10.1). Rózsadomb is built up of Eocene limestone and covered mainly by Eocene-Oligocene marl and Pleistocene clayey detrital blanket. The Triassic dolomite here is deep-seated. Gellért Hill is built up of Triassic dolomite and is partly covered by marl, but at the nearest point to the Danube it occurs as a bare karst surface (Fig. 10.2). Since the current hydrogeological system has an artificially influenced groundwater discharge the primary discharge features – for instance location of the upwelling, temperature, chemical composition of the springs – were summarised based on the evaluation of historic hydrogeological data by Eross *et al.* (2008). This study revealed strong structural control on the spring locations in both areas. In case of the Rózsadomb discharge zone, the distribution of lukewarm (20–35°C) and hot springs (40–65°C) was found to be clearly separated. Furthermore, their chemistry was also different, the hot springs were characterised by higher TDS and CO_2 content (800–1350 and 200–400 mg/l, respectively). In the Gellért Hill discharge zone the springs could be characterised by temporally and spatially uniform temperature (33.5–43.5°C) and chemical composition (1450–1700 mg/L TDS) which clearly differs from the Rózsadomb waters.

10.3 DATA AND METHODS

Recent hydrogeochemical measurements were carried out in order to give an overview about the current distribution of physico-chemical parameters including major and trace components. Thirty fluid samples have been collected from springs (measuring points 3, 4, 5, 19, 23, 26, 27, 28, 30 on Fig. 10.2) and from wells (all other points on Fig. 10.2). The first reconnaissance sampling (10 samples) was accomplished in October 2008 followed by a comprehensive sampling campaign (20 samples) in June 2009. In-situ physico-chemical parameters (electrical conductivity (EC), temperature, pH, dissolved O_2) were recorded during sampling. Water samples were taken for chemical analyses in 1.5 l PET bottles. For nitrate analysis, 20 ml water samples were taken into 0.1 l PET bottles to which 2.8 ml acetonitrile was added. The samples were kept cool before delivery to the laboratories. The measured parameters and analytical methods are summarised in Table 10.1, while Table 10.2 shows the results.

The sampled wells 11, 14 and 16 are not routinely pumped. These wells were therefore pumped for several hours or even days, and the sample was taken when the conductivity and pH were stable, i.e. when the sampled water had reached the formation water characteristics. However, the water temperature could not reach the previously measured values (i.e. those during construction or continuous operation). These data points are therefore assigned with empty symbols on chemical hand temperature diagrams. The original formation temperatures were 74°C (point 11), 49°C (point 16) and 45.5°C (point 14). The composition of minerals was investigated by X-Ray diffraction.

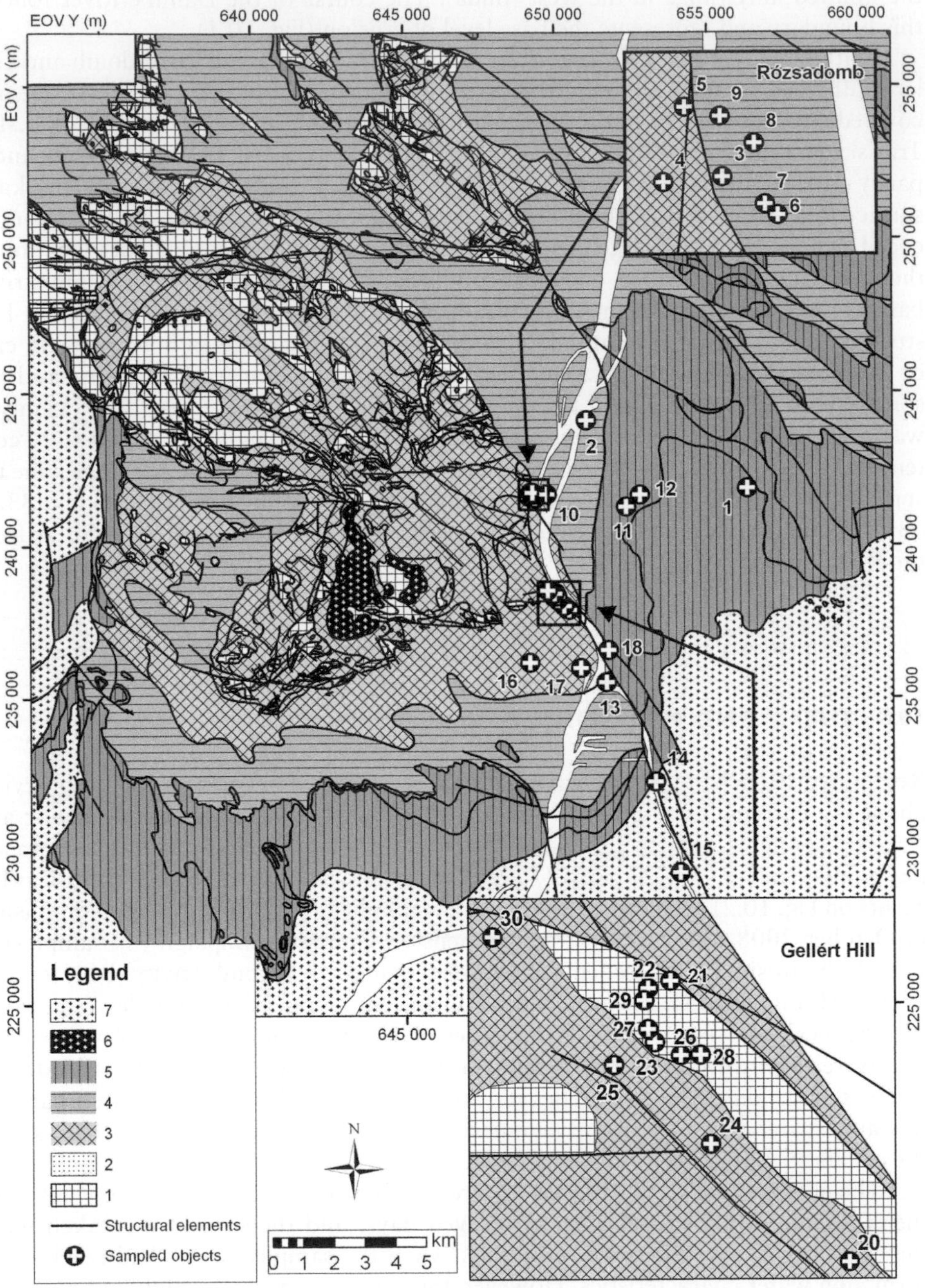

Figure 10.2 Location of the sampled springs and wells indicated on the pre-Pliocene geological map of the Buda Hills, modified after Fodor (unpublished). 1: Triassic carbonates; 2: Bauxitic clay (Eocene?); 3: Eocene to Lower Oligocene carbonate and siliciclastic; 4: Upper Oligocene siliciclastic; 5: Lower and Middle Miocene siliciclastic; 6: Upper Miocene travertine; 7: Upper Miocene siliciclastic and carbonate.

Table 10.1 Summary of the analyzed parameters, methods and laboratories.

Parameters analyzed	*Method*	*Laboratory*
pH, temp., EC, O_2	WTW Multi-Parameter 650i	on site
cations	ICP-AES	MÁFI[1] Laboratory
anions	ion chromatography and ICP-AES	MÁFI[1] Laboratory
trace elements	ICP-MS	MÁFI[1] Laboratory
free and bound CO_2	titrimetry	Wessling Laboratory
dissolved sulphide	UV-VIS spectrophotometry	Wessling Laboratory
methane	GC	Wessling Laboratory
liquid hydrocarbon	GC-MS	ELTE[2] Dept. of Analytical Chemistry
mineral composition (precipitates)	XRD	ELTE[2] Dept. of Mineralogy

[1]Hungarian Geological Institute, [2]Eötvös Loránd University.

10.4 RESULTS

10.4.1 Characteristics of discharging fluids

All investigated waters belong to the Ca + Mg, Na + K cation facies according to Back (1961). However, there are differences in the anion facies: the wells in South Budapest, the waters around the Gellért Hill and some deep wells from the (northern) Pest side (measuring point No. 10 to 30) belong to the Cl+SO_4, HCO_3 anion facies, while all the others to the HCO_3, Cl + SO_4 facies (Table 10.2).

The total dissolved solids (TDS) content of the sampled waters as the function of temperature is displayed on Figure 10.3, where linear increasing trend can be observed between measuring point 3 and 12. These measuring points (1 to 12) will be hereinafter referred as North System that includes the lukewarm and hot waters from the Rózsadomb discharge area and deep wells from the opposite Pest side. Measuring points 13 to 30 will be referred as the South System. To this group belong the waters around the Gellért Hill and deep wells from South Budapest. The waters of the South System form a cluster on the TDS-temperature plot. These waters have narrower temperature range (31.4–46.7°C), and their TDS content is higher (1370–1790 mg/l) than the waters with the highest temperatures (65.5°C) and TDS (1360 mg/l) in the North System.

The same detachment of the two systems can be observed on the Na^+, Mg^{2+}, Ca^{2+}, Cl^-, HCO_3^- and SO_4^{2-} vs. temperature diagrams (Fig. 10.4a–f). Based on these diagrams the South System can be characterised by higher Ca^{2+}, Mg^{2+}, HCO_3^- and SO_4^{2-} contents than the waters of the North System. In case of the South System the high Na^+ and Cl^- content is coupled with lower temperature (31.4–46.7°C), compared with the hot waters (65.5°C) of the North System (Fig. 10.4a–b). The North System shows linear trends of the main components with increasing temperature, while Mg^{2+} is inversely related to temperature. For the South System such linear relation between the main components and temperature cannot be observed, but the data form a cluster.

Table 10.2 Summary of the hydrogeochemical data of the investigated springs and wells.

Object Nr.	*Date of sampling*	*Reference of point elevation (m asl)*	*Temp. (°C)*	*EC (μS/cm) (25°C)*	*pH*	O_2 *(mg/l)*	Na^+ *(mg/l)*	K^+ *(mg/l)*	Ca^{2+} *(mg/l)*	Mg^{2+} *(mg/l)*	Fe^{2+} *(mg/l)*	Cl^- *(mg/l)*
1	24.6.2009	−1288.510	63.90	1245	6.19	–	95	12.4	131.00	38.8	0.03	96.7
2	24.6.2009	−18.850	37.90	1173	6.45	1.82	68.3	8.21	138.00	39	0.01	73.6
3	25.6.2009	102.000	20.80	909	6.94	2.23	21.8	3.15	117.00	43.5	<0.005	32.3
4	25.6.2009	104.000	21.70	932	6.94	2.53	24.7	3.43	118.00	43.6	0.006	30.9
5	07.10.2008	104.000	27.70	1046	6.92	1.43	31.6	4.23	119.00	43.3	0.03	47.2
6	07.10.2008	27.965	51.40	1563	6.35	–	115	13.2	148.00	39.1	0.18	135
7	25.6.2009	11.170	51.40	1403	6.27	–	106	13.1	151.00	38.7	0.07	114
8	25.6.2009	10.280	44.10	1285	6.30	0.66	85.7	10.5	142.00	39.6	0.04	102
9	25.6.2009	76.100	53.50	1494	6.42	–	125	14.5	150.00	37.2	1.67	144
10	07.10.2008	−201.960	60.20	1707	6.26	–	151	17	158.00	36.5	0.04	172
11	02.7.2009	−841.220	38.70	1749	6.17	1.60	158	18.4	151.00	32.6	0.2	180
12	07.10.2008	−1143.640	65.50	1804	6.14	–	172	18.6	154.00	34	0.04	192
13	24.6.2009	−355.200	44.40	1867	6.34	2.26	155	19.3	188.00	55.3	0.11	135
14	25.6.2009	−558.829	35.50	1864	6.63	2.21	152	19.4	185.00	57.4	0.18	168
15	24.6.2009	−1033.275	46.70	1992	6.30	2.20	182	20.5	182.00	62.3	0.08	187
16	25.6.2009	−417.860	30.00	2160	6.37	0.83	203	24.2	196.00	56	10.2	193
17	22.6.2009	−67.645	39.60	1955	6.22	1.97	174	20.7	183.00	54.3	0.06	187
18	08.10.2008	−239.872	45.30	1934	6.32	1.30	151	19.2	188.00	58.3	0.11	146
19	08.10.2008	95.000	39.70	1845	6.71	3.23	145	21	182.00	65.5	0.07	160
20	08.10.2008	91.710	42.50	1785	6.41	2.80	135	17.5	178.00	55.5	0.06	134
21	25.6.2009	68.030	44.60	1750	6.28	0.76	142	18.5	179.00	53.6	0.008	131
22	25.6.2009	86.100	43.10	1767	6.41	0.81	142	19.1	182.00	54.2	0.02	123
23	25.6.2009	104.000	37.10	1659	6.46	0.90	123	17.4	175.00	53.8	0.01	113
24	25.6.2009	101.445	40.10	1653	6.45	2.06	119	16.8	177.00	54.4	0.13	110
25	25.6.2009	102.525	43.20	1650	6.59	0.73	121	16.6	180.00	54	0.01	138
26	30.6.2009	104.000	37.00	1697	6.49	0.62	122	17.1	174.00	53.6	0.09	130
27	30.6.2009	104.000	33.60	1704	6.68	2.45	121	16.8	180.00	54.5	0.17	141
28	08.10.2008	104.000	31.40	1740	6.88	2.47	128	17.2	180.00	57.3	0.09	142
29	08.10.2008	75.020	40.90	1771	6.69	0.86	131	17.5	179.00	56.2	0.03	122
30	08.10.2008	102.000	37.40	1719	6.56	1.47	126	17.2	179.00	55.4	0.55	125

(–: no data available)

HCO_3^- (mg/l)	SO_4^{2-} (mg/l)	NO_3^- (mg/l)	NO_2^- (mg/l)	PO_4^{3-} (mg/l)	S^{2-} (mg/l)	TDS (mg/l)	H_2SiO_3 (mg/l)	free CO_2 (mg/l)	bound CO_2 (mg/l)	CH_4 (mg/l)	Error %	Hydrochemical facies
535	152	<0.1	3.52	0.02	0.31	1110	43.3	210	188	0.04	−2.73	Ca+Mg, Na+K – HCO_3, Cl+SO_4
535	123	<0.1	3.37	0.02	0.44	1020	28.8	130	192	0.05	−1.43	Ca+Mg, Na+K – HCO_3, Cl+SO_4
428	116	7.15	2.66	0.03	0.01	786	13.8	50	149	–	−3	Ca+Mg, Na+K – HCO_3, Cl+SO_4
452	134	11.8	2.23	0.02	<0.01	873	14.6	40	155	–	−0.4	Ca+Mg, Na+K – HCO_3, Cl+SO_4
444	127	1.59	2.84	0.07	<0.01	837	15.8	50	159	–	−2.83	Ca+Mg, Na+K – HCO_3, Cl+SO_4
531	170	<0.1	<0.1	0.02	0.52	1190	38.6	170	193	0.16	−0.26	Ca+Mg, Na+K – HCO_3, Cl+SO_4
535	181	<0.1	4.42	0.03	0.52	1180	39.1	170	193	0.12	−1.32	Ca+Mg, Na+K – HCO_3, Cl+SO_4
511	165	<0.1	3.91	0.02	0.39	1090	30.7	130	183	–	−2.28	Ca+Mg, Na+K – HCO_3, Cl+SO_4
547	176	<0.1	4.14	0.02	0.57	1240	40.0	150	201	0.09	−1.8	Ca+Mg, Na+K – HCO_3, Cl+SO_4
531	206	<0.1	<0.1	0.02	0.76	1320	52.2	190	193	0.16	0.26	Ca+Mg, Na+K – Cl+SO_4, HCO_3
547	195	<0.1	<0.1	0.04	0.76	1340	56.4	290	200	0.53	−1.52	Ca+Mg, Na+K – Cl+SO_4, HCO_3
542	186	<0.1	<0.1	0.03	0.54	1360	57.2	120	192	<0.04	0.93	Ca+Mg, Na+K – Cl+SO_4, HCO_3
606	351	<0.1	3.22	0.02	0.66	1550	34.2	110	209	<0.04	−0.21	Ca+Mg, Na+K – Cl+SO_4, HCO_3
616	337	<0.1	2.98	0.03	0.52	1570	32.7	100	214	<0.04	−2.54	Ca+Mg, Na+K – Cl+SO_4, HCO_3
665	319	<0.1	3.69	0.01	0.47	1660	32.5	180	230	<0.04	−0.98	Ca+Mg, Na+K – Cl+SO_4, HCO_3
653	403	<0.1	3.98	0.02	0.55	1790	47.8	290	236	0.22	−1.32	Ca+Mg, Na+K – Cl+SO_4, HCO_3
583	340	<0.1	4.31	0.02	0.72	1580	34.3	130	208	<0.04	−1.23	Ca+Mg, Na+K – Cl+SO_4, HCO_3
585	366	<0.1	<0.1	0.04	0.55	1550	34.1	130	207	<0.04	−0.15	Ca+Mg, Na+K – Cl+SO_4, HCO_3
531	375	1.1	<0.1	0.07	<0.01	1520	32.9	80	186	–	0.63	Ca+Mg, Na+K – Cl+SO_4, HCO_3
553	337	<0.1	<0.1	0.04	0.37	1440	31.1	150	197	<0.04	−0.2	Ca+Mg, Na+K – Cl+SO_4, HCO_3
547	379	0.44	3.64	0.03	0.96	1490	32.2	160	194	–	−2.11	Ca+Mg, Na+K – Cl+SO_4, HCO_3
571	368	0.64	3.85	0.03	0.65	1500	33.0	110	194	–	−1.63	Ca+Mg, Na+K – Cl+SO_4, HCO_3
547	333	<0.1	3.56	0.01	<0.01	1400	30.0	140	188	–	−1.11	Ca+Mg, Na+K – Cl+SO_4, HCO_3
535	333	<0.1	<0.1	0.02	0.01	1380	30.0	110	189	–	0.38	Ca+Mg, Na+K – Cl+SO_4, HCO_3
559	289	<0.1	4.43	0.03	0.34	1390	30.3	140	197	–	−0.76	Ca+Mg, Na+K – Cl+SO_4, HCO_3
547	299	<0.1	<0.1	0.01	–	1370	29.8	–	–	–	−0.01	Ca+Mg, Na+K – Cl+SO_4, HCO_3
547	306	<0.1	<0.1	0.04	<0.01	1400	30.8	100	192	–	−0.41	Ca+Mg, Na+K – Cl+SO_4, HCO_3
520	332	1.47	<0.1	0.04	<0.01	1410	30.9	90	197	–	0.47	Ca+Mg, Na+K – Cl+SO_4, HCO_3
553	354	4.56	<0.1	0.03	0.57	1450	31.1	100	191	<0.04	−1.37	Ca+Mg, Na+K – Cl+SO_4, HCO_3
542	331	1.1	<0.1	0.08	0.11	1410	32.1	130	185	–	0.2	Ca+Mg, Na+K – Cl+SO_4, HCO_3

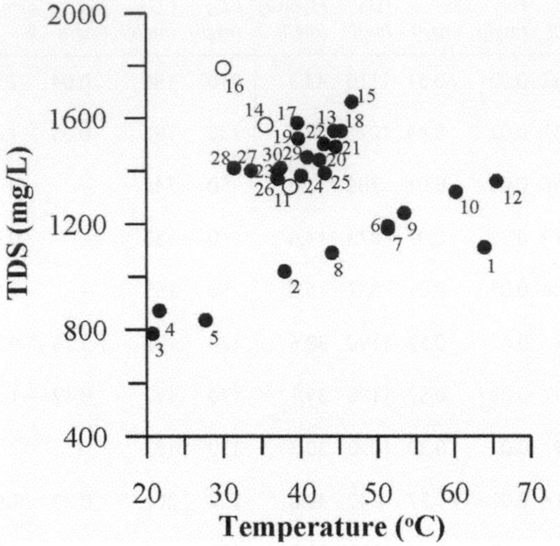

Figure 10.3 Cross-plot illustrating the TDS content of the sampled objects vs. temperature. Empty symbols are the non-operating well data (see data in Table 10.2).

Since the sampled wells have different depths and screened sections, they could provide information about the distribution of parameters with depth. The reference point elevation term used on the diagrams means the elevation (m above sea level, masl.) of the middle of screening section of the well.

Firstly the temperature and TDS of the waters are investigated as a function of the reference point elevation (Fig. 10.5a–b). In case of the South System (dots) the temperature remains the same with increasing depth (about 45°C), only minor decrease (about 15°C) can be observed at the discharge level at (+100) m asl. For the North System (squares) the temperature is also the same with increasing depth, however the temperature decrease at the discharge level is more pronounced, from about 60–65°C to 20°C in the (−200) to (+100) m asl interval (Fig. 10.5a).

On the TDS vs. reference point elevation diagram the same trend can be observed (Fig. 10.5b). Higher TDS (1370–1600 mg/l) concentration occur in the South System. The 1600 mg/l TDS content suddenly decreases to about 1400 mg/l at the discharge level. For the North System ca. 1300 mg/l TDS are characteristic at depth, and decrease moderately to 800 mg/l at the discharge level (Fig. 10.5b). Both the temperature and the TDS content remain the same with increasing depth in both systems. Only the deepest well of the North System (Nr. 1) has a lower TDS (1100 mg/l) than the other deep wells discharging water of similar temperature.

If the main components are plotted against the reference-point-elevation (Fig. 10.5c–h), it can be observed that the deep wells of the North and South Systems have similar Na^+ and Cl^- concentrations (Fig. 10.5c–d), however in case of the North System a larger decrease of the values can be observed at the discharge level. Differentiation of the North and South Systems can be seen regarding the Mg^{2+}, HCO_3^-, Ca^{2+} and SO_4^{2-} concentration (Fig. 10.5e–h) and the TDS (Fig. 10.5b).

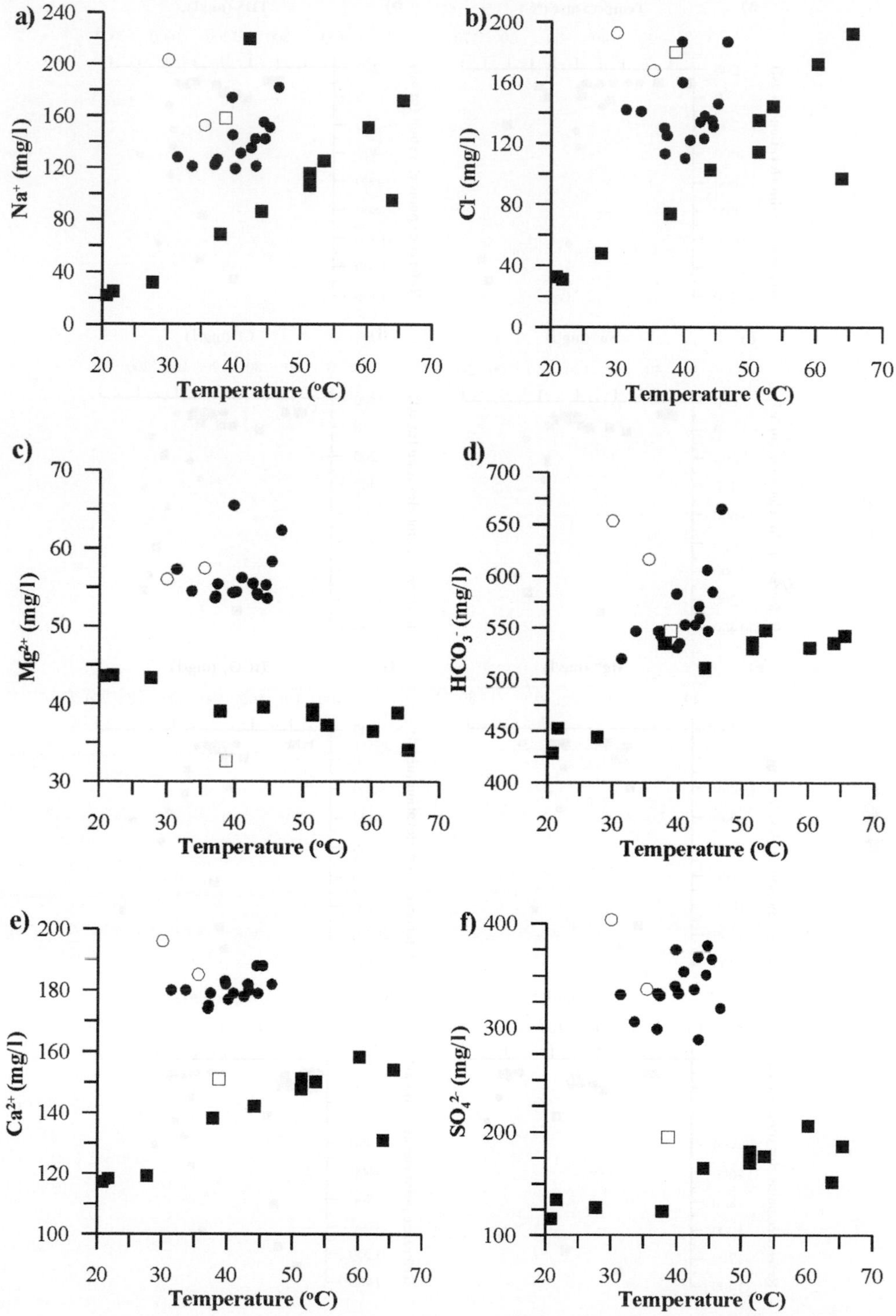

Figure 10.4 a) Na^+, b) Cl^-, c) Mg^{2+}, d) HCO_3^-, e) Ca^{2+} and f) SO_4^{2-} vs. temperature cross-plots of the sampled objects. Squares indicate the North System, dots the South System and empty symbols the non-operating well data (see data in Table 10.2).

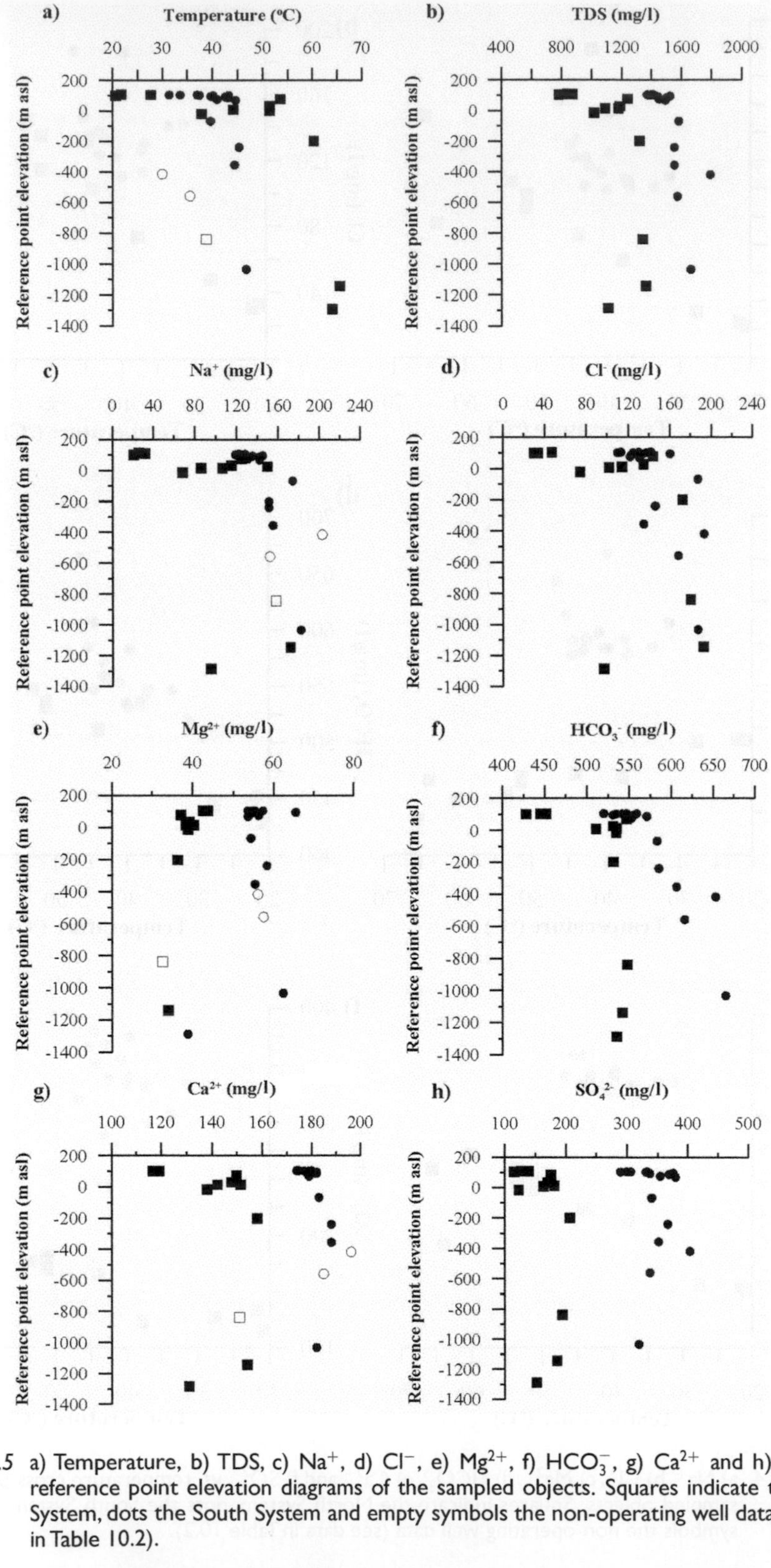

Figure 10.5 a) Temperature, b) TDS, c) Na^+, d) Cl^-, e) Mg^{2+}, f) HCO_3^-, g) Ca^{2+} and h) SO_4^{2-} vs. reference point elevation diagrams of the sampled objects. Squares indicate the North System, dots the South System and empty symbols the non-operating well data (see data in Table 10.2).

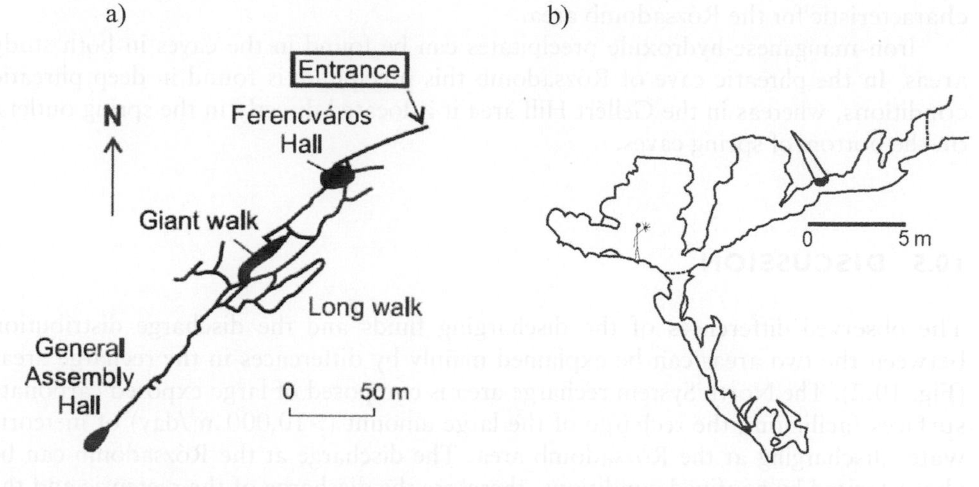

Figure 10.6 a) Map of the Szemlő-hegy Cave (Rózsadomb area) after Horváth (1965). b) Cross section of the Citadella Cave (Gellért Hill area) after Leél-Őssy *et al.* (2007).

10.4.2 Characteristics of caves and recent mineral precipitates

In the 5–6 km^2 area of the Rózsadomb there are more than 60 caves and their total explored length is more than 35 km. The formation of these caves was attributed to the mixing of cold and hot karst waters (Takács-Bolner & Kraus, 1989; Leél-Őssy, 1995; Leél-Őssy & Surányi, 2003). The caves developed mainly in Eocene limestone, the highest levels in Eocene marl, and in some cases the lowest levels reach the Triassic carbonates. The most important, common feature of these caves is their strong structural control (Fig. 10.6a). Along with faults and fractures, cave passages often follow bedding planes and also the contact between the Eocene limestone and the Eocene-Oligocene marl (Leél-Őssy, 1995). As a result, multi-level complicated, maze-like cave systems were developed. At the discharge zone of the Rózsadomb area, at the actual base level there is a phreatic cave, which is the modern analogue of the already dry caves found at higher levels.

The caves of the Gellért Hill are much smaller and most of them developed in Triassic dolomite. Compared to the multi-level, maze-like pattern of the Rózsadomb caves, these caves are usually fracture-related isometric spherical cavities up to 12 m high (Fig. 10.6b) (Korpás *et al.*, 2002). At the present discharge level the springs discharge from enlarged fractures in the dolomite.

Dry caves are usually rich in minerals, especially the abundance and variety of calcite and gypsum. These two minerals are widespread in the caves located at the base level and partly filled with water. There is an important recently observed difference between the two study areas in the occurrence of these precipitates. In case of the Gellért Hill area intensive calcite precipitation can be observed on the water surfaces (as calcite rafts), and extensive accumulation of calcite "mud" can be found on the bottom of the spring caves, while in the Rózsadomb area only a thin film of calcite can be observed in the lukewarm springs. The gypsum can be found as a crust on the cave walls in the Gellért Hill spring caves above the water table, whereas crystals are characteristic for the Rózsadomb area.

Iron-manganese-hydroxide precipitates can be found in the caves in both study areas. In the phreatic cave of Rózsadomb this precipitate is found in deep phreatic conditions, whereas in the Gellért Hill area it is located directly in the spring outlets, on the bottom of spring caves.

10.5 DISCUSSION

The observed differences of the discharging fluids and the discharge distribution between the two areas can be explained mainly by differences in the recharge areas (Fig. 10.2). The North System recharge area is composed of large exposed carbonate surfaces facilitating the recharge of the large amount (>10,000 m^3/day) of meteoric water discharging at the Rózsadomb area. The discharge at the Rózsadomb can be characterised by confined conditions, therefore the discharge of the meteoric and the upwelling hydrothermal fluids is structurally controlled and has an important consequence on the mixing of these waters (Fig. 10.7). Mixing can only occur through

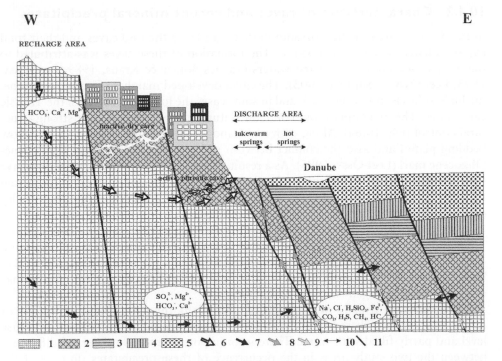

Figure 10.7 Conceptual model for the Rózsadomb discharge area. 1: Triassic carbonate; 2: Eocene to Lower Oligocene carbonate and siliciclastic; 3: Upper Oligocene siliciclastic; 4: Lower and Middle Miocene siliciclastic; 5: Upper Miocene siliciclastic and limestone; 6: local-intermediate flow system; 7: regional flow system; 8: regional + basinal fluids; 9: basinal fluids; 10: complex rock-water interaction; 11: structural elements. Location: see Fig. 10.2.

structures. Since the recharge area of the South System can be characterised by a limited surface of exposed carbonates (Fig. 10.2) the meteoric fluid contribution is also limited to the discharge area, and the upwelling hydrothermal waters dominate and no lukewarm spring can be found there (Fig. 10.8). The natural discharge rate of the system (3200 m^3/day) is lower compared to the Rózsadomb (18 000 m^3/day) also indicating differences in the recharge areas. The differences of the fluid chemistry between the North and South Systems might be explained by regional geological differences.

Based on the hydrogeochemical investigations, the contribution of basinal fluids to the discharging waters of Buda Thermal Karst (in form of Na^+, K^+, Cl^-, H_2SiO_3, CO_2, H_2S, CH_4, and liquid hydrocarbons) has to be considered. The similar Na^+, Cl^-, and CO_2 concentration of the waters in both systems suggest a common basinal origin. However, differences were also identified between the two systems regarding the basinal components, e.g. no hydrocarbon was found in the South System. The basinal fluid contribution is probably the result of complex fluid-rock interactions, and might be controlled by structural elements and pressure conditions. The present study suggests the overpressure derived from tectonic compression (Tóth & Almási, 2001; Bada *et al.*, 2006) as a driving force of the basinal fluid contribution.

The differences in the recent mineral precipitates can be deduced both from natural and anthropogenic factors. Calcite precipitation needs an open water table, where CO_2 degassing can occur. Firstly, in case of the Rózsadomb area most of the natural springs were substituted by boreholes. Calcite precipitation could only be observed in one lukewarm spring. The spring outlets of the Gellért Hill are characterised, however, by

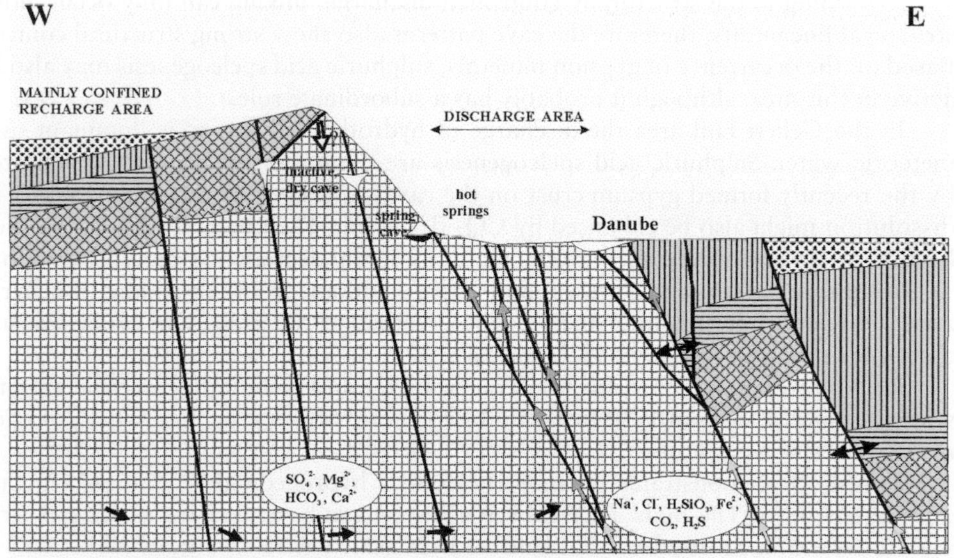

Figure 10.8 Conceptual model for the Gellért Hill discharge area. 1: Triassic carbonate; 2: Eocene to Lower Oligocene carbonate and siliciclastic; 3: Upper Oligocene siliciclastic; 4: Lower and Middle Miocene siliciclastic; 5: Upper Miocene siliciclastic and limestone; 6: local-intermediate flow system; 7: regional flow system; 8: regional + basinal fluids; 9: basinal fluids; 10: complex rock-water interaction; 11: structural elements. Location: see Fig. 10.2.

more natural conditions such as spring caves. There is a possibility of CO_2 degassing to allow calcite precipitation. Secondly, the Ca^{2+} and HCO_3^- concentrations are higher in the waters of the Gellért Hill (South System) than in the waters of the Rózsadomb area (North System).

The occurrence of iron-manganese-hydroxide precipitates may denote the mixing of anoxic deep hydrothermal waters and oxygen-rich meteoric waters of Rózsadomb as these precipitates were found in deep phreatic conditions. In the Gellért Hill caves they may indicate the oxidation zone, as they are found in the spring outlets.

10.6 SUMMARY

The North System comprises all the waters around the Rózsadomb and deep wells on the opposite Pest side. The South System includes waters around the Gellért Hill and deep wells in South Budapest. Regarding the main components, the South System is characterised by higher Ca^{2+}, Mg^{2+}, HCO_3^-, SO_4^{2-}, and TDS concentrations compared to the North System. These higher values correspond to lower temperatures within a narrower range (30–47°C). The differences between the two systems can also be identified on the distribution of parameters with depth. However, with regard to Na^+ and Cl^- there are no differences between the two systems. At greater depth the parameters are characterised by near constant values, meanwhile significant differences occur in the proximity of the discharge level, thus suggesting that the processes inducing changes in the parameters operate in the vicinity of the discharge zone.

According to the structurally controlled discharge, mixing can only occur along structural lineaments, therefore the cave patterns also show strong structural control. Based on the occurrence of gypsum minerals, sulphuric acid speleogenesis may also be active in this area, although it probably has a subordinate role.

In the Gellért Hill area the discharge of hydrothermal waters is dominant over meteoric water. Sulphuric acid speleogenesis are the dominant process as indicated by the recently formed gypsum crust on the cave walls close to the water table. The dissolution might also be enhanced by CO_2 degassing. The H_2S and CO_2 have rather basinal origin and associated with the maturation and degradation of hydrocarbons.

Several processes can be present and active simultaneously at a regional discharge zone. However, the predominance of one or other of the individual processes may determine the main character of the caves.

Regional karst aquifer systems are qualitatively and quantitatively stable groundwater resources. They are affected only by long-term dewatering and local pollution near to the discharge zone.

This study demonstrates the importance of a hydrogeological approach in cave and karst studies. The Buda Thermal Karst system can be considered as the type area and in same time the modern analogue for hypogenic karsts.

ACKNOWLEDGEMENTS

This study was accomplished within the framework of the collaboration between Shell International E&P and the Eötvös Loránd University.

REFERENCES

Alföldi L. (1982) A layered thermal-water twin flow system. *J Hydrol* 56: 99–105.

Back W. (1961) Techniques for mapping of hydrochemical facies. *USGS Professional Paper*, 424-D, 380–382.

Bada G., Horváth F., Dövényi P., Szafián P., Windhoffer G., Cloetingh S. (2007) Present-day stress field and tectonic inversion in the Pannonian basin. *Global Planet Change* 58(1–4): 165–180.

Eross A., Mádl-Szőnyi J., Csoma É.A. (2008) Characteristics of discharge at Rose and Gellért Hills, Budapest, Hungary. *Central Euro Geol* 51(3): 267–281.

Horváth J. (1965) A Szemlőhegyi-barlang 1961–62. évi felérése (Mapping of Szemlő-hegy Cave in 1961–62) [in Hungarian]. Karszt és Barlang I: 21–30.

Klimchouk A.B. (2007) Hypogene speleogenesis: hydrogeological and morphogenetic perspective. Special Paper no. 1, National Cave and Karst Research Institute, Carlsbad, NM.

Korpás L., Fodor L., Magyari Á., Dénes G., Oravecz J. (2002) A Gellért-hegy földtana, karszt- és szerkezetfejlodése (Geology, karst and structural evolution of Gellért Hill) [in Hungarian]. Karszt és Barlang I–II: 57–93.

Leél-Őssy C., Leél-Őssy S., Adamkó P. (2007) A Citadella-kristálybarlang (Citadella Cave) [in Hungarian]. Karszt és Barlang I–II: 67–78.

Leél-Őssy S. (1995) A Rózsadomb és környékének különleges barlangjai (Particular caves of the Rózsadomb Area) [in Hungarian]. Földt Közlöny 125(3–4): 363–432.

Leél-Őssy S., Surányi G. (2003) *Peculiar hydrothermal caves in Budapest, Hungary.* Acta Geol Hung 46(4): 407–436.

Takács-Bolner K., Kraus S. (1989) The results of research into caves of thermal water origin. Karszt és Barlang Special Issue 31–38, Hungarian Speleological Society, Budapest.

Tóth J. (1963) A theoretical analysis of groundwater flow in small drainage basins. *J Geophys Res* 68: 4795–4812.

Tóth J. (1999) Groundwater as a geologic agent: an overview of the causes, processes, and manifestations. *Hydrogeol J* 7: 1–14.

Tóth J. (2009) Springs seen and interpreted in the context of groundwater flow-systems. *GSA Annual Meeting*, Portland, OR, 18–21 October 2009.

Tóth J., Almási I. (2001) Interpretation of observed fluid potential patterns in a deep sedimentary basin under tectonic compression: Hungarian Great Plain, Pannonian Basin. *Geofluids* 1(1): 11–36.

Zuber A., Witczak S., Rozanski K., Sliwka I., Opoka M., Mochalski P., Kuc T., Karlikowska J., Kania J., Jackowicz-Korczyński M., Dulinski M. (2005) Groundwater dating with ^3H and SF_6 in relation to mixing patterns, transport modelling and hydrochemistry. *Hydrological Processes* 19: 2247–2275.

Chapter 11

Geochemistry of thermal waters of the Sikhote-Alin ridge, Russia

Ivan V. Bragin, George A. Chelnokov & Maksim G. Blokhin

Far East Geological Institute, Russian Academy of Sciences, Vladivostok, Russian Federation

ABSTRACT

Along the Sikhote-Alin ridge of the Russian Far East, 30° to 50°C thermal waters emerge in a series of sodium bicarbonate and sodium bicarbonate-sulfate type springs. Atmospheric nitrogen and oxygen are the dominant dissolved gases. Rare earth element data constrain understanding of water-rock interaction occurring in the source region. Thermal waters are of meteoric origin and the water exchange time is low according to the hydrogen and oxygen isotopic data and geochemical modelling. A conceptual model of groundwater evolution of Sikhote-Alin thermal waters is proposed.

11.1 INTRODUCTION

Low-temperature thermal waters of the continental margin of the Russian Far East are widespread along the coast of Japan and Okhotsky Seas. All thermal water systems occur in the Sikhote-Alin ridge (Fig. 11.1). Most of them were studied from the 1930s until 1960s (Makerov, 1938). Some are now spas: Annensky, Tumninsky ("warm spring"), Amgunsky and Chistovodny and others are used by locals for self-treatment: Khucin and St. Helen's springs at Amgunsky and "Hot spring" at the Chistovodny thermal area. Latest investigations (Bragin *et al.*, 2007) have shown the geochemical properties of waters. However, information about geochemistry and rare earth elements (REE) is limited. This chapter presents new data on geochemistry, isotopes, and REEs in the thermal springs of the Sikhote-Alin ridge.

11.2 SPRINGS CHARACTERISTICS

The Annensky spa thermal area is located on the East of the Khabarovsky krai, in the lowest part of the Amur river, 80 km from the Okhotsky sea shore. According the borehole data, thermal waters occur in the contact zone between effusive and tuffaceous sedimentary rocks of upper Cretaceous age in the Bolbinskaya and Tatarkinskaya suites. The rocks are covered with alluvial sediments (thickness up to 5 to 8 m) consisting of clay, poorly sorted sand and gravel. The water is low-temperature (49.5 to 54.5°C), weakly mineralized (TDS 0.20 to 0.35 g/l), alkaline (pH 8.5 to 9.4) and

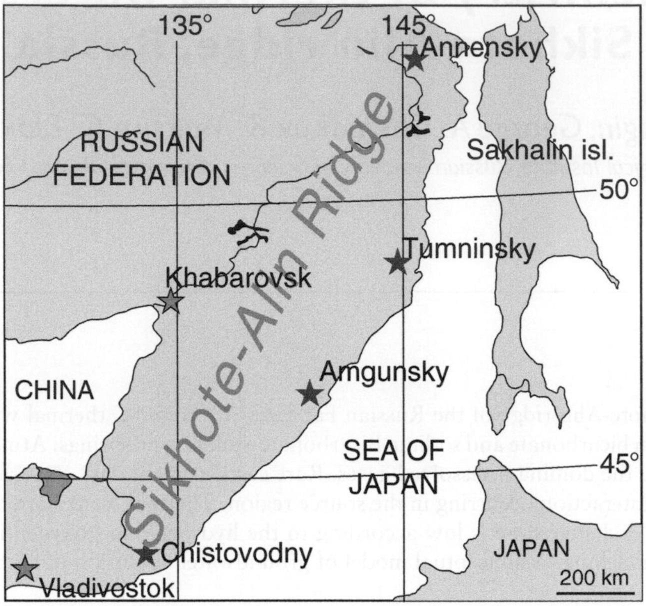

Figure 11.1 Sikhote-Alin ridge – location.

the chemical classification of the groundwater is a Na-HCO_3-SO_4 type with a high concentration of silica (25 to 35 mg/l) and fluoride (up to 7 mg/l).

The Tumninsky spa thermal area is located on South-East of the Khabarovsky krai, 9 km from the Tumnin River, and 40 km away from the Tatarsky channel (Japan sea). The geological structure is defined by the contact zone of granites and andesitic basalts of the Eocene Kuznecskaya suite (Sidorenko, 1971). Waters from borehole No. 9 (depth 460 m) are low-temperature (43.0 to 46.0°C), alkaline (pH 8.7 to 9.7), low-mineralised (TDS 0.15 to 0.25 g/l) and the chemical classification of the groundwater is a Na-HCO_3-SO_4 type.

The Chistovodny geothermal area is located in Lazo area of the Primorsky krai, 70 km away from the Japan Sea shore and is known from a group of springs and boreholes (Chudaev, 2003). Host rocks consist of weathered granites of upper Cretaceous age, broken by Paleocene dykes of aplites and diorite porphyries. Bedrock is covered by Quaternary alluvial deposits with thickness of 3 to 7 m. The geothermal field is presented by several outlets, notably "Chistovodny spa" and "Hot Spring". Waters are low-temperature (30°C) low-mineralized (TDS 0.14 to 0.16 g/l), low-alkaline (pH 7.0 to 8.4) of Na-HCO_3 type with high content of silica (up to 60 mg/l) and fluoride (up to 4 mg/l).

The Amgunsky geothermal area is located in the Terney area of the Primorsky region, 10 km from the Japan Sea shore. This geothermal field is presented by three well-known outlets: Banny, St. Helen's, and Khucin. The geological structure is defined by the contact of granite intrusions with rhyolites, tuffs and ignimbrites of Mesozoic and Cenozoic age (Chudaev, 2003). Waters are weakly mineralised (TDS 0.1 to 0.2 g/l), alkaline (pH 8.5 to 9.0) of Na-HCO_3 type with up to 22 mg/l silica.

Table 11.1 Spring characteristics.

	Annensky	*Tumninsky*	*Amgunsky*	*Chistovodny*
pH	9.2	9.3	9.1	8.9
mg/l				
TDS	235	195	183	167
Na^+	61.0	32	33.4	19.62
K^+	1.17	0.35	0.7	0.33
Ca^{2+}	2.0	2.11	2.0	5.42
Mg^{2+}	0.01	0.01	0.08	0.007
HCO_3^-	112.8	78.1	70.7	63.4
SO_4^{2-}	25.4	7.1	13.6	5.7
Cl^-	4.0	1.4	3.6	2.4
F^-	2.7	0.8	0.9	3.9
SiO_2	46.8	35.2	31.1	18.6
°C				
T, °C (surface)	49.9	44.8	34.5	27.5
T, °C quartz geothermometer (Fournier, 1977)	98.7	86.1	80.9	60.7
T, °C Na/K geothermometer (Arnorsson *et al.*, 1983)	123.7	95.9	128.7	116.8

11.3 ANALYTICAL RESULTS

From the southern to northern thermal areas of Sikhote-Alin an increase in water temperature is reflected in quartz geothermometry (Bragin *et al.*, 2007). The temperature increase is paralleled by increases in most dissolved species (Tab. 11.1). These trends may reflect enhanced reaction rates in higher temperature waters combined with longer flow paths as the higher temperature waters penetrate to greater depths than the lower temperature waters. Oxygen and hydrogen isotope ratios were measured in thermal waters from the whole Sikhote-Alin folded area. The δD and $\delta^{18}O$ values for thermal waters in the Annensky and Tumninsky areas correlate with data of Chudaev (2003) on Chistovodny, Amgunsky and for precipitations in Primorye, Southern Sikhote-Alin (Yushakin, 1968). On a δD vs. $\delta^{18}O$ diagram the Sikhote-Alin thermal waters lie close to the global meteoric water line (Craig, 1961) as shown in Fig. 11.2. REEs were measured using an Agilent 7500C inductively-coupled plasma mass spectrometer (ICP MS). The REE data, normalized to North American Shale Composite (NASC), show the common rising trend caused by greater solubility of heavy REE (Fig. 11.3).

In contrast to the small REE concentrations in waters (0.02 to 20×10^{-9} g/l), the high pCO_2 Lastochka waters, also from the Sikhote-Alin region, have high REE concentrations, probably reflecting low alkalinity of studied thermal waters. Some show negative anomalies on Ce and Eu. The Eu anomaly may be caused by Eu-deficient plagioclase. Whereas Ce could be precipitated under oxidizing conditions.

Micro-analyses of the granite, which is host to the Tumninsky springs, showed the possible source of some REEs to be monazite (Fig. 11.4) (La 12.49%, Ce 28.89%, Nd 12.9%, Sm 2.32%, Gd 2.14%). Other REEs are probably derived from plagioclase.

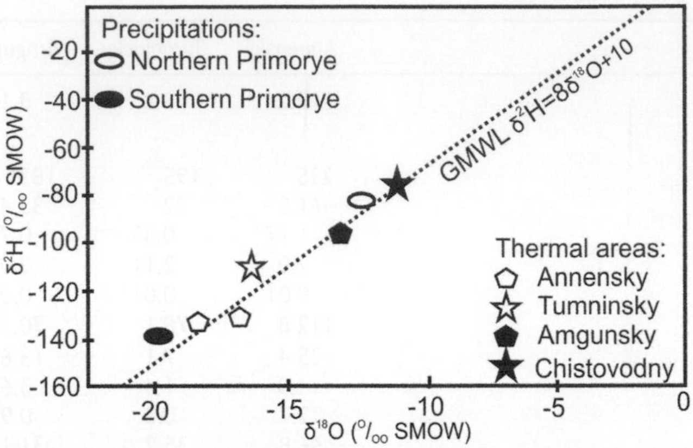

Figure 11.2 Deuterium and Oxygen-18 isotopes.

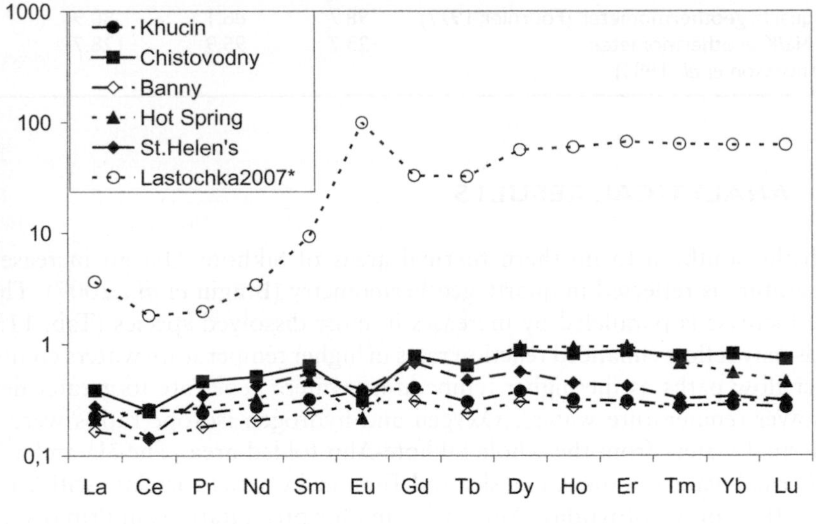

Figure 11.3 Concentration of REE in thermal and high pCO_2 waters (Lastochka) of Sikhote-Alin ridge normalized to NASC. Lastochka data obtained by Tchepkaia (2007).

11.4 WATER-ROCK INTERACTION MODELLING

Calculations using the computer code WATEQ4F (Ball & Nordstrom, 1991) show that all the thermal waters are slightly supersaturated with calcite and chalcedony and strongly undersaturated with fluorite and albite (Bragin *et al.*, 2007).

Temperatures calculated by quartz (Fournier, 1977) and Na-K (Arnorsson *et al.*, 1983) geothermometers are shown in the Tab. 11.1. The Na-K geothermometer shows higher temperatures than quartz, possibly caused by mixing of fresh cold groundwater, which lowers the quartz temperature but not Na-K geothermometer temperature.

A conceptual model (as an example) for the Tumninsky geothermal area (Fig. 11.5) is proposed wherein meteoric groundwaters penetrate deep into the permeable rock

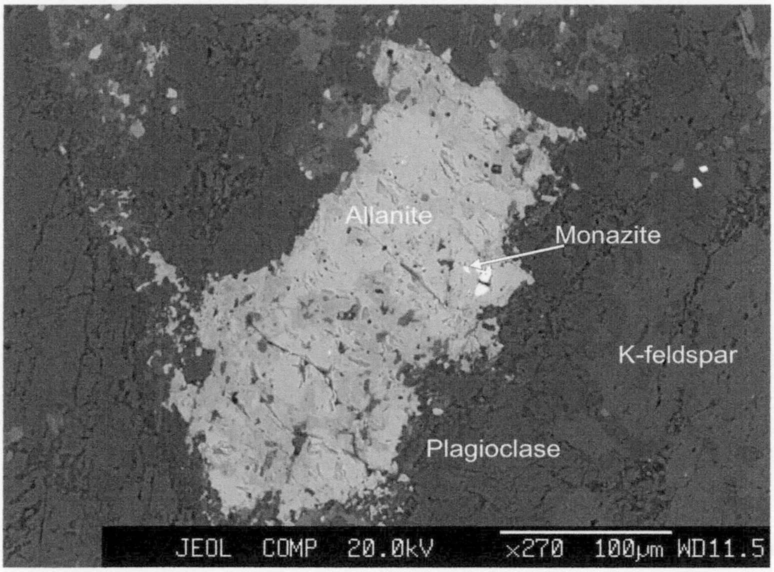

Figure 11.4 Monazite in granites of Tuminsky spring host rocks.

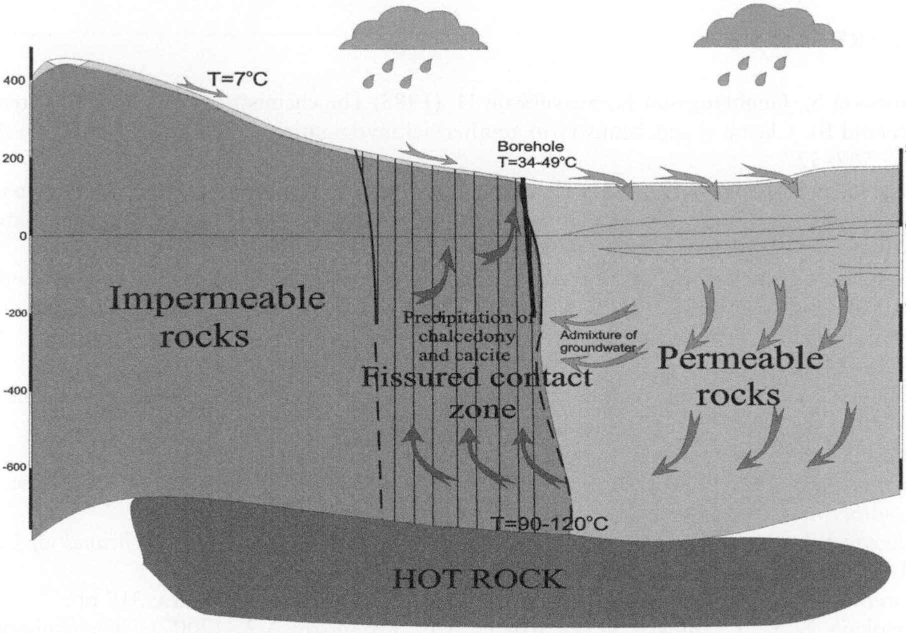

Figure 11.5 Conceptual model of evolution of thermal waters of Sikhote-Alin ridge.

to become heated to 90 to 120°C, and then rise through the contact zone or some fissured zone. Rising water cools and precipitates chalcedony and calcite accounting for equilibrium with those two minerals, and outflows at the surface.

11.5 CONCLUSIONS

The following conclusions can be drawn from the investigations of the thermal waters of the Sikhote-Alin ridge:

1 Data on stable isotope ratios ($\delta^{18}O$, δ^2H) show that the waters are of meteoric origin (rainwater),
2 measurement showed low REE concentrations caused by low alkalinity; some waters have Ce and Eu anomalies which should be studied further,
3 analytical results coupled with thermodynamic and geothermal modelling help create a conceptual model of evolution of studied waters.

ACKNOWLEDGEMENT

This research is supported by grants of Far East Branch of Russian Academy of Sciences No. 10-3-B-08-194.

REFERENCES

Arnorsson S., Gunnlaugsson E., Svavarsson H. (1983) The chemistry of geothermal waters in Iceland III. Chemical geochemistry in geothermal investigations. *Geochim Cosmochim Acta* 47: 567–577.

Ball J.W., Nordstrom D.K. (1991) User's manual for WATEQf4 with revised thermodynamic data base and test cases for calculating speciation of minor, trace and redox elements in natural waters. *U.S. Geol. Surv. Open File Rep.*, 91–183.

Bragin I.V., Chelnokov G.A., Chudaev O.V., Chudaeva V.A. (2007) Low-temperature geothermal waters of continental margin of Far East of Russia. *Proceedings of the 12-th international symposium on water-rock interaction WRI-12*. Taylor & Francis Group, London, UK, 481–484.

Chudaev O.V. (2003) Composition and origin of the recent hydrothermal systems of the Far East Russia. Vladivostok: Dalnauka, 85–107.

Craig H.I. (1961) Isotopic variations in meteoric waters. *Science* 133: 1702–1703.

Fournier R.O. 1977. Chemical geothermometers and mixing model for geothermal systems. *Geothermics* 5: 41–50.

Makerov Y.A. (1938) Mineral waters of the Far East. *Journal of Far Eastern Branch of USSR Academy of Sciences*, No. 38. Vladivostok: Dalgiz, 18–35.

Sidorenko A.V. (1971) *Hydrogeology of the USSR*, No 23. Moscow: Nedra, 310 pp.

Tchepkaia N.A., Chelnokov G.A., Kiselev V.I., Karabtsov A.A. (2007) Geochemistry of Na-HCO_3 groundwater and sedimentary bedrocks from the central part of Sikhote-Alin mountain region (Far East of Russia). *Applied Geochemistry* 1764–1776.

Yushakin E.P. (1968) Report of the examination of mineral springs of Primorye. Open File Report, Primorskaya Geological survey.

Chapter 12

Arsenic in groundwater in Western Anatolia, Turkey: A Review

Orhan Gunduz[1], *Alper Baba*[2] & *Handan Elpit*[1]

[1]*Dokuz Eylul University, Department of Environmental Engineering, Izmir, Turkey*
[2]*Izmir Institute of Technology, Department of Civil Engineering, Izmir, Turkey*

ABSTRACT

Western Anatolia in Turkey is an area of complex geology with active tectonics and high geothermal potential. This natural setting serves as a suitable environment for the presence of high levels of arsenic in subsurface waters. High arsenic concentrations in groundwater have been detected in many provinces of Western Anatolia including but not limited to İzmir, Kütahya, Çanakkale, Afyon, Manisa, Aydın and Denizli with values ranging from 20 to 560 ppb, exceeding the national and international drinking water quality criteria of 10 ppb. On the other hand, arsenic concentrations in geothermal fluids are about three times higher then the corresponding concentrations in groundwater. Considering the potential of contamination of regional groundwater reserves with geothermal fluids, levels in hot waters of Western Anatolia demonstrate additional problems. Based on these fundamentals, this study discusses the potential sources and concentrations of arsenic in water resources of the region with particular emphasis on local geologic and tectonic properties of the Western Anatolian Plate.

12.1 INTRODUCTION

Occurrence of arsenic in groundwater and its impacts on millions of individuals worldwide is a serious environmental health issue. The sources of As are both natural and anthropogenic, and affect different regions of the world both at local as well as at regional scales (Bundschuh *et al.*, 2010). Considering its toxic effects on human health, the presence of elevated levels of arsenic in groundwater resources used in drinking water supply has been an active research field throughout the world (Van Halem *et al.*, 2009). In this regard, case studies from Bangladesh, India, Nepal, El Salvador, Ecuador, Honduras, Mexico, Chile, China, Canada, Argentina, Peru, Taiwan, United States, Bolivia and Turkey have been documented with regards to the detection of natural arsenic levels in groundwater, the occurrence and distribution mechanisms, the human health effects and treatment techniques (Jean *et al.*, 2010). In many of these locations, arsenic is naturally found in the subsurface strata within volcanic and sedimentary formations as well as in areas of geothermal systems related to tectonic activity. It is also known that discharge of geothermal waters may result in high arsenic concentrations in surface and subsurface waters. In geothermal springs of Yellowstone National Park, arsenic concentrations exceed 1 mg/l (Ball *et al.*, 1998) and reach 0.36 mg/l in the Madison River down gradient from the springs' area (Nimick, 1994). Over 100 000 kg of geothermally derived arsenic is estimated to leave the western boundary of the Park each year, affecting water quality within a large region (Nimick *et al.*, 1998). Arsenic concentrations at the geothermal field El Tatio, in North Chile, are reported to be as high as 27 mg/l (Romero *et al.*, 2003).

Western Anatolia in Turkey is an area of complex geology with active tectonics and a high geothermal potential. This natural setting serves as a suitable environment for the presence of high levels of arsenic in subsurface waters. Based on these fundamentals, this study presents a general overview of arsenic presence in western Anatolia.

12.2 GEOLOGICAL SETTINGS IN WESTERN ANATOLIA

Turkey is one of the most seismically active regions in the world. Its geological and tectonic evolution has been dominated by the repeated opening and closing of the Paleozoic and Mesozoic oceans (McKenzie, 1972; Dewey & Sengor, 1979; Jackson & McKenzie, 1984). It is located within the Mediterranean Earthquake Belt, whose complex deformation results from the continental collision between the African and Eurasian plates (Bozkurt, 2001). The border of these plates constitutes seismic belts marked by young volcanics and active faults, the latter allowing circulation of water as well as heat. The distribution of hot springs in Turkey roughly parallels the distribution of the fault systems, young volcanism and hydrothermally altered areas (Simsek *et al.*, 2002). There are a total of about 1000 thermal and mineral water spring groups in the country (MTA, 1980; Simsek *et al.*, 2002) (Figure 12.1). The activity in Western Anatolia is believed to be a result of tensional forces that resulted from rigid behaviour during the Neogene and Quaternary, and the development of extended near-coastal graben areas (Baba & Ármannsson, 2006).

12.3 ARSENIC LEVELS IN WESTERN ANATOLIA

The primary mechanism for the presence of numerous trace elements in the earth's crust including, but not limited to, arsenic, antimony, boron, nickel, lead and zinc, is the dominant volcanic structure in the geological formations of Turkey, particularly in western Anatolia. They are also found as impurities in the ores of other minerals including coal (Karayigit *et al.*, 2000; Baba *et al.*, 2009), reaching concentrations as high as 6413 mg/l arsenic. These trace elements are dissolved in all geothermal waters and in some fresh groundwater resources in many areas in western Anatolia. The concentrations of arsenic in rocks and ores of western Anatolia are given in Table 12.1. Arsenic levels as high as 4% are observed in mineral deposits particularly in the Kütahya-Emet area, which is known to contain the world's largest boron deposits. In this area, arsenic is typically found in the boron minerals as discussed by Helvaci (1986), Helvaci & Orti (1998) and Helvaci & Alonso (2000).

The high arsenic levels (Table 12.1) are mostly related to the alterations in volcanic formations. In Western Anatolia, arsenic is observed in the alteration zones of volcanic formations in addition to some sedimentary rocks. Based on the tectonic characteristics (see Figure 12.1) and the geological structure, many parts of Turkey are likely to have arsenic contained in geological formations within which groundwater may have

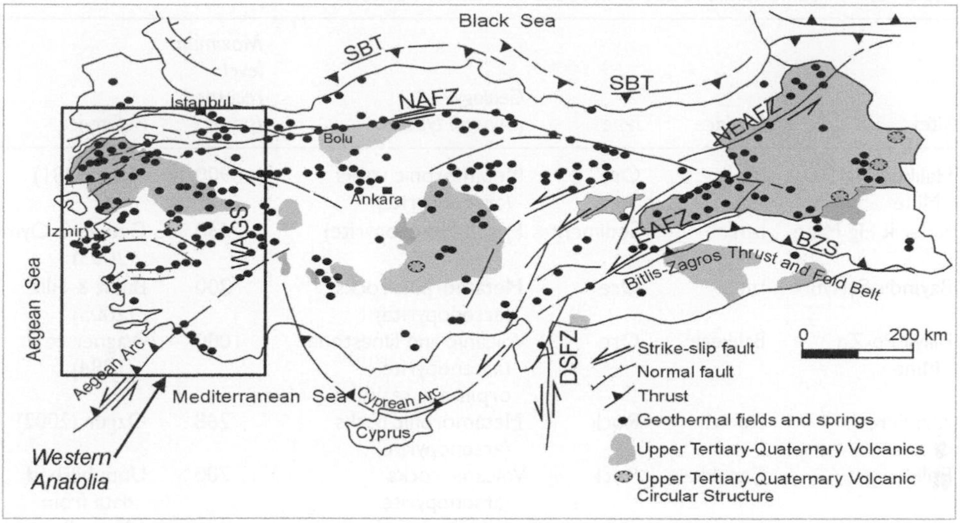

Figure 12.1 Tectonic map of the eastern Mediterranean region showing structures developed during the Miocene to Holocene time and distribution of geothermal areas around Turkey (compiled from; Simsek *et al.*, 2002 and Yigitbas *et al.*, 2004). (SBT, Southern Black Sea Thrust; NAFZ, North Anatolian Fault Zone; NEAFZ, Northeast Anatolian Fault Zone; EAFZ, Eastern Anatolian Fault Zone; WAGS, Western Anatolian Graben System; DSF, Dead Sea Fault Zone; BZS, Bitlis-Zagros Suture) (Baba & Armannsson, 2006).

correspondingly high concentrations. Most of the rocks are altered and fractured due to the effects of active faults. Basement rocks are composed of Oligocene volcanic rocks such as andesite, dacite, rhyodacite, basalt, tuff and agglomerate. Several mineral deposits including numerous industrial metals as well as some precious metals have been found in the alteration zones or fractured parts of these volcanic rocks, where arsenic is typically seen as an impurity (Baba, 2010).

The most common arsenic minerals in western Anatolia are arsenopyrite, orpiment and realgar. In Kutahya and Balıkesir regions, where large boron deposits are present, dominant arsenic minerals are orpiment and realgar. On the other hand, arsenopryite dominates in Izmir region.

Due to the neotectonic structure and volcanism, various altered rock types may affect the quality of water resources. Thus, arsenic found in groundwater is typically geogenic in origin and has strong links to the local regional geology. In particular, arsenic is an indicator parameter for hot water reserves of Western Anatolia. It is found in almost all geothermal waters, and is used as a tracer (together with lithium and boron) for the detection of contamination in surface and subsurface waters with geothermal fluids (Gunduz *et al.*, 2010).

Arsenic levels are extremely high in many geothermal fields such as Hamamboğazı in Uşak, Kızıldere in Denizli and Balçova-Narlıdere in İzmir (Table 12.2). Although ingestion of geothermal waters is not a typical practice in Turkey as it is in some other parts of the world, these high levels serve as potential contamination sources for local groundwater and surface waters that are used for drinking water purposes, as

Table 12.1 Arsenic concentrations in rocks in Western Anatolia.

Site	*Province*	*Type*	*Geology (mineral types)*	*Maximum level recorded (ppm)*	*Reference*
Halıköy Hg-Sb Mine	İzmir	Ore	Metamorphic rocks (arsenopyrite)	8900	Akar (1981)
Kalecik Hg Mine	İzmir	Sediment	Fylsch (arsenopyrite)	9660	Gemici & Oyman (2003)
Bayındır-Sarıyurt	İzmir	Ore	Metamorphic rocks (arsenopyrite)	200	Bulut & Filiz (2005)
Balya Pb-Zn Mine	Balıkesir	Ore	Volcanic and limestone (arsenopyrite, orpiment, realgar)	1000	Wagner *et al.* (1984)
Kızıldere	Denizli	Rock	Metamorphic rocks (arsenopyrite)	268	Ozgur (2002)
Etili	Çanakkale	Rock	Volcanic rocks (arsenopyrite, orpiment, realgar)	700	Unpublished data from Alper Baba
Doğancılar	Çanakkale	Rock	Volcanic rocks (arsenopyrite, orpiment, realgar)	3000	Wagner *et al.* (1984)
Soğukpınar	Çanakkale	Rock	Volcanic rocks (arsenopyrite, orpiment, realgar)	6000	Wagner *et al.* (1984)
Çan	Çanakkale	Coal	Coal deposit (arsenopyrite)	6413	Baba *et al.* (2009)
Emet	Kütahya	Rock	Lacustrine sediment and volcanic rocks (orpiment, realgar)	500	Aydin *et al.* (2003)
Gökler coal mine	Kütahya	Coal	Coal deposit (arsenopyrite)	3854	Karayigit *et al.* (2000)
Emet	Kütahya	Rock	Lacustrine sediment and volcanic rocks (orpiment, realgar)	3900	Dogan & Dogan (2007)
İğdeköy-Emet	Kütahya	Ore	Lacustrine sediment and volcanic rocks (orpiment, realgar)	40 000	Colak *et al.* (2003)
Simav Sb Mine	Kütahya	Ore	Metamorphic rocks (arsenopyrite)	660	Gunduz *et al.* (2010)
Dulkadir	Kütahya	Rock	Lacustrine sediment and volcanic rocks (orpiment, realgar)	4197	Atabey (2009)
Emet	Kütahya	Rock	Lacustrine sediment and volcanic rocks (orpiment, realgar)	19 487	Atabey (2009)
Kırka Borate Mine	Kütahya	Ore	Lacustrine sediment and volcanic rocks (orpiment, realgar)	>2000	Helvaci & Alonso (2000)
Alaşehir Hg Mine	Manisa	Waste rock	Metamorphic (arsenopyrite)	1164	Gemici (2008)

Table 12.2 Arsenic concentrations in geothermal fields in Western Anatolia.

Geothermal Field	*Province*	*Maximum level recorded (ppb)*	*Reference*
Heybeli	Afyon	1249	Gemici & Tarcan (2004)
Çan	Çanakkale	100	Baba & Deniz (2008)
Tuzla	Çanakkale	136	Baba *et al.* (2009)
Karaılıca	Çanakkale	88	Baba & Deniz (2008)
Kestanbol	Çanakkale	100	Baba & Ertekin (2007)
Alibeyköy	Çanakkale	290	Yilmaz *et al.* (2009)
Kızıldere	Denizli	1500	Ozgur (2002)
Balçova-Narlıdere	İzmir	1420	Aksoy *et al.* (2009)
Seferihisar	İzmir	172	Tarcan & Gemici (2003)
Dikili	İzmir	480	Personal comm. with Alper Baba
Simav	Kütahya	594	Gunduz *et al.* (2010)
Gediz	Kütahya	300	Dogan & Dogan (2007)
Yoncalı	Kütahya	950	Dogan & Dogan (2007)
Salihli	Manisa	315	Tarcan *et al.* (2005)
Alaşehir	Manisa	939	Bulbul (2009)
Sart	Manisa	198	Ozen (2009)
Kursunlu	Manisa	3455	Ozen (2009)
Hamambogazi	Uşak	6936	Davraz (2008)

Table 12.3 Arsenic concentrations in groundwater resources in Western Anatolia

Site	*Province*	*Source*	*Maximum level recorded (ppb)*	*Reference*
Bigadiç	Balıkesir	Spring	337	Gemici *et al.* (2008)
Ayvacık	Çanakkale	Well	282	Baba (2010)
Çan	Çanakkale	Spring	71	Baba et *al.* (2009)
Etili	Çanakkale	Well	150	Unpublished data Alper Baba
Menderes plain	İzmir	Well	463	Simsek *et al.* (2008)
Nif mountain	İzmir	Spring	294	Simsek *et al.* (2008)
Balçova	İzmir	Well	170	Aksoy et *al.* (2009)
Aliağa	İzmir	Spring	120	Unpublished data Orhan Gunduz
Simav plain	Kütahya	Well	562	Gunduz et *al.* (2010)
Hisarcık	Kütahya	Spring	152	Atabey (2009)
Emet	Kütahya	Spring	634	Oruc (2004)
İğdeköy	Kütahya	Spring	9300	Dogan et *al.* (2005)
Göksu-Sarıkız	Manisa	Well	59	Personal comm. with IZSU* officials
Eşme	Uşak	Well	50	Local newspaper article (2006)

*İzmir Municipality Water and Sewerage Administration

discussed in detail by Aksoy *et al.* (2009) and Gunduz *et al.* (2010). The concentrations are two-to-three orders of magnitude higher than the national (ITASHY, 2005) and international standards (EPA, 2003; WHO, 2004).

Similar to geothermal waters, arsenic levels are also high in many freshwater areas in western Anatolia (Table 12.3). High arsenic concentrations exceeding the standards are observed in provinces such as Balıkesir, Çanakkale, İzmir and Kütahya. The majority of these high concentrations occur in spring and shallow groundwaters that are in direct contact with alteration zones with high arsenic concentrations or are influenced by geothermal fluids. It should also be noted that these high levels represent carcinogenic risks to people who ingest these waters. There is a need for a state-wide arsenic survey based on hot and cold water samples taken from surface and subsurface water resources. Such an inventory will provide the necessary spatial and temporal extent required for a detailed review.

12.4 CONCLUSIONS

Due to its neotectonic structure and the influence of volcanism, the Anatolian Plate contains various altered rock types that contain high concentrations of arsenic and other trace elements. These rocks demonstrate a strong potential to influence the quality of water resources as a result of water-rock interaction in geological formations. Thus, high arsenic levels of geogenic origin are observed on a wide spatial extent in Western Anatolia. Values in the order of milligrams per litre are common in some parts of this area. These values are several orders of magnitude higher than the national and international standard values and demonstrate a significant health risk for people consuming these waters.

ACKNOWLEDGEMENTS

The authors acknowledge the financial support provided by The Scientific and Technological Research Council of Turkey (TUBITAK) through the project number 109Y029.

REFERENCES

Akar A. (1981) Enrichment and removal of arsenic from arsenic containing antimonite ore in Ödemis–Halıköy-Emirli Mine. *7th Scientific and Technical Congress on Mining in Turkey*. pp. 239–274 (in Turkish).

Aksoy N., Simsek C., Gunduz O. (2009) Groundwater contamination mechanisms in a geothermal field: A case study of Balcova, Turkey. *Journal of Contaminant Hydrology* 103(1–2):13–28

Atabey E. (2009) *Arsenic and its effects*. Publication of General Directorate of Mineral Research and Exploration (MTA), Ankara (in Turkish).

Aydin A.O., Gulensoy H., Akincioglu A., Sakarya A. (2003) Influence of arsenic in colemanite on production of boric acid and borax. *Journal of BAU Institute of Natural Sciences* 5(1):51–58 (in Turkish).

Baba A. (2010) High arsenic levels in water resources resulting from alteration zones: A case study from Biga Peninsula, Turkey. *Proceedings of AS2010: The third international congress on Arsenic in the Environment*, 17–21 May, 2010, Taiwan.

Baba A., Ármannsson H. (2006) Environmental impact of the utilization of a geothermal area in Turkey, *Energy Source*, 1: 267–278.

Baba A., Deniz O. (2008) Assessment of the geothermal potential of Biga Peninsula, its area of use and environmental impacts. The Scientific and Technological Research Council of Turkey (TÜBITAK) Project No: ÇAYDAG-104Y082 (in Turkish).

Baba A., Deniz O., Ozcan H., Erees S.F., Cetiner S.Z. (2009) Geochemical and radionuclide profile of Tuzla geothermal field, Turkey. *Environmental Monitoring and Assessment* 145:361–374.

Baba A., Ertekin C. (2007) Determination of the source and age of the geothermal fluid and its effects on groundwater resources in Kestanbol (Çanakkale-Turkey). *Proceedings CDROM of GQ07: Securing Groundwater Quality in Urban and Industrial Environments*, 7th International Groundwater Quality Conference, Fremantle, Western Australia.

Baba A., Save D., Gunduz O., Gurdal G., Bozcu M., Sulun S., Ozcan H., Hayran O., Ikiisik H., Bakırcı, L. (2009) The assessment of the mining activities in Çan Coal Basin from a medical geology perspective. Final Report. The Scientific and Technological Research Council of Turkey (TÜBITAK) Project No: ÇAYDAG-106Y041, Ankara (in Turkish).

Ball J.W., Nordstrom D.K., Jenne E.A., Vivit D.V. (1998) Chemical Analysis of Hot Springs, Pools, Geysers, and Surface Waters from Yellowstone National Park, Wyoming and its Vicinity. USGS Open-File Report 98:182.

Bozkurt E. (2001) Neotectonics of Turkey – A synthesis. *Geodinamica Acta* 14:3–30.

Bulbul A. (2009) Hydrogeological and hydrogeochemical assessment of cold and hot water systems of Alasehir (Manisa). PhD Thesis, Dokuz Eylul University Graduate School of Natural and Applied Sciences, Izmir, Turkey. (in Turkish).

Bulut M., Filiz S. (2005) Hydrogeology, hydrochemistry and isotopic properties of Bayındır Geothermal Field (Izmir, Western Anatolia, Turkey). *MTA Journal* 131: 63–78 (in Turkish).

Bundschuh J., Bhattacharya P., Hoinkis J., Kabay N., Jean J.S., Litter M.I. (2010) Groundwater arsenic: From genesis to sustainable remediation, *Water Research* 44: 5511.

Colak M., Gemici U., Tarcan G. (2003) The effects of colemanite deposits on the arsenic concentrations of soil and groundwater in Igdeköy-Emet, Kütahya, Turkey. *Water, Air, and Soil Pollution* 149: 127–143.

Davraz A. (2008) Hydrogeochemical and hydrogeological investigations of thermal waters in Uşak area (Turkey). *Environmental Geology* 54:615–628.

Dewey J.F., Sengor A.M.C. (1979) Aegean and surrounding regions: complex multi-plate and continuum tectonics in a convergent zone. *Geol. Soc. America Bull.* Part 1. 90:84–92.

Dogan M., Dogan A.U. (2007) Arsenic mineralization, source, distribution, and abundance in the Kutahya region of the western Anatolia, Turkey. *Environmental Geochemistry and Health*, 29:119–129.

Dogan M., Dogan A.U., Celebi C., Baris Y.I. (2005) Geogenic arsenic and a survey of skin lesions in Emet Region of Kutahya, Turkey. *Indoor-Built Environment* 14(6):533–536.

EPA (2003) Office of Water National Primary Drinking Water Standards. Environmental Protection Agency.

Gemici U. (2008) Evaluation of the water quality related to the acid mine drainage of an abandoned mercury mine (Alaşehir, Turkey). *Environmental Monitoring and Assessment* 147:93–106.

Gemici U., Oyman T. (2003) The influence of the abandoned Kalecik Hg mine on water and stream sediments (Karaburun, İzmir, Turkey). *Science of the Total Environment* 312: 155–166.

Gemici U., Tarcan G. (2004) Hydrogeological and hydrogeochemical features of the Heybeli Spa, Afyon, Turkey: Arsenic and the other contaminants in the thermal waters. *Bulletin of Environmental Contamination and Toxicology* 72/6:1104–1114.

Gemici U., Tarcan G., Helvaci C., Somay A.M. (2008) High arsenic and boron concentrations in groundwaters related to mining activity in the Bigadic borate deposits (Western Turkey). *Applied Geochemistry* 23:2462–2476.

Gunduz O., Simsek C., Hasozbek A. (2010) Arsenic pollution in the groundwater of Simav Plain, Turkey: Its impact on water quality and human health. *Water, Air and Soil Pollution*, 205:43–62.

Helvaci C. (1986) Stratigraphic and structural evolution of the Emet borate deposits, Western Anatolia, DEU Research Paper No: MM /jeo-86 ar 008, 28 p. (in Turkish)

Helvaci C., Orti F. (1998) Sedimantology and diagenesis of Miocene colemantite-ulexite deposits (Western Anatolia, Turkey). *Journal Sedimentary Research* 68(5):1021–1033.

Helvaci C., Alonso R.N. (2000) Borate Deposits of Turkey and Argentina; A Summary and Geological Comparison. *Turkish Journal of Earth Sciences* 9:1–27.

ITASHY (2005) *Regulation on waters for human consumption.* Official Gazette dated 17/02/2005, No.25730, Ankara. (in Turkish)

Jackson J., McKenzie (1984) Active tectonics of the Alpine-Himalayan belt between western Turkey and Pakistan. *Geophysical Journal Royal Astronomy Society* C 77:185–264.

Jean J-S., Bundschuh J., Bhattacharya P. (2010) *Arsenic in Geosphere and Human Diseases*, CRC Press/Balkema, 595p.

Karayigit A.I., Spears D.A., Booth C.A. (2000) Antimony and arsenic anomalies in the coal seams from Gokler coalfield, Gediz, Turkey. *International Journal of Coal Geology* 44:1–17.

McKenzie D.P. (1972) Active tectonics of the Mediterranean region. *Geophysical Journal International*, 30(2):109–185.

MTA (1980) Hot waters, springs and mineral water inventory of Turkey. General Directorate of Mineral Research and Exploration Report No. 6833, 78 s. Ankara (in Turkish).

Nimick D.A., (1994) Arsenic hydrogeochemistry in an irrigated river valley: a reevaluation. *Ground Water* 36: 743–753.

Nimick D.A., Moore J.N., Dalby C.E., Savka M.W. (1998) The fate of geothermal arsenic in the Madison and Missouri rivers, Montana and Wyoming. *Water Resour Res* 34: 3051–3067.

Oruc N (2004) Arsenic levels in Emet-Kütahya drinking water and its relation to boron deposits. *Proceedings of 2nd International Congress on Boron*, Eskisehir (in Turkish).

Ozen T. (2009) Hydrogeological and hydrogeochemical assessment of Salihli geothermal field. PhD Thesis, Dokuz Eylul University Graduate School of Natural and Applied Sciences, Izmir, Turkey (in Turkish).

Ozgur N. (2002) Geochemical signature of Kızıldere geothermal field, Western Anatolia, Turkey. *International Geology Review* 44(2):153–163.

Romero L., Alonso H., Campano P., Fanfani L., Cidu R., Dadea C., Keegan T., Thornton I., Farago M. (2003) Arsenic enrichment in waters and sediments of the Rio Loa (Second region, Chile). *Applied Geochemistry* 18 (9): 1399–1416.

Simsek C., Elci A., Gunduz O., Erdogan B. (2008) Hydrogeological and hydrogeochemical characterization of a karstic mountain region. *Environmental Geology* 54(2):291–308.

Simsek S., Yildirim N., Simsek Z.N., Karakus H. (2002) Changes in geothermal resources at earthquake regions and their importance. *Proceedings of Middle Anatolian Geothermal Energy and Environmental Symposium* pp. 1–13.

Tarcan G., Gemici U. (2003) Water chemistry of the Seferihisar geothermal area, Izmir, Turkey. *Journal of Volcanology and Geothermal Research* 126:225–242.

Tarcan G., Gemici U., Aksoy N. (2005) Hydrogeological and geochemical assessments of the Gediz Graben geothermal areas, Western Anatolia, Turkey. *Environmental Geology* 47(4): 523–534.

Van Halem D., Bakker S.A., Amy G.L., van Dijk J.C. (2009) Arsenic in drinking water: a worldwide water quality concern for water supply companies. *Drinking Water Engineering and Science*, 2:29–34.

Wagner G.A., Pernicka E., Seeliger T.C., Oztunalı Ö., Baranyi I., Begemann F., Schmitt-Strecker S. (1984) Geological investigations of Early Metallurgy of Northwestern Anatolia. *MTA Journal* 101–102: 92–127 (in Turkish).

Yigitbas E., Elmas A., Sefunc A., Ozer N. (2004) Major neotectonic features of eastern Marmara region, Turkey: development of the Adapazari-Karasu corridor and its tectonic significance. *Geological J* 39(2):179–198.

Yilmaz S., Baba B., Baba A., Yagmur S., Citak M. (2009) Direct quantitative determination of total arsenic in natural hot waters by anodic stripping voltammetry at the rotating lateral gold electrode. *Current Analytical Chemistry* 5(1):29–34.

WHO (2004) *Guidelines for Drinking Water Quality* Third Edition, World Health Organization, Vol. 1. Geneva.

Part 2

Groundwater Management

Chapter 13

Regional spatial-temporal assessment of the sustainability of groundwater exploitation in the south of Portugal

Tibor Stigter[1]*, José Paulo Monteiro*[2]*, Luís Nunes*[2]*,*
Luís Ribeiro[1] *& Rui Hugman*[2]

[1] *Geo-Systems Centre/CVRM – Instituto Superior Técnico, Lisbon, Portugal*
[2] *Geo-Systems Centre/CVRM – Universidade do Algarve, Faro, Portugal*

ABSTRACT

Groundwater was the main source for public supply in the Algarve, in the south of Portugal, until the end of the 20th century, after which it was replaced by surface water supplied by large reservoirs. The large drought that hit the region in 2004 and 2005 revealed the problems related to a water supply strategy based on a single source and stressed the need of an integrated water resource management (IWRM) scheme. Following a qualitative and quantitative screening of groundwater resources for integration into the public water supply system of the region, current work aims to address the regional quantification of groundwater availability in those aquifers that are available for public water supply. For this purpose the strengths and flaws of simple analytical water balance formulae to calculate annual aquifer storage volumes will be discussed. These calculations are validated with the help of a numerical finite-element groundwater flow model, and it is shown that their application is suitable for calculating available storage volumes in regions where groundwater models do not exist, as long as a conceptual hydrogeological model is developed.

13.1 INTRODUCTION

Groundwater was the main source for public supply in the Algarve, in the south of Portugal, until the end of the 20th century, after which it was replaced by surface water supplied by large reservoirs. The large drought that hit the region in 2004 and 2005 revealed the problems related to a water supply strategy based on a single source. It is well-known that in semi-arid regions such as the Algarve, the seasonal and annual variations in rainfall are extreme. Moreover, the intensity and frequency of occurrence of extreme droughts is expected to increase significantly in the future (Giorgi, 2006; Santos & Miranda, 2006; Stigter *et al.*, 2011). Recent research has been carried out on climate scenarios and their impacts on groundwater resources and dependent ecosystems in the Central Algarve (as well as the in the Ebre Delta in Spain and the Atlantic Sahel in Morocco), in the scope of the CIRCLE-Med project CLIMWAT (Stigter *et al.*, 2009b, 2011). For the Central Algarve, although mean annual rainfall is expected to decrease only slightly in the short-term, i.e. up to 2050, significant shifts in seasonal distribution and interannual variability are predicted. Rainfall will be more concentrated in winter seasons, with large reductions in spring and autumn. Calculations show that this in fact will lead to a slightly higher fraction of rainfall contributing to recharge. However, the interannual variability in both rainfall and recharge will increase. In the long-term, 2070–2100, the work of Stigter *et al.* (2011) shows that a significant reduction in both rainfall and recharge is predicted by climate models, considering the A1b CO_2 emission scenario and downscaled data of the EU-funded ENSEMBLES project (Van der Linden & Mitchell, 2009). Global warming will also increase crop water demand and thus irrigation requirements, seriously jeopardizing the sustainability of water use in the region in the long-term, particularly when considering ecological water demands for groundwater dependent ecosystems.

The conjunctive use and management of multiple water sources for different water-consuming activities, as part of the more complex concept of integrated water resource management, will be essential both in the near and distant future. These multiple water sources include surface water, groundwater and alternative resources such as treated wastewater for irrigation. Within this scope, a qualitative and quantitative screening of groundwater sources for integration into the public water supply system of the Algarve region was performed by Stigter *et al.* (2009a), as part of the OPTEXPLOR project, a R&D project financed by the Algarve Water Utility with the aim to create a decision support system based on an optimization model (Nunes *et al.*, 2009; Vieira *et al.*, 2011). Current work aims to address the regional quantification of groundwater availability and exploitation sustainability in those aquifers that are available for public water supply, as well as their dependence on factors such as the spatial and temporal distribution of recharge, aquifer heterogeneity and the location of the pumping wells. In this chapter the strengths and flaws of simple analytical water balance formulae to calculate annual aquifer storage volumes are discussed. Calculations are validated with the help of a numerical finite-element groundwater flow model for the largest and most productive aquifer system. It is shown that such simple water balance calculations can be applied to calculate available storage volumes in regions where groundwater models do not exist, as long as a conceptual hydrogeological model is developed.

13.2 GROUNDWATER RECHARGE VERSUS CONSUMPTION

The present state of development of the hydrogeology of Algarve allows the definition of 17 aquifer systems with regional importance, shown in Figure 13.1 (Almeida *et al.*, 2000). The most productive aquifers are karstified limestones and dolomites. The six most important aquifers for public water supply are characterised in Table 13.1. Other aquifer systems are mainly exploited for irrigation. Due to its large area and significant recharge, as well as the high degree of karstification, the aquifer system M5, known as the Querença-Silves aquifer, constitutes the most important groundwater reservoir.

The estimation of aquifer recharge is a crucial and continuously ongoing task. Stigter *et al.* (2009a) provide an overview of some of the applied methods. During the implementation of regional flow models in the Algarve, the accuracy of recharge estimates was improved, due to: i) improved accuracy of the geometric representation of the lithological outcrops; ii) improved accuracy and spatial resolution of rainfall, through the implementation of kriging with external drift on an orthogonal grid with a

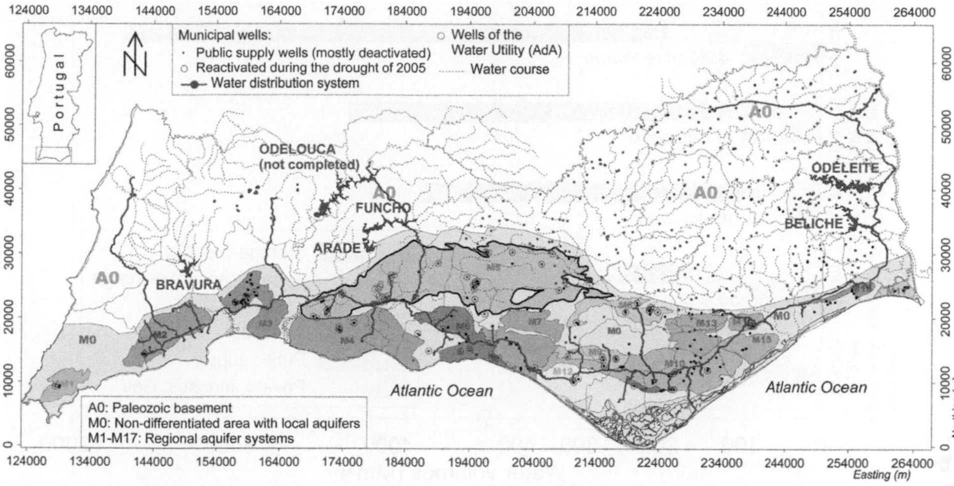

Figure 13.1 Location and geometry of the aquifer systems in the Algarve, highlighting the Querença-Silves aquifer (M5); also shown is the location of the municipal wells and surface water reservoirs.

Table 13.1 Characterization of aquifer systems with regional expression in the Algarve.

Aquifer system		*Main aquifer lithology*	*Area (km²)*	*Recharge (hm³ yr⁻¹)*
M2	Almádena – Odeáxere	lmst, dlmt	63.49	16.6
M3	Mexilhoeira Grande – Portimão	lmst, dlmt, sand	51.71	8
M5	Querença – Silves	lmst, dlmt	317.85	100
M8	S. Brás de Alportel	lmst, dlmt	34.42	5.5
M9	Almansil – Medronhal	lmst, dlmt	23.35	6.5
M14	Malhão	lmst, dlmt	11.83	3

lmst = limestone; dlmt = dolomite

resolution of 1 km^2, with elevation proving to be the most representative auxiliary variable (Nicolau, 2002). Recent research has further taken into account parameters, such as daily precipitation, soil texture, moisture content and vegetation cover, allowing for a deeper insight into the processes controlling recharge and its temporal evolution (Oliveira *et al.*, 2008). Figure 13.2 presents the estimated mean annual recharge (MAR) volumes for the 17 defined aquifer systems (M1–M17), the six most relevant aquifers for public water supply, and the aquifer system M5. Roughly estimated total storage capacities are also presented, considering an aquifer thickness of 100 m and effective porosity of 10%. Though these estimates are extremely simplified, they are also conservative and allow for a good perception of their magnitude as compared to surface water storage, also presented in Figure 13.2. The total estimated storage of the aquifer systems is about 20 times higher than that of the surface reservoirs including the Odelouca reservoir, currently in the phase of completion. It is also 50 times the MAR volume. The key question here is what fraction of storage is exploitable in a

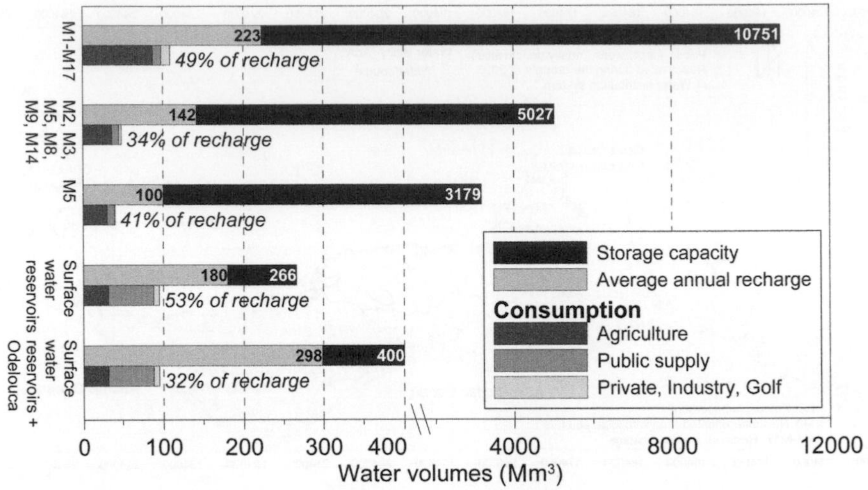

Figure 13.2 Storage capacity, mean annual recharge and water consumption volumes for groundwater and surface water in the Algarve. For groundwater the three categories refer to: all the aquifer systems (M1-M17), the main aquifers for public supply (M2, M3, M5, M8, M9, M14), and M5; for water consumption, the labels indicate total volumes as a percentage of mean annual recharge.

sustainable way, particularly on a short-term (i.e. yearly) basis. In other words, how much of the storage can be captured in dry years without having negative consequences for water quality or dependent ecosystems?

Figure 13.2 also presents current groundwater consumption volumes, and their distribution among users. The numbers are based on a detailed study of available data, provided by the Regional Water Utility, the Portuguese Ministry of Agriculture, and Do Ó & Monteiro (2006). Agriculture is by far the main consumer of groundwater, with a mean yearly total of approximately 90 hm^3 withdrawn from the 17 aquifer systems, 40% of MAR. Total consumption amounts to half of MAR, but for the six main aquifer systems for public supply this percentage is less, namely 34% of MAR. Though groundwater is the main source for irrigation, irrigation with surface water is gaining importance, allocating increasing water volumes.

13.3 SUSTAINABLE YIELD ANALYSIS

Safe yield was initially defined by Sophocleous (1997) as the attainment and maintenance of a long-term balance between the amount of groundwater withdrawn annually and the annual amount of recharge. Subsequently, the emphasis shifted to sustainable yield (e.g. Sophocleous, 2000; Custodio, 2002), which reserves a fraction of safe yield for ecological demands. This fraction depends on factors such as climate (variability), hydrogeological setting, location of wells, and the presence of groundwater dependent ecosystems. The concept of sustainable yield (or volume) can be studied by analysing different groundwater recharge/capture/discharge scenarios. Capture is defined by Lohman *et al.* (1972) as the sum of the increase in recharge and decrease in discharge, caused by abstractions due to pumping. Capture predominantly results in a decrease of groundwater discharge and a removal of water from storage.

The analysis starts with the definition of the so-called "safe storage volume" (S_{safe}), below which undesirable effects may occur as a result of overexploitation, such as the drying up of groundwater dependent streams and wetlands or the intrusion of seawater. The water volume resulting from one year of MAR is considered to adequately represent the safe storage volume. Considering a simple black box model, the hypothetical evolution of aquifer storage is calculated for different discharge scenarios, using the following equations:

$$S_t = (1 - f)(S_{t-1} + Rn_{\{(t-1),t\}} - W_{\{(t-1),t\}} + Ra_{\{(t-1),t\}}) \tag{13.1}$$

$$Q_{\{(t-1),t\}} = f(S_{t-1} + Rn_{\{(t-1),t\}} - W_{\{(t-1),t\}} + Ra_{\{(t-1),t\}}) \tag{13.2}$$

$$Rn_{\{(t-1),t\}} = \frac{P_{\{(t-1),t\}}}{\overline{P}} \times \overline{Rn} \tag{13.3}$$

where: S_t and S_{t-1} are the aquifer storage at time t and $t-1$, respectively, with a discrete time step of one hydrological year, $P_{\{(t-1),t\}}$ is precipitation between hydrological years $t-1$ and t, $Rn_{\{(t-1),t\}}$ is natural recharge, $W_{\{(t-1),t\}}$ is withdrawal, $Ra_{\{(t-1),t\}}$ is artificial recharge (irrigation return flow), and $Q_{\{(t-1),t\}}$ is groundwater discharge for the same period; f is the fraction of surplus contributing to discharge. Natural recharge is calculated as a ratio of observed to mean annual precipitation times MAR (Equation 13.3). Surplus is defined as the storage at the beginning of the preceding year plus the difference between natural and artificial recharge and abstractions throughout the year. Equation 13.2 indicates that a higher surplus will result in a higher discharge. For the first year, the value for initial aquifer storage (S_0) was set to half the volume of natural recharge of the preceding year. Though this may seem arbitrary, it showed a good fit when compared to observed tendencies, and sensitivity analysis showed that calculated surplus in subsequent years rapidly became independent from S_0. The hypothetical scenarios of water storage are generated by varying parameter f between 0 (no outflow) and 1 (100% outflow).

The curves defining each of the six scenarios, 0, 20, 40, 60, 80 and 100% groundwater outflow, have been calculated for the QS aquifer and drawn in Figure 13.3, where water storage S is plotted as a fraction of S_{safe}, the volume obtained by one year of MAR in the aquifer. A period of eight (hydrological) years is considered, starting in October 2001, when the multi-municipal public water supply system (MPWSS) was fully operational and groundwater consumption was comparable to the present-day picture. A conceptual hydrogeological model of the QS aquifer system is presented in Figure 13.4, indicating MAR as a fraction of rainfall, based on data from Oliveira *et al.* (2008), the location of wells and springs, as well as the boundary condition of the Arade Estuary. On average 45 hm^3 of water is annually pumped from the QS aquifer, but in the dry year of 2005 abstractions exceeded 65 hm^3.

In the scenario of 0% groundwater discharge, naturally unrealistic, all surplus water is stored in the aquifer. In the opposite, equally unrealistic scenario of 100% outflow, no surplus water exists and available water volumes are 0% of safe

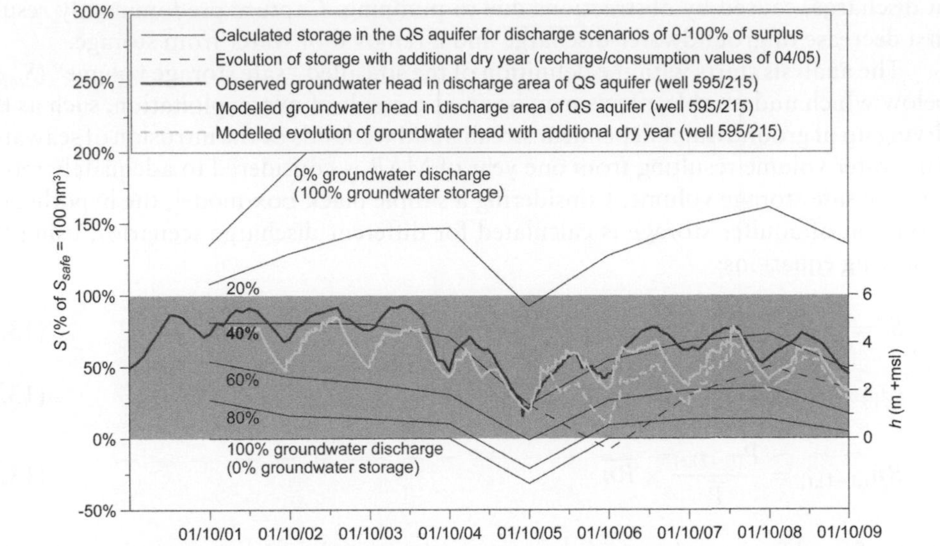

Figure 13.3 Scenarios of the evolution of water storage as a percentage of safe storage volume in the Querença-Silves aquifer system, as well as modeled and observed time series of groundwater levels in the discharge area. Dashed lines indicate potential evolution with a second consecutive dry year.

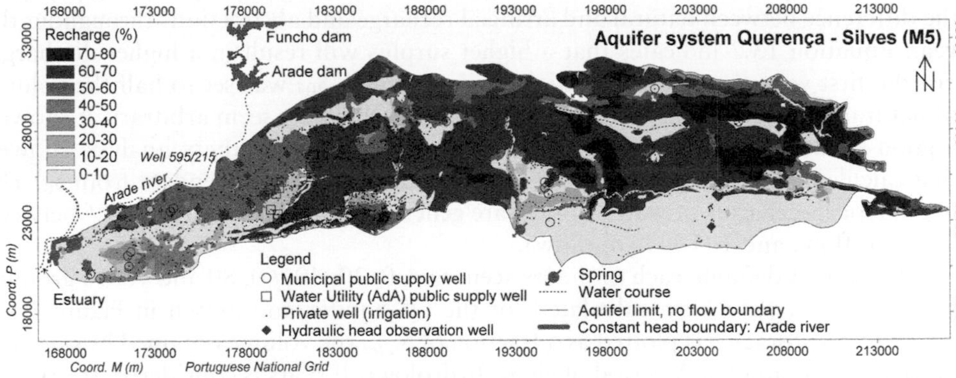

Figure 13.4 Conceptual hydrogeological model of the Querença-Silves aquifer system, indicating mean annual recharge (MAR) as a fraction of rainfall (based on data from Mendes Oliveira, 2008), location of wells, springs and the boundary condition of the Arade Estuary.

storage. The remaining four discharge scenarios are all hypothetically realistic, but the question is which one, if any, represents a more or less truthful simulation of reality. Naturally, the equations greatly simplify the actual behaviour of the systems. For instance, groundwater discharge depends on the spatial and temporal distribution of recharge, whereas the impact of abstractions also depends on the location of the pumping wells. Moreover, in reality discharge is not a fixed percentage of the available water volume in the aquifer; it will be higher in wet years and lower in dry years.

In order to interpret the accuracy of the simple analytical calculations, they can be compared to advanced numerical simulations, as well as observed groundwater head time series, which are an indicator of aquifer storage volume. Time series are shown in Figure 13.3 for well 595/215 located in the discharge area (see Figure 13.4), near the western boundary of the aquifer system. The choice of the axis limits for the head time series plot may be debatable, but is based on observations of the historical time series, where the head of 6 m proved to be a representative indication of 100% safe storage volume. On the other hand, 0 m represents the limit below which gradient inversion occurs, resulting in zero discharge at the springs and seawater intrusion. First, it can be seen that the yearly trends are correctly portrayed by this simple analysis. Second, it appears that the 40% surplus discharge scenario ($f = 0.4$), most correctly follows the observed water level trend. In fact, the relation between piezometry and this f-curve, which indicates a 60% storage of annual surplus in the aquifer, is surprisingly accurate. This may be related to the fact that this karst aquifer has a well-defined geometry and clear boundary conditions.

The development of steady-state and transient groundwater flow models for the QS aquifer system is described and discussed by Monteiro *et al.* (2006, 2007) and Stigter *et al.* (2009a). The conceptual flow model was translated to a finite element mesh with 11663 nodes and 22409 triangular finite elements. Transmissivity values were optimised by inverse calibration of the model and allowed for a significant improvement of the simulation reliability of the observed regional flow pattern (Stigter *et al.*, 2009a, 2011). For the transient model, the spatial distribution of the storage coefficient was calibrated by trial-and-error for a model run from 2002 and 2006, using available piezometric data of eight wells in the official monitoring network of the Regional Water Basin Administration (RWBA), and then validated for 2006–2009. Estimated discharge rates were compared to measurements performed by the RWBA at springs of the Arade estuary that forms the imposed boundary condition in the model (Figure 13.4). Direct calibration based on measured spring discharge was not possible, as it only constitutes 25–35% of the total discharge from the aquifer (Stigter *et al.*, 2011).

The results of a model for the 2001–2009 period are shown in Figure 13.3 for the groundwater head time series of well 595/215 and in Figure 13.5 for the aquifer discharge hydrograph. The latter plot clearly shows the relation between observed discharge at the springs of the Arade Estuary and total aquifer discharge at the Arade boundary condition. It can be noted that model variants including boundary conditions for the smaller springs in the central and eastern sector revealed a negligible impact on the regional flow pattern and water balance.

The trend and amplitude of oscillation of the water level in well 595/215 are correctly simulated by the model. During calibration of the transient model it was clearly noticed that groundwater pumping from private wells in 2005 started earlier than usual, namely in January (rather than in May). To simulate the larger drawdowns in that year, 13 hm^3 had to be added to the annual 31 hm^3 considered in the model for irrigation. This fact clearly indicates the "double-negative" aspect of droughts, i.e. lower recharge and higher (uncontrolled) pumping. Total abstraction in this year was 190% of recharge. When simulating a second consecutive dry year following 2004/2005, with the same extraction and recharge rates (190% of recharge), it is observed that both the

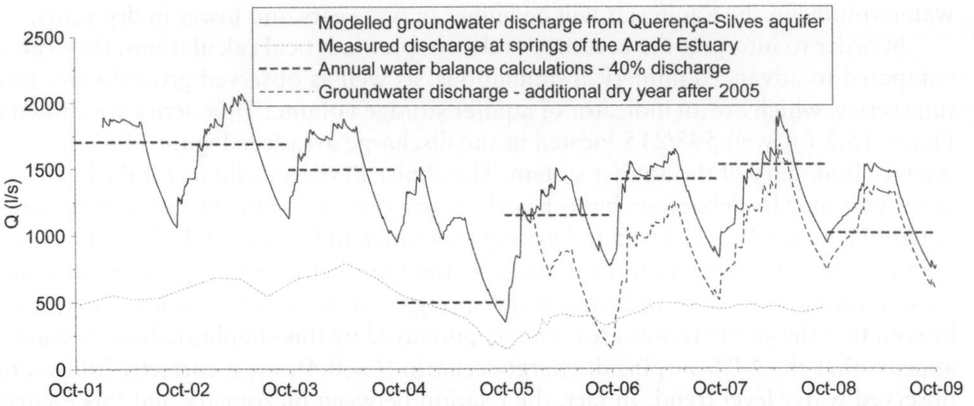

Figure 13.5 Groundwater discharge from the Querença-Silves aquifer system between 2001 and 2009, as output from the transient numerical model, compared to the annual water balance calculations.

groundwater head near the aquifer boundary and the discharge drop to values close to 0 (Figure 13.4 and 13.5). Nevertheless, their recovery in subsequent years is rapid and in 2009 the impact of the additional dry year is hardly noticed, though complete recovery would take some more years. This means that the overexploitation on an annual basis, as defined by groundwater withdrawals exceeding recharge rates, is not necessarily unsustainable in the short-term, especially in aquifers with high resilience.

When considering the 40% outflow scenario, total storage can be calculated by extrapolating the results for the six main public supply aquifers, based on known (estimated) abstractions and recharge rates. Figure 13.6 provides the results, as % of maximum storage (140 hm^3). The figure also gives an idea of potential changes in storage and discharge volumes for a 20 hm^3 higher annual abstraction or 25% lower recharge scenario. It is clear from Figure 13.6 and the analytical formulas behind it, that higher abstractions have a significant effect on both storage and discharge. It is therefore not an easy or straightforward task to optimise abstraction rates and determine sustainable yields. Moreover, much of the increase in water demands, either for crops (irrigation) or for public supply (including tourism) occurs in the dry season, when outflow is reduced, so that abstraction has a higher impact on storage. Increasing storage and reducing discharge will be essential challenges in the future.

13.4 FINAL CONSIDERATIONS

The results presented in this paper show that the application of annual water budget calculations to determine water storage volumes may be a valid tool in regions where groundwater models and monitoring data are unavailable. Naturally, there are limitations involved, as storage depends on factors such as the spatial and temporal distribution of recharge, aquifer heterogeneity and the location of the pumping wells. Nevertheless, if the conceptual hydrogeological model of a region is known and the

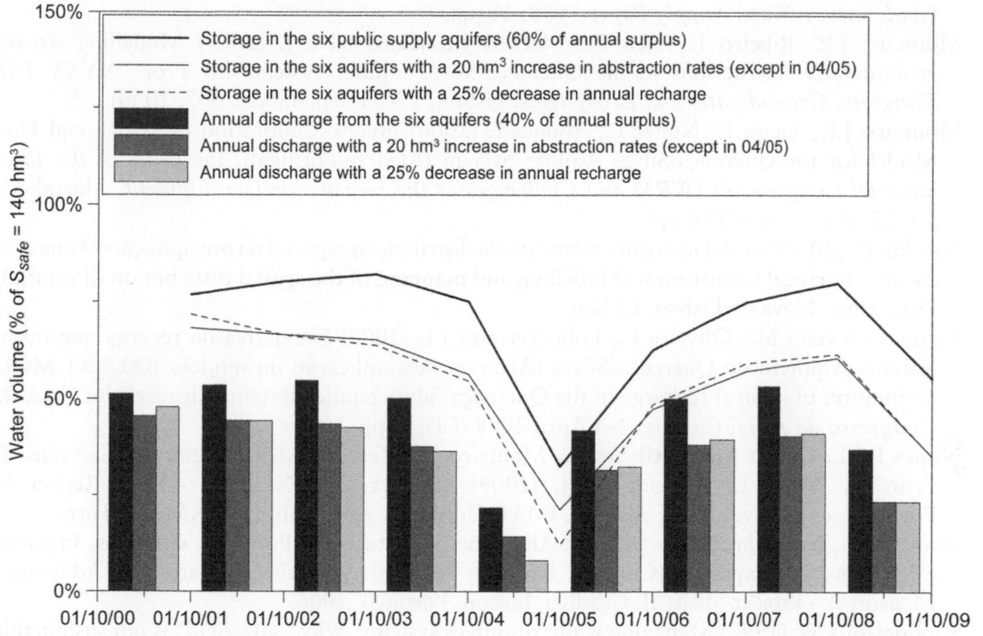

Figure 13.6 Evolution of annual storage and discharge volumes in the six main public supply aquifers, using f = 0.4 (40% of surplus is outflow). Also shown are higher consumption scenarios (dashed lines).

wells are distributed equally over the area, or located in the discharge sector of the aquifer, the method is expected to provide accurate indications of storage, particularly in semi-arid and arid regions where surface runoff is negligible. The method can also be adjusted to include infiltration from rivers, though estimates can be difficult to obtain and less reliable. Validation of the method should be performed, if water level monitoring data are available under similar climate and hydrogeological conditions as those of the aquifer to be studied. The method is not suitable for analysing seasonal variations of storage, although, theoretically, it could be adjusted by increasing the temporal resolution. However, this would require estimates of the water budget components on a seasonal or monthly basis.

REFERENCES

Almeida C. (1985) Hidrogeologia do Algarve Central, Ph.D., Univ. de Lisboa, Lisbon.

Custodio E. (2002) Aquifer overexploitation: what does it mean? *Hydrogeol J* 10: 254–277.

Do Ó A., Monteiro J.P. (2006) Estimação da procura real de água no Algarve por sectores. In *Proc. V Congresso Ibérico sobre Gestão e Planeamento da Água*, Faro, 4–8 Dec., 10 pp.

Giorgi F. (2006) Climate change hot-spots. *Geophys Res Let* 33, L08707, doi:10.1029/2006GL025734.

Lohman S.W., Bennett R.R., Brown R.H., Cooper Jr. H.H., Drescher W.J., Ferris J.G., Johnson A.I., McGuinness C.L., Piper A.M., Rorabaugh M.I., Stallman R.W., Theis C.V.

(1972) Definitions of selected ground-water terms: revisions and conceptual refinements. *US Geol. Survey, Water Supply Paper* 1988, 21 pp.
Monteiro J.P., Ribeiro L., Reis E., Martins J., Matos Silva J. (2007) Modelling stream-groundwater interactions in the Querença-Silves aquifer system. In: *Proc. XXXV IAH Congress, Groundwater and Ecosystems*, Lisbon, 17–21 September 2007, 10 pp.
Monteiro J.P., Vieira J., Nunes L., Younes F. (2006) Inverse Calibration of a Regional Flow Model for the Querença-Silves Aquifer System (Algarve-Portugal), in: *Proc. of the International Congress on IWRM and Challenges of the Sustainable Development*, Marrakech, 23–25 May 2006 (CD 6 pp).
Nicolau R. (2002) Modelação e mapeamento da distribuição espacial da precipitação – Uma aplicação a Portugal Continental (Modelling and mapping of the spatial distribution of rainfall). PhD, Univ. Nova de Lisboa, Lisbon.
Mendes Oliveira M., Oliveira L., Lobo Ferreira J.P. (2008) Estimativa da recarga natural no Sistema Aquífero de Querença-Silves (Algarve) pela aplicação do modelo BALSEQ_MOD. Estimation of natural recharge in the Querença-Silves aquifer system (Algarve). In: *Proc. 9.º Congresso da Água*, Cascais, 2–4 April 2008 (CD 15 pp).
Nunes L.M., Cunha M.C., Ribeiro L., Monteiro J.P., Teixeira M.R., Stigter T., Guerreiro P., Brito S., Vieira J., Nascimento J. (2009) "Project OPTEXPLOR – Final Report (in Portuguese)." University of Algarve, IMAR-University of Coimbra, CVRM-IST, Faro.
Santos F.D., Miranda P. (Eds.) (2006) Alterações climáticas em Portugal. Cenários, Impactos e Medidas de Adaptação (Climate change in Portugal. Scenarios, Impacts and Adaptation Measures) – Project SIAM II, Gradiva, Lisbon, Portugal, 2006.
Sophocleous M. (1997) Managing water resources systems: Why "safe yield" is not sustainable. *Ground Water* 35:561–570.
Sophocleous, M. (2000): From safe yield to sustainable development of water resources – the Kansas experience, *J. Hydrol.*, 235, 27–43.
Stigter T., Ribeiro L., Samper J., Fakir Y., Pisani B., Li Y., Nunes J.P., Tomé S., Oliveira R., Hugman R., Monteiro J.P., Silva A.C.F., Tavares P.C.F., Shapouri M., Cancela da Fonseca L., El Mandour A., Yacoubi-Khebiza M., El Himer H. (2011) Assessing and Managing the Impact of Climate Change on Coastal Groundwater Resources and Dependent Ecosystems – Final Report. CIRCLE-Med Project. Instituto Superior Técnico, Lisbon, 187 pp.
Stigter T.Y., Monteiro J.P., Nunes L.M., Vieira J., Cunha M.C., Ribeiro L., Nascimento J., Lucas H. (2009a). Screening of sustainable groundwater sources for integration into a regional drought-prone water supply system. *Hydrology and Earth System Sciences* 13: 1–15.
Stigter T., Ribeiro L., Oliveira R., Samper J., Fakir Y., Monteiro J.P., Nunes J.P., Pisani B., Tavares P.C.F. (2009b) Assessing and managing the impact of climate change on coastal groundwater resources and dependent ecosystems: the CLIMWAT project. *International Conference on Ecohydrology and Climate Change*, Tomar, 10–12 September, 2009.
Van Der Linden P., Mitchell J.F.B. (Eds) (2009) ENSEMBLES: Climate Change and its Impacts: Summary of research and results from the ENSEMBLES project, Met Office Hadley Centre, Exeter, UK.
Vieira J., Cunha M.C., Nunes L.M., Monteiro J.P., Ribeiro L., Stigter T., Nascimento J., Lucas H. (2011) Optimization of the Operation of Large-Scale Multisource Water-Supply Systems, *Journal of Water Resources Plan Manag* 137:150–158, doi:10.1061/(ASCE)WR.1943-5452.0000102.

Chapter 14

Quality and quantity status and risk assessment of groundwater bodies in the karst areas of Croatia

Ranko Biondić[1], *Božidar Biondić*[1], *Josip Rubinić*[2] & *Hrvoje Meaški*[1]

[1] *University of Zagreb, Faculty of Geotechnical Engineering, Varaždin, Croatia*
[2] *University of Rijeka, Faculty of Civil Engineering, Rijeka, Croatia*

ABSTRACT

Almost half of Croatia is karst. It is difficult to separate groundwater from surface waters in karst areas. Karst areas require an integrated water quality and quantity assessment as well as water resources risk assessment for each groundwater body (GWB). In heterogeneous karst conditions there are considerable difficulties in delineating GWBs as well as in the assessment of groundwater quality and quantity status according to the European Union Water Framework Directive. The fact that groundwater monitoring in Croatia is in a developing stage is an additional problem, so that initial characterisation and groundwater risk assessment have been based upon data gathered during the period 2000–2007. This paper presents the methodological approach which was applied on the Croatian karst, and a summary of the findings.

14.1 INTRODUCTION

Karst aquifers developed in the south-western and southern part of Croatia occupy an area of approximately 26 750 km^2, i.e. about 50% of the country (Fig. 14.1). The rocks are mostly karstified carbonate rocks that belong to the macro-structural unit *"Dinarides"*, which extends from Slovenia through Croatia and Bosnia and Herzegovina to Montenegro. The volume of groundwater contained in the karst aquifers in Croatia represents almost half the total available amounts of water in the country. For karst areas and economically developed coastal areas these aquifers are the only sources of drinking water. Part of the karst area lies within the Adriatic Sea catchment and part in the Black Sea catchment.

Basic characteristics of Dinaric karst aquifers are large catchments with high rainfall (up to 4000 mm per year), low retention capacity of karstified ground, rapid subsurface flows, periodic flooding of karst fields, appearances of large karst springs, multiple discharges and sinks within the same catchment, long drought periods, exposed rock and significant intrusion of sea water in coastal and island aquifers. Karst aquifers are highly vulnerable and special protection measures are required to preserve the quality and quantity of the groundwater. Only 2% of the total groundwater reserves are used for public water supply and almost 25% are held in a numerous surface water reservoirs and used for hydro power plants.

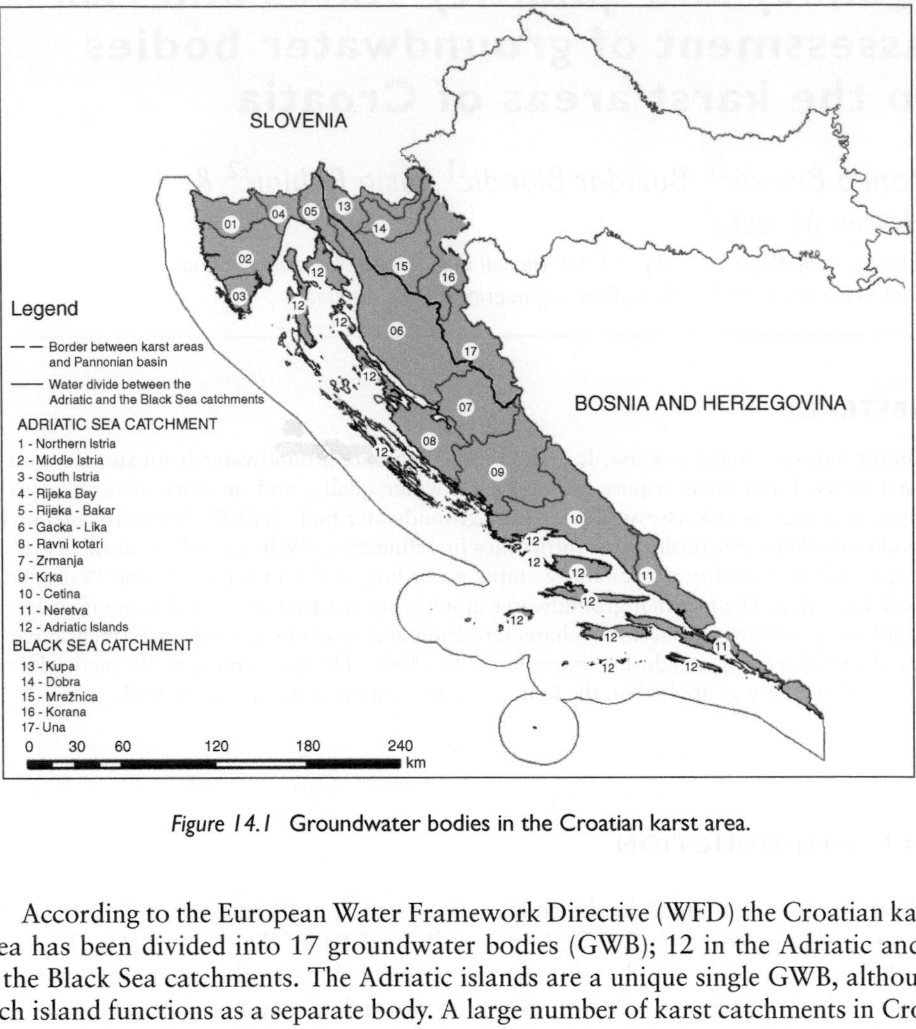

Figure 14.1 Groundwater bodies in the Croatian karst area.

According to the European Water Framework Directive (WFD) the Croatian karst area has been divided into 17 groundwater bodies (GWB); 12 in the Adriatic and 5 in the Black Sea catchments. The Adriatic islands are a unique single GWB, although each island functions as a separate body. A large number of karst catchments in Croatia extend into the neighbouring countries of Slovenia, Bosnia and Herzegovina, as transboundary aquifers. Towards the south of Croatia, the number of transboundary GWB units increases, so in the southern part of Croatia discharge zones depend on catchment areas in Bosnia and Herzegovina.

14.2 CHARACTERISATION OF GROUND WATER BODIES

WFD is a strategic document for the organization of management and protection of water resources in the European Union to prevent degradation of water resources, as well as to enable effective protection, improvement of aquatic ecosystems and sustainable use of highly sensitive natural resources. Risk assessment of GWBs is a basic requirement for further monitoring of water resources. Assessments of impacts and pressures were carried out using the European approach described by COST project

620 (EU COST 620, 2004), which consists of three interconnected phases: analysis of natural vulnerability, hazard and of risk.

Analysis of the natural vulnerability of karst aquifers is based on geological structure, estimation of the degree of karstification, slope analyses and rainfall. Different classes of natural vulnerability were separated out to produce a *Vulnerability map*, which can be used to determine groundwater source protection zones for water supply sources.

Hazard analyses include the development of databases for point and diffuse (agriculture) pollutants to the karst aquifers. The final products are the *Classified maps of hazards*, which present the locations and weighted values for each pollutant. Special attention is paid to the impact of agricultural activities, because the chemical analyses of groundwater in some regions indicates significant agricultural impact on some groundwater resources. The *Risk map* is achieved by overlapping the Vulnerability and Classified Hazard maps.

Qualitative status assessment (QUAL) has been carried out for each GWB, based on chemical analyses of waters from 55 karst springs that are included in the *National monitoring network*, and on detailed analyses of individual water supply springs for the period of 2000–2007.

The following basic parameters were used for QUAL (WFD, 2000): dissolved oxygen, pH, specific electrical conductivity (SEC), nitrate and ammonia, and additional groundwater parameters such as free CO_2, water temperature (T), orthophosphate, turbidity, Fe, Mn, mineral oils, As, Cd, Pb, Hg, chloride, sulphate, trichloroethylene and tetrachloroethylene.

The first step of the qualitative analysis is to determine the reference value (REF) prescribed by national legislation according to the use of the groundwater in each GWB. According to the WFD, as well as supporting technical reports (e.g. Towards a Guidance on Groundwater Chemical Status and Threshold Values) two basic types of criteria must be taken into consideration. These are environmental criteria and usage, with the strictest criteria applied for public water supply. There is at least one public water-supply station or one is planned in almost all the GWBs. For ecosystems, and other types of use, no specific REF values for the water quality parameters were set, so that the maximum allowable concentration (MAC) for drinking water prescribed by the Regulations on the quality of drinking water was adopted.

The next step in the qualitative status assessment is to determine the natural background level (BL) and the threshold value (TV). BL is determined only for the quality parameters analysed in the quality assessment analyses of the GWBs, because there was not enough data for the statistical analysis of other parameters. At all observation points $BL < REF$ was recorded. In this case the TV can be set and implemented in several ways ($TV = REF$; $TV = 0.75 * REF$; $TV = (BL + REF)/2$. For the purposes of qualitative analysis in the karstic area of Croatia the statement $TV = REF$ was used.

Specific characteristics of karst aquifers in Croatia and elsewhere in the world include high groundwater velocities and relatively short residence times, as well as rapid water quality changes in short time intervals. It is interesting to note exceptional quality of water in karst springs particularly during the long summer dry periods. This indicates the important role of epikarst and the unsaturated zone of karst aquifers, which prevent contaminants infiltrating to the saturated karst aquifer. Heavy rainfall after a long dry period causes strong and relatively short-term pollution of karst groundwaters.

During these events several water quality parameters exceed the MAC. Another major problem, especially of coastal karst aquifers, is occasional sea water intrusion deep into the coastal aquifers during the summer dry periods and management of groundwater abstraction to minimize sea water ingress.

An illustrative parameter for QUAL in karst groundwater in Croatia is the concentration of nitrate. In the central part of the Croatian karst area, is mountainous, nitrate concentrations are low and have a constant value without any meaningful trends. This is an indicator of untouched pristine GWB that is characterised by sparsely populated land. In the Dinaric border areas, on the Istrian Peninsula, the situation is quite different and nitrate concentrations are in most cases still within MAC (on average 46 mg NO_3/l), they are close to the MAC value of 50 mg NO_3/l. This is because of the intensive agricultural activities on the Istrian Peninsula.

MAC of pH for drinking water is in the range from 6.5 to 9.5. The results of all analyses during the characterization of GWBs lie within this range, but in the border areas of the mountainous Dinaric lands pH is decreasing. This reflects microbiological and chemical denitrification processes occurring in natural systems.

The major problem of coastal and island karst aquifers in Croatia is a periodic or permanent marine influence on freshwater systems. During the first half of the 20th century numerous springs were captured in coastal areas and on islands for water supply needs, and in that time water quantities were adequate for local population needs. However, population growth and tourism development have caused increased demands for drinking water and existing springs have become increasingly exploited. Increased fresh water exploitation caused gradual increases of salinity in these springs, and nowadays many of them are already out of service. This problem is especially significant on the islands where fresh water gradually becomes brackish, and today only three large islands (Cres, Krk, Vis) have their own fresh groundwater resources. Water supplies on other islands are either associated with mainland supply systems with undersea water pipelines, or desalination of salt water.

QUAL analyses of karst groundwater in Croatia highlighted two GWBs that have significant water quality problems. These are the GWBs *South Istria* and *Ravni kotari* (Fig. 14.2). *South Istria* has problems with nitrate (with concentrations above 100 mg NO_3/l in some abstraction wells) and chloride that is increased during the summer dry periods, due to uncontrolled exploitation of water for agriculture purposes and uncontrolled agricultural production (Biondic *et al.*, 2009). GWB *Ravni kotari* has the biggest problem with sea water intruding beneath the land causing higher chloride concentrations in some wells. Fresh water resources in the Adriatic islands are also influenced by sea water, some of them temporarilly and some permanently (smaller islands), but because there are fewer big water abstraction sites and good water quality their QUAL status is assessed as good.

Risk assessment was also performed under QUAL, based on analyses of trends for selected parameters. The boundary condition of *at risk* is 75% of the allowable reference value at the end of the reporting period, i.e. end of 2015. With this methodology GWB *Central Istria* is *at risk*", and this includes a large part of the Istrian Peninsula. Significant improvement of the qualitative status of the groundwater is needed. The risk assessment also includes the fact that most of transboundary GWBs have their recharge areas in the neighbouring countries (Slovenia and Bosnia and Herzegovina). The problem lies in differing water policies within these countries and within Croatia,

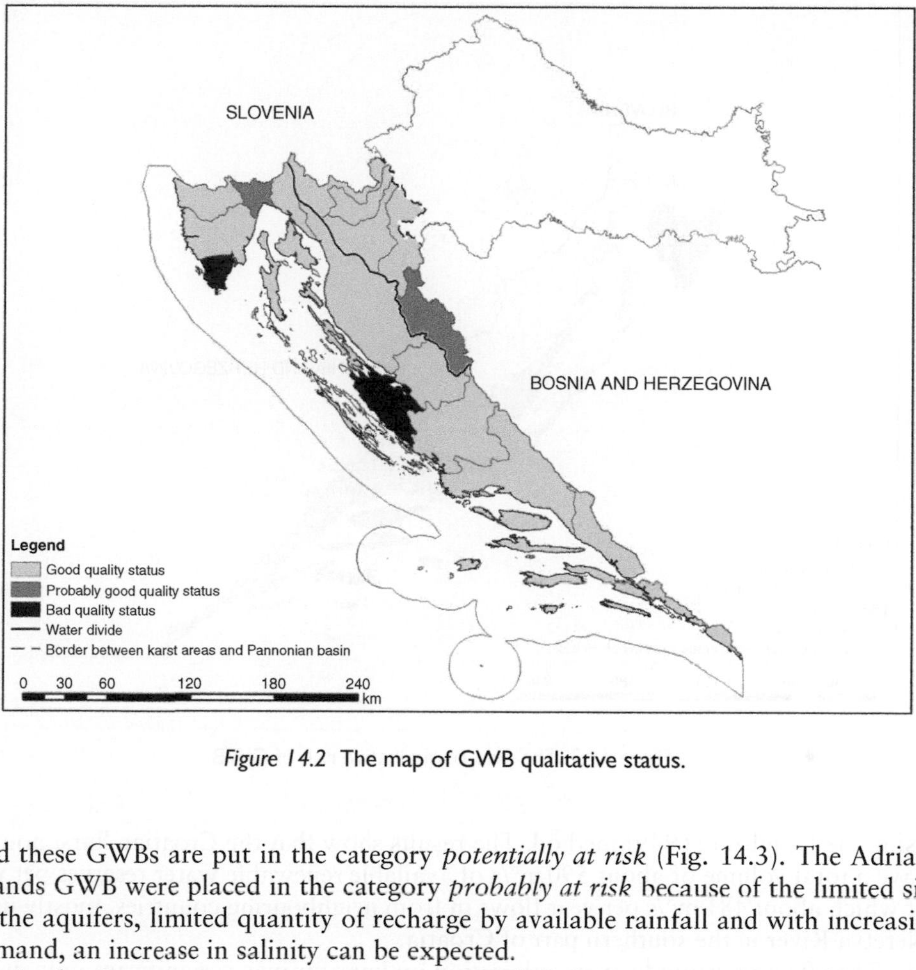

Figure 14.2 The map of GWB qualitative status.

and these GWBs are put in the category *potentially at risk* (Fig. 14.3). The Adriatic islands GWB were placed in the category *probably at risk* because of the limited size of the aquifers, limited quantity of recharge by available rainfall and with increasing demand, an increase in salinity can be expected.

In the Quantitative status assessment (QUAN) four comparative analyses (that include effects of anthropogenic change in recharge, groundwater flow and discharge) have been applied:

1 analyses of GWBs water balance, with particular attention to the effects of exploitation of groundwater on viability of surface flows,
2 surface flow analyses,
3 impact analyses of quantitative relations in GWBs with terrestrial ecosystems, and
4 analyses of sea water intrusion into coastal aquifers.

For the purpose of the QUAN analyses the data period of 1961–1990, with 567 stations, has been used. For the spatial coverage of boundary areas, data from 85 stations in neighbouring countries have also been included.

Estimation of mean annual discharges have been made from groundwater models of Turc (1954) and Langbein (1949), and compared with the measured hydrologic data for several tested GWBs (Horvat & Rubinić, 2006). The results were obtained

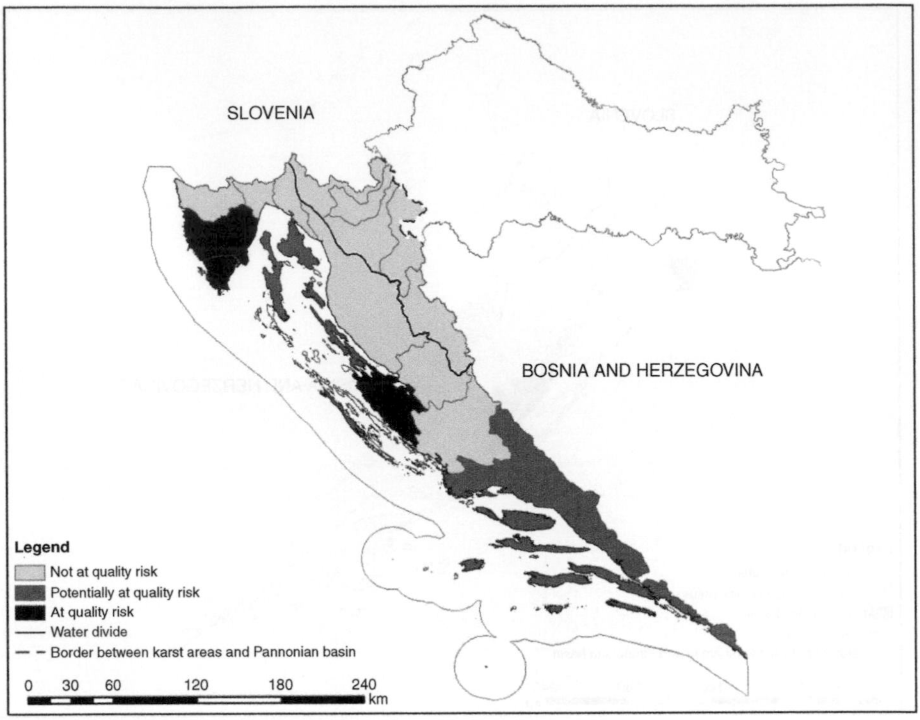

Figure 14.3 The map of qualitative risk of GWB.

using the Langbein (1949) method. The results show that the Croatian karst aquifers have a total volume of about 590 m^3/s of available renewable water reserves per year, of which about 481 m^3/s per year flows in from neighbouring countries, mostly in the Neretva River in the southern part of Croatia.

The effect of groundwater exploitation on karst springs is significant only during the summer dry periods when the maximum abstraction takes place.

Most of the waters from the karst catchments are used for hydro power plant. Reservoirs built in high zones of the karst areas arrest flood waters; thereby also reducing the risk of flooding of karst fields and valleys, where the largest rivers drain to the Adriatic Sea or to the Black Sea catchments. The reservoirs have caused significant damage to the environment as they were built. However, because of projected river flow losses, nowadays they have a valuable function, as they tend to increase the discharge of karst springs and streams during the summer dry periods. For the period 1961–1990 the estimated average runoff from the Croatian karst areas toward the Adriatic coast was about 420 m^3/s, adding inflow of 435 m^3/s from Bosnia and Herzegovina. Natural discharge (springs, rivers) into the sea is about 210 m^3/s, while hydropower facilities discharge into the sea about 200 m^3/s. Hydrologically uncontrolled groundwater discharge into the sea amounts to 445 m^3/s (52%). Therefore, the Dinaric karst area, which is drained toward the Adriatic Sea, provides enormous amounts of fresh water, which nowadays freely drain into the sea, and can be used in the

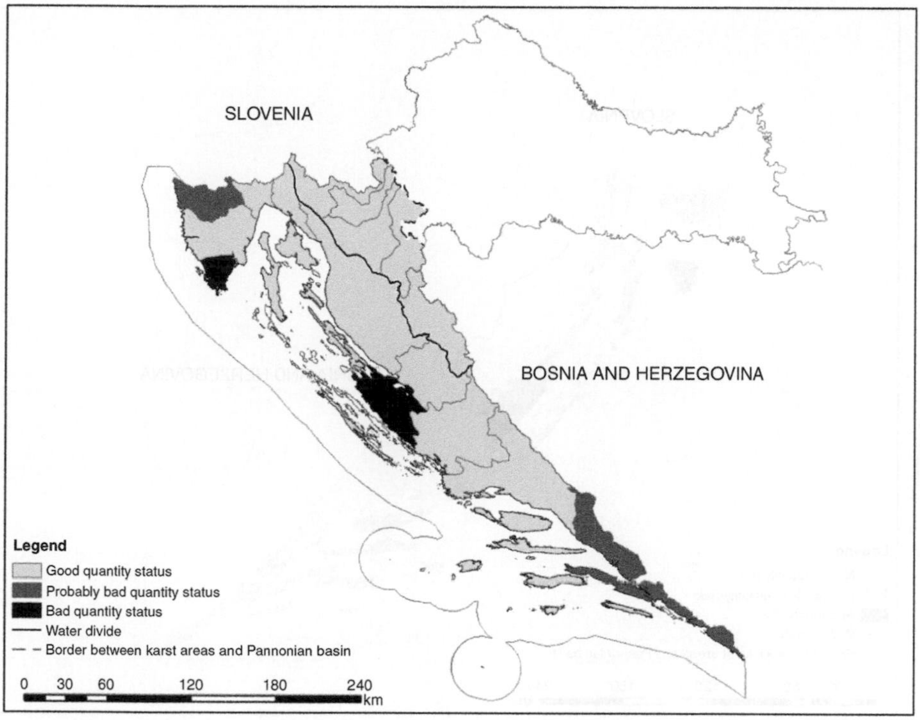

Figure 14.4 Quantitative status of GWBs.

first place for Croatian needs (water supply of islands), but also for eventual commercialisation of a part of those reserves in the future water market in the Mediterranean area.

Previous hydrological analysis show about a 10% reduced water balance in the Dinaric karst in Croatia during the thirty years observation period. The same is probably the case in the neighbouring countries where waters also drain to the Adriatic and Black Sea catchments. However, the water regime is not under threat by current abstraction rates. Problems occur only locally, during summer dry periods, when the exploitation is maximised and the natural recharge of karst aquifer minimised. This is particularly related to the maintenance of important ecosystems along karst rivers. However, water reserves are renewable and the first rainfalls after dry periods compensate for water deficits created by long term droughts.

In several major karst springs in Croatia the analyses of environmental tracers was used to determine the age of groundwater, groundwater flow directions, dominant recharge areas of those springs, the dynamics of groundwater base-flow and overflow components and potential contaminant transport mechanisms. These results were used for the GWB delineation and partly in quantity and quality status analyses. Because of its importance such research should be systematically implemented throughout the karst area and the results used as the additional data for allowing better assessment of the GWB status and risk in the next period of assessment.

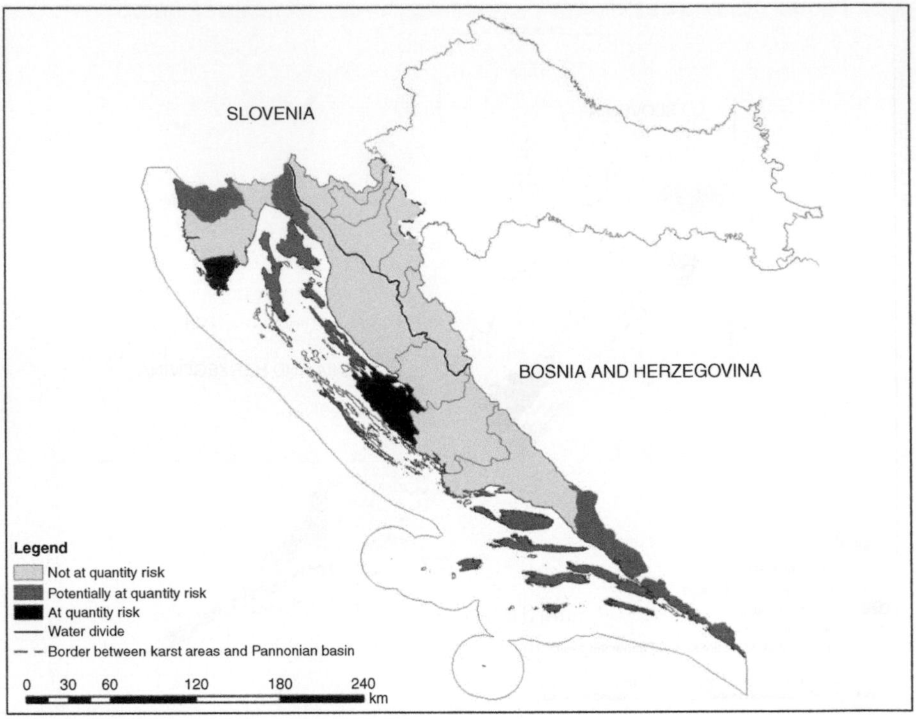

Figure 14.5 Quantitative risk of GWBs.

QUAN analysis has shown that two GWBs have *poor status* (*South Istria* and *Ravni Kotari*), and two others, due to insufficient data and based on expert judgement, belong to the category *probably poor status* (*North Istria* and *Neretva*) (Fig. 14.4). Except these GWBs with *poor status* or *probably poor status* the assessment of quantitative risk extends this list to GWB *Rijeka – Bakar* and GWB *Adriatic Islands* due to sea water intrusion and its effect on groundwater, which causes occasionally higher salinisation of water supply springs (Fig. 14.5).

14.3 CONCLUSIONS

In the Croatian Dinaric karst region 17 GWBs has been delineated, of which 12 are in the Adriatic, and 7 in the Black Sea catchments. Analyses of pressures and impacts, qualitative and quantitative status of groundwater, groundwater dependent ecosystems and risk assessment have been carried out for each GWB. Special attention is given to the monitoring network extension, because for further characterisation of risk or potential risk there are insufficient data.

The assessment of the qualitative and quantitative status of GWBs in the Croatian Dinaric karst area shows generally good status of GWBs, which confirms that the

Adriatic region has high-quality water resources in sufficient quantities for development, but also for the future commercialisation in the Mediterranean region. Problems with water are registered in two GWBs, but with the possibility of alternative sources for water supply. Status *at risk* due to the high content of nitrates and occasional intrusion of sea water has two GWBs *South Istria* and *Ravni Kotari*. Several GWBs have the status *potentially at risk*: *Rijeka – Bakar*, *Adriatic Islands* and *Neretva* due to periodic sea water intrusion, and GWB *Cetina* and *Neretva* due to transboundary conditions and inability to manage transboundary recharge areas.

REFERENCES

COST 620 (2004) Vulnerability and risk mapping for the protection of carbonate (karst) aquifers. EUR 20912 EN, Final report. Directorate-General Science, Research and Development, Brussels, Belgium.

Biondić R., Biondić B., Rubinić J., Meaški H. (2009) Assessment of status and risk of groundwater bodies in the karst area of Croatia – Final report. Varaždin, technical report, unpublished.

Horvat B., Rubinic J. (2006) Annual runoff estimate- an example of karstic aquifers in the transboundary region of Croatia and Slovenia. *Hydrological Sciences Journal*. 51:2 314–324.

Langbein W.B. (1949) Annual runoff in the United States. *US Geol. Survey Circular* 52, Washington D.C. USA.

Turc L. (1954) Le bilan d'eau des sols, relation entre les precipitations, l'evaporation et l'ecoulement. *Ann Agron* 5: 491–596; 6: 5–131.

Water Framework Directive (WFD) 2000/60/EC (2000) Water Framework Directive of the European Parliament and of the Council establishing a framework for Community action in the field of water policy.

Chapter 15

WEAP-MODFLOW as a Decision Support System (DSS) for integrated water resources management: Design of the coupled model and results from a pilot study in Syria

Jobst Maßmann[1], Johannes Wolfer[1], Markus Huber[2], Klaus Schelkes[1], Volker Hennings[1], Abdallah Droubi[3] & Mahmoud Al-Sibai[3]

[1] *Federal Institute for Geosciences and Natural Resources (BGR), Hannover, Germany*
[2] *Geo:Tools, München, Germany*
[3] *Arab Center for the Studies of Arid Zones and Dry Lands (ACSAD), Damascus, Syria*

ABSTRACT

Within the framework of an Arab-German technical cooperation project, an easy to use and inexpensive Decision Support System (DSS) for integrated water resources management was applied, further developed and disseminated. This DSS mainly consists of two pre-existing components: the *W*ater *E*valuation *A*nd *P*lanning software WEAP and the 3-D groundwater flow model MODFLOW. As a novelty, a linkage between these two components has been developed, which is essential to determine water balances in arid/semi-arid regions. The calibrated DSS provides the capacity to investigate, compare and evaluate various water management scenarios, considering the interactions between surface water and groundwater. Future constraints, such as changes in demography, economy, climate, land use, irrigation efficiency, or return flow, can easily be taken into account.

Successful applications of the WEAP-MODFLOW DSS in several countries Morocco, Tunisia, Palestine, Syria and Jordan, prove its strength to investigate the impacts of climate change, change in water demand and supply, waste water reuse and artificial recharge on the water balance.

15.1 INTRODUCTION

The situation of the water supply in parts of the Middle East is characterised by water scarcity and, at the same time, by increasing demand caused by population growth as well as expanding economy and agriculture. Furthermore, climate change models predict even more severe conditions in the water sector, associated with rising temperatures and decreasing precipitation (Bates *et al.*, 2008). The decision makers have to respond to the most urgent questions: how will the water balance change in time and which action is required to achieve a sustainable water supply? An integrated approach is obligatory. In addition to the development of the demand and availability of water resources, the whole social, economic, cultural and environmental framework has to

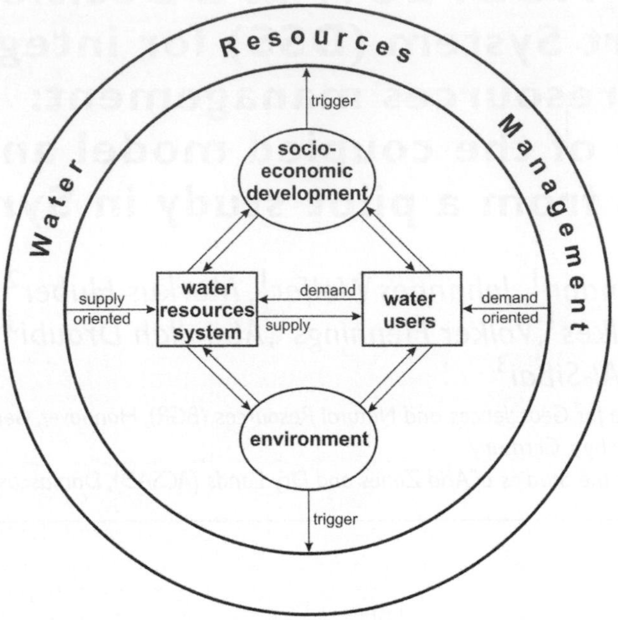

Figure 15.1 Concept of Integrated Water Resources Management after Koudstaal *et al.* (1992).

be taken into consideration, as depicted in Figure 15.1. To understand such a complex interacting system and the outcome of any changes, a Decision Support System (DSS), based on computational models, has been applied.

Within the technical cooperation project "Management, Protection and Sustainable Use of Groundwater and Soil Resources in the Arab Region", ACSAD and BGR, supported by the Stockholm Environment Institute (SEI), have jointly worked on the improvement, application and dissemination of a DSS software for integrated water resources management (Droubi *et al.*, 2008a). This system has successfully been applied in several countries in the Arab region, as Morocco (Fakir *et al.*, 2010), Tunisia, Palestine, Syria and Jordan.

This chapter describes the development of an adequate DSS for water balancing on the river basin scale in arid/semi-arid regions by coupling two pre-existing software codes. It presents the need and advantage of this approach with a case study as example. In contrast to the report by Droubi *et al.* (2008b), wherein the same case study is presented in detail, the basic idea of the coupling and the scenario development and evaluation are the focus points of this chapter.

15.2 WEAP-MODFLOW DECISION SUPPORT SYSTEM (DSS)

Following the classical definition from Sprague & Carlson (1982), a DSS is "an interactive computer-based support system that helps decision makers to utilise data and models to solve unstructured problems". In general, it consists of three main components: a user-interface for dialog generation and providing the interface between the user and the system, a model management subsystem and an information management subsystem (database). Typically, the problem solving process is accomplished in an interactive manner, whereby the experience and judgment of decision makers are essential, because the DSS does not create new approaches automatically but helps to evaluate man-made approaches. In terms of integrated water resources management, the main tasks of a DSS are the analysis of the present water management system and the forecasting of its future behaviour.

The DSS, presented here, is based on a combination of the two pre-existing software codes: MODFLOW and WEAP. MODFLOW, developed by the U.S. Geological Survey (Harbaugh *et al.*, 2000), numerically solves the three-dimensional groundwater flow equation for a porous medium by using the finite-difference method. It is based on Darcy's law for laminar flow and the conservation of the water volume in porous media. The water evaluation and planning system WEAP (Sieber & Purkey, 2011) was developed by the SEI (Yates *et al.*, 2005). Based on a schematic representation of the hydraulic system, WEAP calculates the groundwater and surface water balances on the catchment level.

Since the Middle East is largely characterised by an arid/semi-arid climate, groundwater is the most important, and in many areas the only available freshwater resource. Consequently, there is the need to predict the interaction between surface water and groundwater, resulting in local and regional development of the groundwater resources in terms of water table drawdown, storage volume and flow. Because WEAP cannot spatially analyse a groundwater system, a coupling between WEAP and a groundwater flow model, such as MODFLOW, is essential.

In order to ensure that WEAP results address MODFLOW cells correctly and vice versa, a linkage was developed, which acts like a dictionary between the two models (Fig. 15.2). The hydrogeological and numerical parameters are stored in the MODFLOW sub-model, but all time-dependent values are handled by WEAP. The user of the DSS controls the whole model by the graphical user interface of WEAP only and MODFLOW is run by WEAP in the background. For each time step, results of the one model are transferred as input data to the other. Groundwater recharge, abstraction rates, and river stages are calculated by WEAP. These data act as boundary conditions for MODFLOW, which calculates hydraulic heads, storage volumes and flows in the groundwater system. These values are used in turn by WEAP. Thus, river-groundwater interactions, spring discharge or recharge as well as management constraints regarding the groundwater head or discharge can be considered (Al-Sibai *et al.*, 2009). The calibrated DSS provides the capacity to investigate, compare and evaluate various water management scenarios. Future constraints, such as changes in demography, economy, climate, landuse, irrigation efficiency, or return flow, can easily be taken into account. The results are visualised as graphs, maps, and tables. They depict the impacts of the scenarios on the water balance in a whole watershed or in detail, e.g. in terms of hydraulic heads, flow rates, water storage, demand satisfaction, irrigation amounts and the development of the resources. Due to the coupling with MODFLOW, the reactions and dynamics of the groundwater system, discretized in time and space, can be displayed and evaluated within WEAP.

The WEAP-MODFLOW DSS has been improved continuously. Recent developments focus on the integration of the simple particle tracking model MODPATH (Pollok, 1994), the optimisation of abstraction rates and pumping allocation with

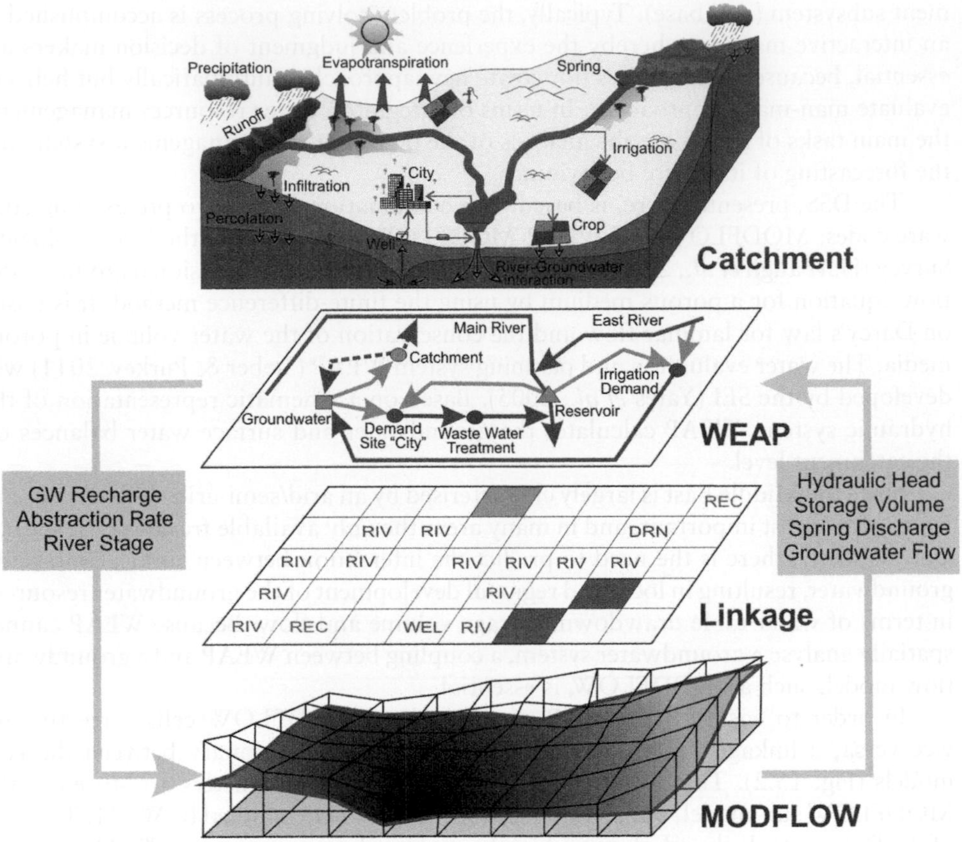

Figure 15.2 Schematic setup of the WEAP-MODFLOW linkage.

consideration of water quality, drawdown and cost (Nouiri *et al.*, 2010), and an additional soil-water balance model called MABIA (Sahli & Jabloun, 2005). MABIA is based on the FAO-56 dual crop coefficient approach (Allan *et al.*, 1998). It provides the use of real world field data as well as FAO reference parameters.

15.3 ZABADANI BASIN, SYRIA

The Zabadani Basin is located in the Antilebanon Mountains in the NW of Damascus, Syria. It covers an area of about 140 km^2. Geomorphologically and hydrogeologically it can be divided into three NNE-SSW trending blocks, as depicted in Fig. 15.3 the Chir Mansour Mountain range in the west, reaching up to 1884 m a.s.l., characterised by faulting, intensive karstification and high transmissivities; 2) the Zabadani graben, ranging from 1080 to 1400 m a.s.l., with moderate transmissivities; 3) the Cheqif mountain range in the E, reaching up to 2466 m a.s.l., with minor karstification and high transmissivities. The Barada River, representing the only perennial stream in the

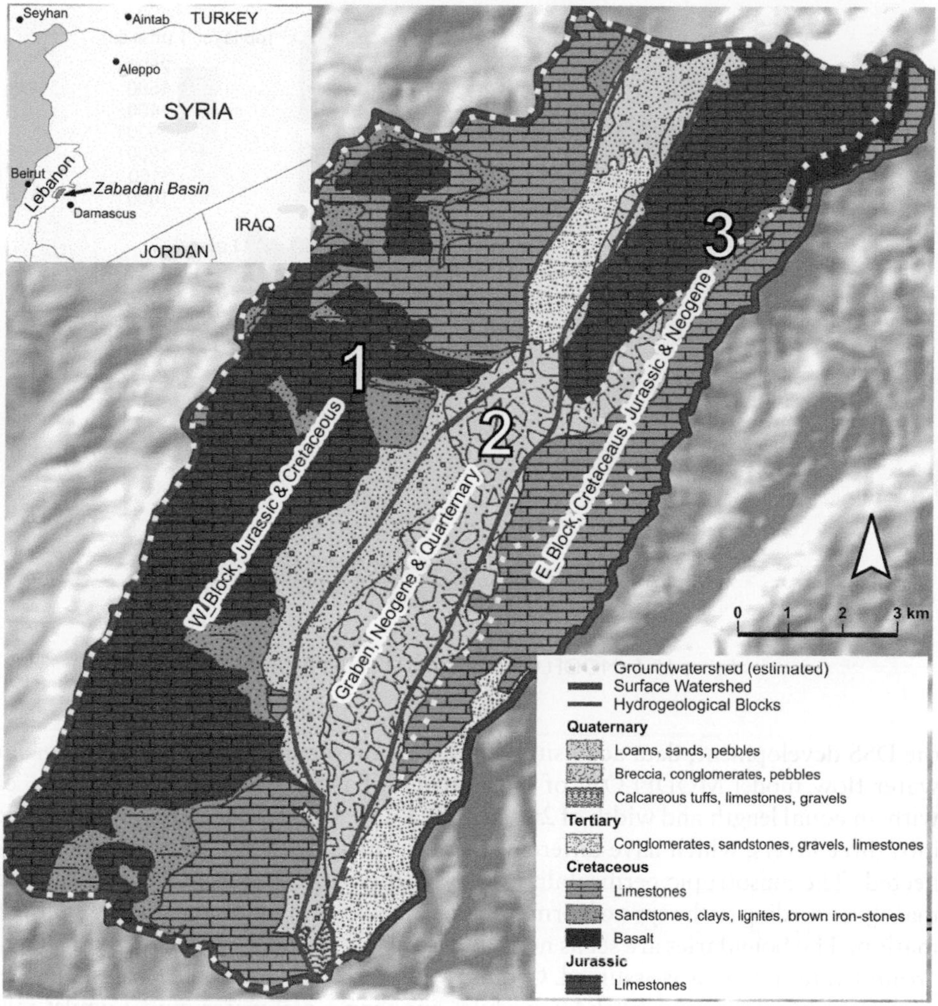

Figure 15.3 Location and geology of the Zabadani Basin, Syria.

region, rises in this basin. The mean annual rainfall is about 700 mm. About 48 000 people live permanently in the area; however, during summer time the population increases significantly due to tourism.

15.3.1 Setup of a DSS for IWRM in the Zabadani Basin

A water competition exists in the Zabadani Basin between the local drinking water suppliers, the Damascus water supply authority, as well as agricultural and tourist demand. Since the beginning of the project, a steering committee has been set up, integrating all relevant stakeholders of involved ministries, the municipality and water suppliers into

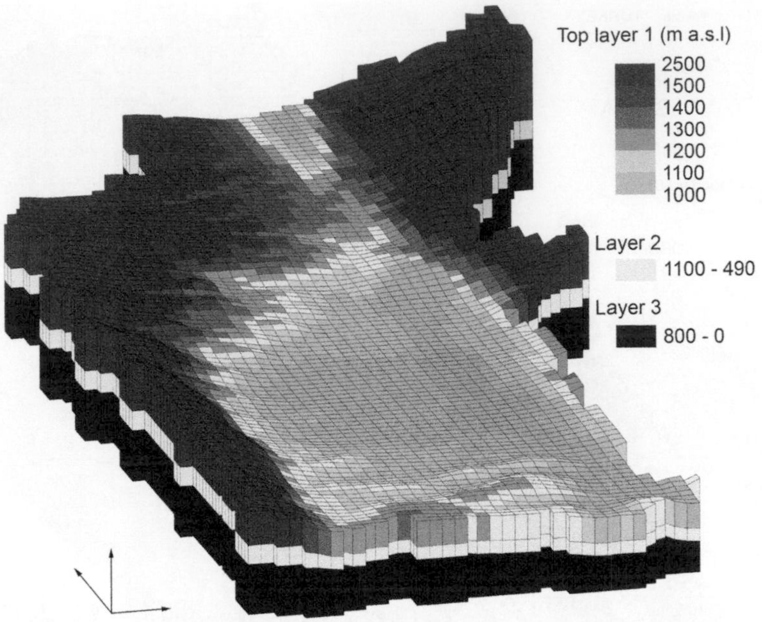

Figure 15.4 MODFLOW model grid of the Zabadani Basin.

the DSS development, data acquisition and scenario planning. The numerical groundwater flow model MODFLOW of the Zabadani Basin consists of 10 044 cells, each with an equal length and width of 200 m (Fig. 15.4). The regional aquifer is subdivided into three layers, which have different hydraulic properties but are hydraulically connected. The anisotropic permeabilities vary in the range between 0.01 to 60.0 m/day, mainly according to the type of formation, density of lineaments and dipping of the formation. The boundaries are set as no flow Neumann boundaries, except in the S where groundwater inflow was assumed. Groundwater recharge and irrigation demands were determined by an internal WEAP module, applying a soil-water model whilst taking into account 48 landuse classes. This module was validated with the external models SWAP (Kroes & Van Dam, 2003) and CROPWAT (Clarke *et al.*, 1998). Groundwater abstractions from well fields for domestic use and from rural wells for irrigation are considered. Furthermore, surface water-groundwater interactions at the Barada River and Barada spring are modeled by Cauchy boundary conditions.

The WEAP software was used to build a planning and evaluation model, which then was linked to the MODFLOW groundwater flow model as component of the DSS. Within WEAP, the basin was divided into 11 sub-catchment, based on the locations of the major drinking water well fields and surface watersheds. In addition to areal data, climate data was also assigned at sub-catchment level. In the next step, each sub-catchment was further subdivided into respective landuse classes. Irrigation pattern, crop coefficient, leaf area index, root zone conductivity and soil water capacity values were assigned to them. The basis for the landuse mapping was provided by aerial photographs, geological information (Kurbanov *et al.*, 1968) and data from ministries

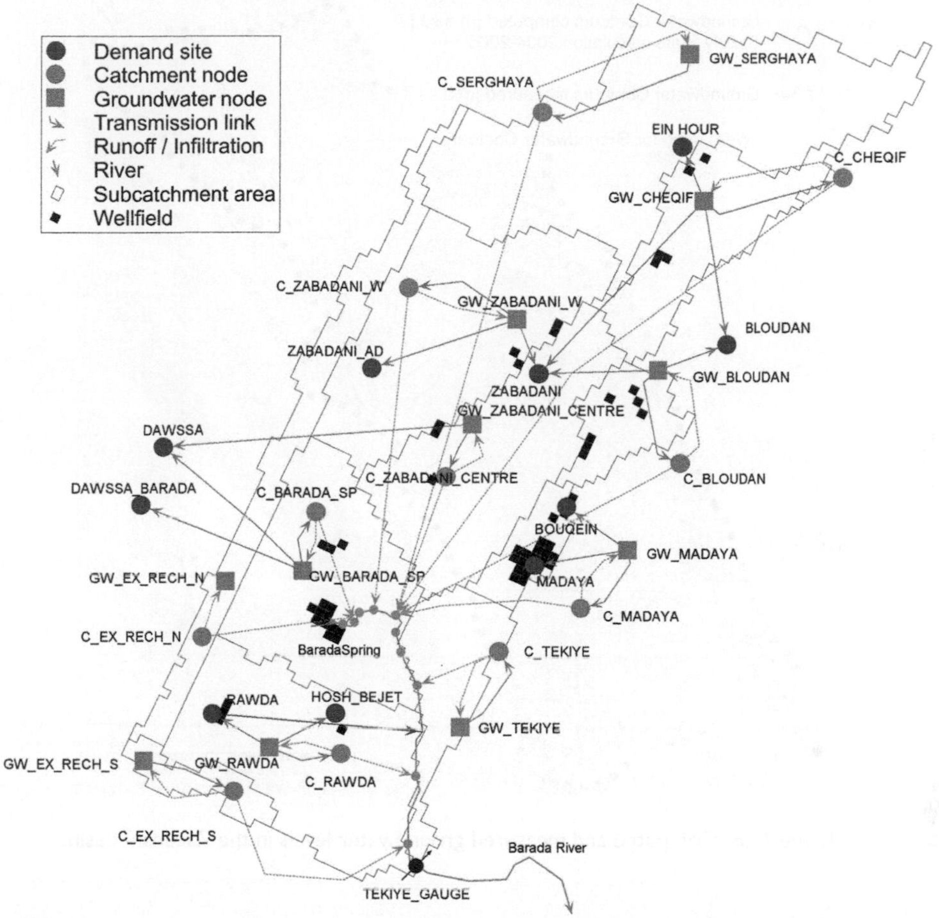

Figure 15.5 WEAP schematic of the Zabadani Basin.

and local farmers. Figure 15.5 depicts the WEAP schematic with integrated nodes as demand and supply sites and their links. Additional details on the WEAP-MODFLOW DSS of the Zabadani Basin are presented in Droubi *et al.* (2008b).

15.3.2 Calibration

The hydrological year 2004/2005 was a year with precipitation resulting in a full recovery of the groundwater table after winter rains and snowmelt. The precipitation in this year was slightly above the long-term average. Furthermore, in this time period the most complete data series were available. Therefore, data from this year were used for steady-state and transient calibration of the hydraulic properties and the groundwater inflow from the south in order to fit the Barada spring discharge and the measured groundwater heads.

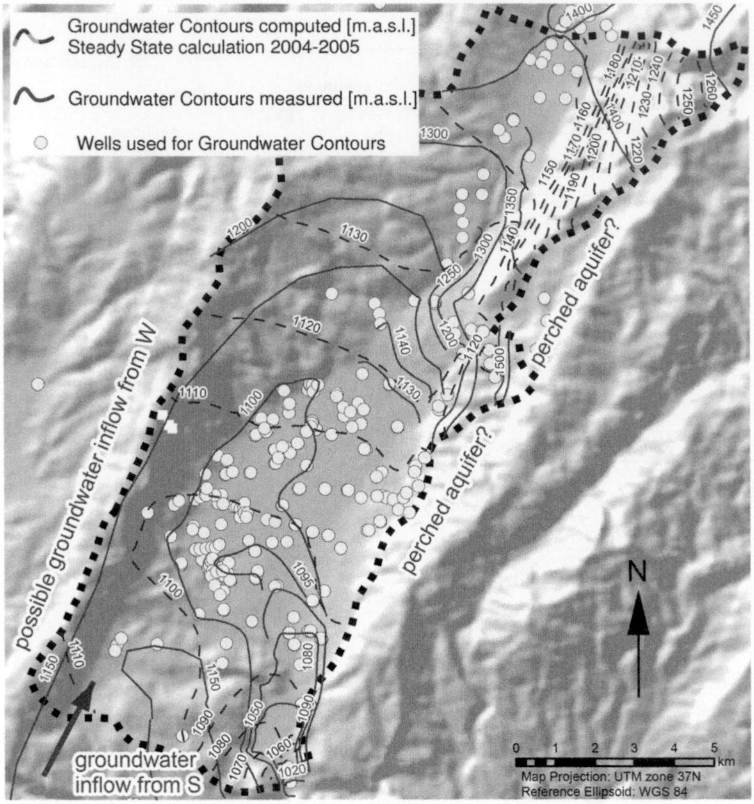

Figure 15.6 Computed and measured groundwater levels in the Zabadani Basin.

The calibrated model is able to predict exactly the yearly discharge of the Barada spring, only the monthly values differ, because the combination of rapidly and slowly moving flow components, typical for karst, cannot be predicted by such a continuum model of an equivalent porous medium. In order to get realistic monthly results, another model approach should be used, which takes into account conduits as preferential flow paths. One possibility should be the usage of an additional MODFLOW package, as presented in Shoemaker *et al.* (2007) and reviewed in Zheng (2009) or the application of a FEM based model, which is the best choice for highly heterogeneous aquifers, because it could consider complex geometries and conduits. If rapid groundwater flow is considered, it is necessary to reconsider the linkage concept to ensure numerical stability. The rough approach of a continuum model has been used, because there is not enough information available about the karst aquifer. Neither the location and alignment, nor the hydraulic properties of the conduits are known.

The computed and measured/estimated heads are shown in Figure 15.6. The regional groundwater flow is directed from the north and NW respectively to the SE. The Barada spring and Barada River constitute the natural drain of the domain. In the plain area of the Zabadani valley, the residuals between measured and computed

Table 15.1 Definition of planning scenarios as applied for the Zabadani Basin, Syria. Changes are defined in comparison with the reference scenario.

Scenario O	Reference Scenario
Storyline	Precipitation and temperature based on the average of the period 1961–1990 Constant demand based on the year 2004/2005
Data source	Steering committee (water supplier, municipalities, ministries)
Scenario A	Demand Change
Storyline	Due to the socio-economic development, the water demand changes Population growth and urbanization are assumed
Change	• Municipal water demand is tripled until 2020 • Rural domestic demand is doubled until 2020
Data source	Steering committee (water supplier, municipalities, ministries)
Scenario B	Drought Cycle
Storyline	Considering 4 consecutive drought years
Change	• Halving of the precipitation in the years 2011–2014
Data source	Historic precipitation distribution recorded at the Damascus station
Scenario C	Climate Change
Storyline	Climate change leads to decreasing precipitation, increasing temperature and decreasing lateral groundwater inflow
Change	• Linear decrease in precipitation up to 16% in 2050 • Linear increase in temperature up to 2.7° in 2050 • Linear decrease in lateral groundwater inflow up to 16% in 2050
Data source	Zganjar *et al.* (2011), based on Bates *et al.* (2008) (the results of 16 modeling groups and the climate change scenarios A2, A1B, B1 are considered)

heads are within 5 m, to the north and to the margins they increase significantly. This was the best match that was obtainable using the available data and knowing the model constraints. Since the main abstraction and water competition area is located within the plain area, the model is considered as a valuable and fairly accurate tool to model groundwater flow, head and spring discharge for this region.

Summarising it can be stated that the linked WEAP-MODFLOW DSS is able to calculate realistic groundwater, surface and soil water balances as well as hydraulic heads, and can be used to investigate annual water balances. In order to improve the model results, additional measurements and tracer methods are needed to characterise the aquifer more precisely and to distinguish between the conduit and matrix flow.

15.3.3 Scenario planning

Scenario planning is used to analyse possible evolutions of a complex system. Founded on certain assumptions, the scenarios do not represent one expected future but many possible evolutions. By choosing reasonable assumptions, the range of possible future developments can be outlined.

Based on the initial model setup of the Zabadani Basin, three scenarios were investigated for the planning period 2005–2050 (Table 15.1). The scenarios have been

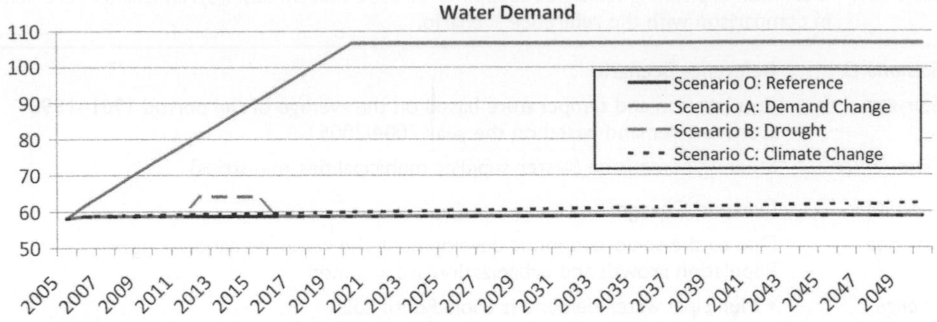

Figure 15.7 WEAP-MODFLOW DSS: Annual agricultural and domestic water demand.

developed by the project steering committee. They deal with realistic assumptions on domestic and agricultural demand as well as influences of climate change and consecutive drought years. Additionally, a reference scenario was defined without any parameter changes, based on the monthly average of precipitation and temperature of the period 1961–1990.

The results do not represent a realistic future. This is due to the fact that the weather is characterised by high annual variations, which are not considered. However, the impact of some expected changes can be worked out clearly and investigated, which is the main goal of this study.

In the reference year 2004/2005 the total annual water demand in the Zabadani Basin is about 58 Mm^3. Almost half of the water is used domestically, the remainder for irrigation. As depicted in Figure 15.7, the highest impact on the water demand is scenario A (demands change). The total demand is approximately doubled through the increase of domestic water demand. After the year 2020 no further change has been considered because no reliable assumption can be made about the urban and regional development. In the scenarios B (drought) and C (climate change) only the agricultural water demand is increased. In order to investigate the expected higher probability of droughts due to climate change (Bates *et al.*, 2008), four consecutive drought years are considered in scenario B. The precipitation is halved in these years based on historical climate records in Damascus. As a result, an additional amount of 6 Mm^3 is needed for irrigation, whereby an increase of the irrigated area is not taken into account. In the climate change scenario, a linear increase of temperature and decrease of precipitation until the year 2050 is applied. The result is an increase of the annual irrigation demand up to 4 Mm^3 in 2050.

15.3.4 Results

Almost the entire water demand is satisfied by groundwater and there is no surface water flowing into the basin. As a consequence, the most important question for the future security of water supply is the knowledge about the aquifer. In scenario A (demand change), the demand is satisfied by additional pumping. The groundwater recharge is not directly influenced, in contrast to scenarios B and C, where a decrease

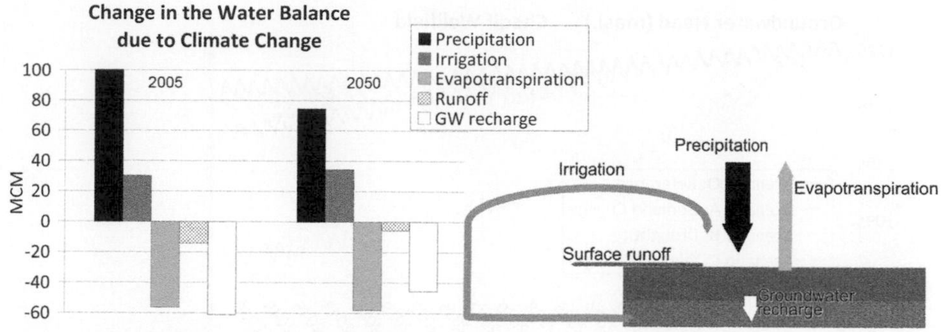

Figure 15.8 WEAP-MODFLOW DSS results: Water balance in the Zabadani Basin in the year 2005 and 2050 as determined by scenario C, climate change.

in precipitation and an increase in temperature lead to lower natural recharge and a higher irrigation demand (Fig. 15.8). Since the water for irrigation is taken from wells, the groundwater storage is double affected by climate change. As depicted in Figure 15.8, climate change has a remarkable influence on the water balance in the year 2050: The groundwater recharge is nearly decreased by one third and the runoff is more than halved.

The integrated WEAP-MODFLOW approach provides the capacity to investigate the impact of these scenarios on the aquifer in detail. As an example, the groundwater table for the reference scenario is depicted in Figure 15.9. At two specific MODFLOW cells, representing the Chaqif well field and the Zabadani graben, line plots show the temporal development of the groundwater table for all scenarios.

The groundwater abstractions influence the local flow regime essentially. For scenario A (demand change) extreme drawdowns up to 38 m are predicted at the Chaqif well field. At the selected MODFLOW cell within the plain Zabadani graben, where wells are absent and the influence of climate change is stronger, the drawdowns are more moderate but still crucial. For both locations, scenario A (demand change) shows the most severe consequences, but climate change also leads to distinct drawdowns. Furthermore, the groundwater table reacts quite fast to drought and a long time period is needed for recovering.

In order to study the influence of the scenarios on the entire aquifer, the temporal changes in groundwater storage for the whole study area are shown in Figure 15.10. Principally, they present the same trend as the local drawdowns (Fig. 15.9). All scenarios lead to a negative groundwater balance within the basin. The decreasing precipitation in scenarios B and C leads to decreasing groundwater recharge and increasing irrigation requirement. Even in the reference scenario the aquifer is overexploited.

Scenario B (drought) illustrates that the entire groundwater system reacts rapidly to a change in recharge. Consecutive years of droughts have the potential to affect the water balance in a short time period more substantially than the expected changes due to climate change or water demand. However, in the long run, the expected change of water demand (scenario A) leads to the most extreme decrease in groundwater storage.

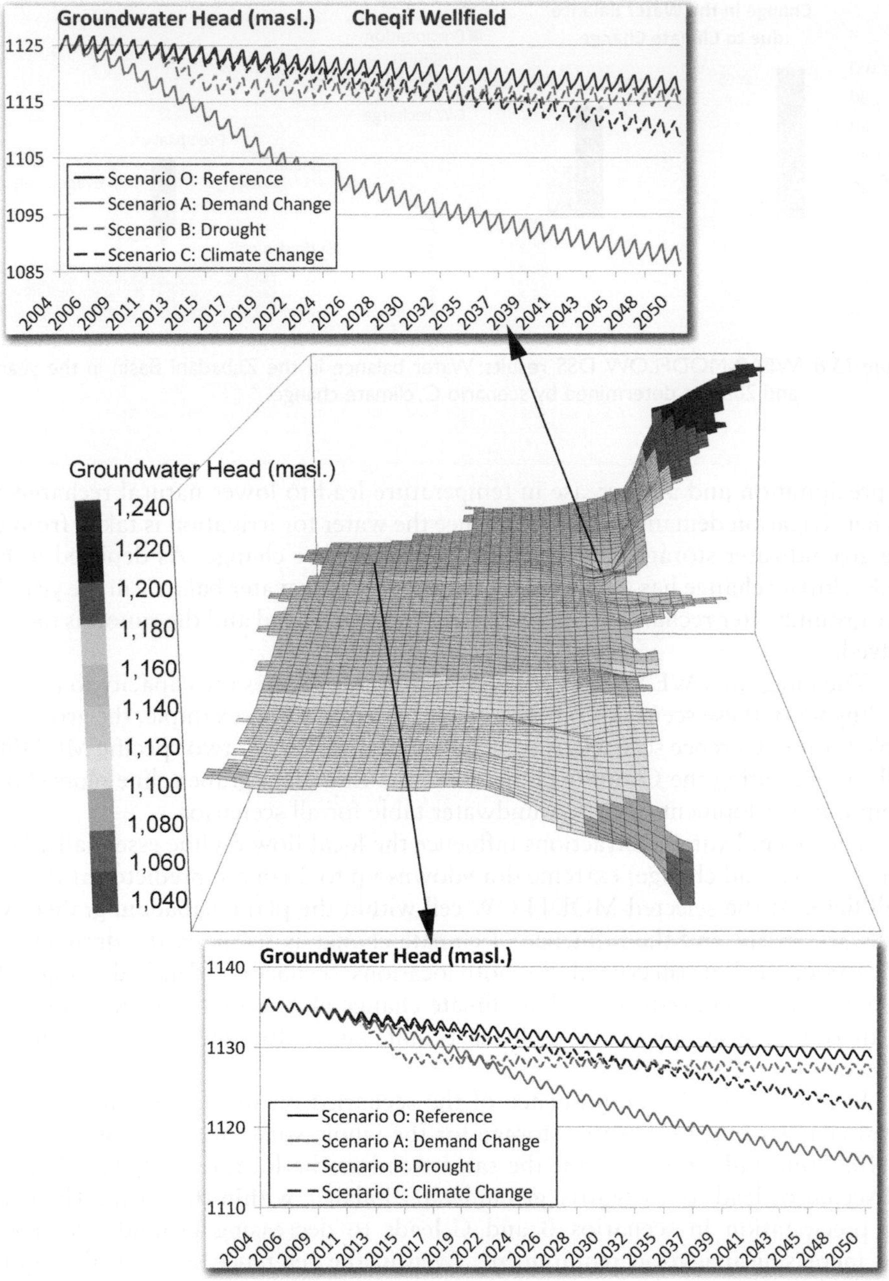

Figure 15.9 WEAP-MODFLOW DSS results: contour plot: Groundwater head of the uppermost aquifer in October 2050, reference scenario; line plots: Groundwater Head at two specific cells.

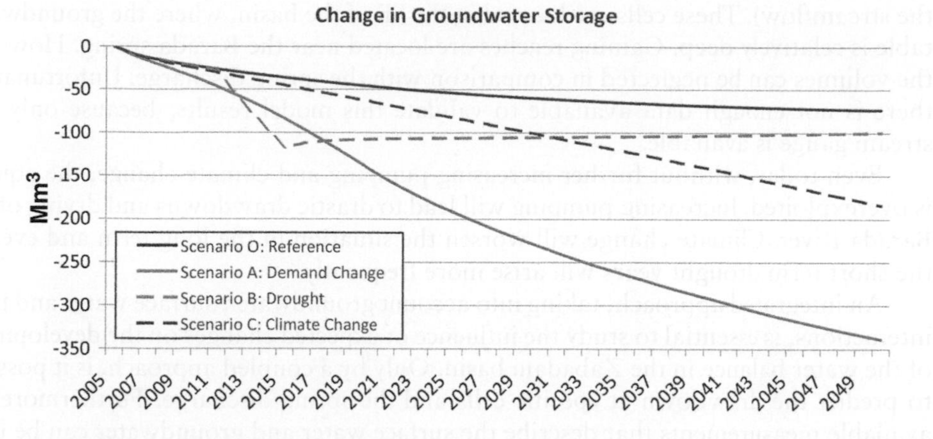

Figure 15.10 WEAP-MODFLOW DSS result: Development of the groundwater storage in comparison to the hydrological year 2004/2005.

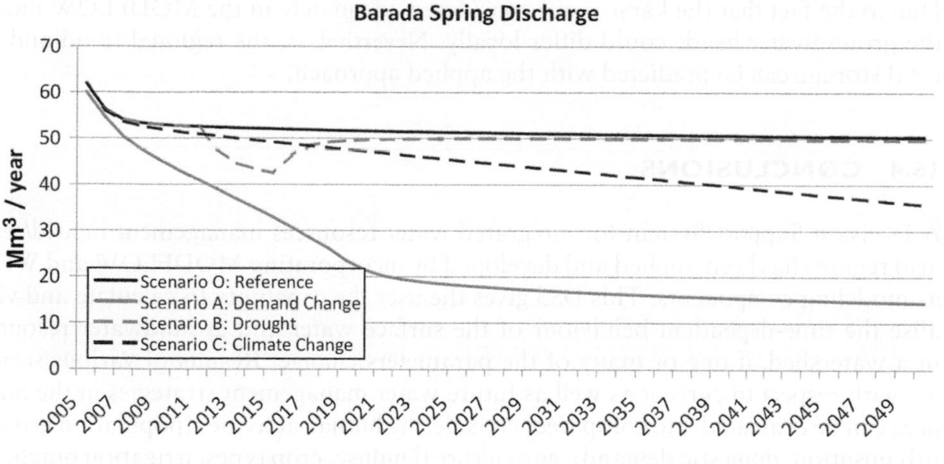

Figure 15.11 WEAP-MODFLOW DSS result: Development of the discharge at the Barada spring.

Among the influence of well extractions on the aquifer, the river-groundwater interaction is important in the Zabadani Basin. Even if local variations exist, all scenarios result in a decline of the water table within the planning period. This decline also causes a decrease in the Barada spring discharge, as shown in Figure 15.11.

In scenario A, a drastic decrease in the Barada spring discharge is predicted, which could lead to problems downstream, where the water is used for irrigation. It is obvious that continued over-pumping will dry up the Barada spring and thus the Barada River. The WEAP-MODFLOW approach is capable of quantifying stream-aquifer interaction in each MODFLOW cell and each time step. Indeed, at the Barada River this exchange is only marginal. In the early summer months, when the precipitation is low but the spring discharge is still remarkable, the Barada has losing reaches (<1% of the streamflow). These cells are located in the SE of the basin, where the groundwater table is relatively deep. Gaining reaches are located near the Barada spring. However, the volumes can be neglected in comparison with the spring discharge. Unfortunately, there is not enough data available to validate this model results, because only one stream gauge is available.

Even today, without further increasing pumping and climate change, the aquifer is overexploited. Increasing pumping will lead to drastic drawdowns and drying of the Barada River. Climate change will worsen the situation in the long term and even in the short term drought years will arise more frequently.

An integrated approach, taking into account groundwater, surface water and their interactions, is essential to study the influence of expected changes on the development of the water balance in the Zabadani basin. Only by a coupled approach, is it possible to predict the drawdown at specific cells and the spring discharge. Furthermore, all available measurements that describe the surface water and groundwater can be used for the calibration and validation process, which is essential in regions where the data density is low. Thus, the quality of the model can be improved in comparison to a pure WEAP model. The usage of WEAP makes it easy to build up and handle scenarios. No man-made changes in the groundwater model are needed to investigate new scenarios. Due to the fact that the karst is not considered adequately in the MODFLOW model, the groundwater heads could differ locally. Nevertheless, the regional trend and the total storage can be predicted with the applied approach.

15.4 CONCLUSIONS

A Decision Support System for integrated water resources management in arid/semi-arid regions has been applied and developed by incorporating MODFLOW and WEAP as modeling components. This DSS gives the user the capability to calculate and visualise the time-dependent behaviour of the surface water and groundwater resources in a watershed, if one or many of the parameters change. Results of various scenarios with respect to current as well as future water management strategies in the target area can be evaluated and compared, considering human activities (population growth, urbanisation, domestic demand), agriculture (landuse, crop types, irrigation practices), climate impacts (climate change, droughts), network characteristics (transmission link losses and limits, well field characteristics, well depths), and additional resources (artificial recharge, waste water reuse). They support the decision making process among relevant stakeholders and decision makers.

The WEAP-MODFLOW DSS is already established in several institutions within the Arab region in Morocco, Tunisia, Palestine, Syria and Jordan. In Syria, a pilot study has been successfully applied in the Zabadani Basin. Based on recommendations of a local steering committee, different water planning scenarios were investigated, considering climate change, consecutive drought years and increasing rural and municipal water demand. The expected change in domestic water demands represents the most severe scenario, because it will lead to extreme over-pumping of the aquifer. Climate change, which could accompany a higher probability of drought years, could worsen the situation even in the short term, because the groundwater system reacts rapidly to changes in recharge. A decline in groundwater storage is predicted for all investigated scenarios, along with extreme drawdowns locally and a decrease in spring discharge. Consequently, changes in the water management are needed in order to obtain a sustainable water balance.

As in many Arab regions, the surface water-groundwater interaction is important in the Zabadani Basin. Spring discharge, local drawdowns and river-groundwater interaction can only be investigated by a coupled approach. Furthermore, the model accuracy can be improved, because measurements of the surface water bodies and the groundwater can be used for the calibration. Thus, the integration of a groundwater flow model is essential for a reasonable DSS, which should help to detect and evaluate water relevant problems in order to find solutions for the future.

ACKNOWLEDGEMENTS

The authors would like to acknowledge the German Federal Ministry for Economic Cooperation and Development (BMZ), which has funded this project.

REFERENCES

Allen R.G., Pereira L.S., Raes D., Smith M. (1998) Crop evapotranspiration – Guidelines for computing crop water requirements – FAO Irrigation and drainage paper 56. FAO – Food and Agriculture Organization of the United Nations Rome.

Al-Sibai M., Droubi A., Abdallah A., Zahra S., Wolfer J., Huber M., Hennings V., Schelkes K. (2008) Incorporate MODFLOW in a Decision Support System for Water Resources Management. *Proceedings of Int. Conference: MODFLOW and More 2008*, Colorado, USA.

Bates B.C., Kundzewicz Z.W., Wu S., Palutikof J.P. (Eds) (2008) Climate Change and Water. *Technical Paper of the Intergovernmental Panel on Climate Change*, IPCC Secretariat, Geneva.

Clarke D., Smith M., El-Askari K. (1998) CropWat for Windows: User Guide. Prepared by Land and Water Development Division of FAO, Institute of Irrigation and Development Studies (IIDS) of Southampton University & National Water Research Center (NWRC) of Egypt. http://www.fao.org/nr/water/ infores_databases_cropwat.html.

Droubi A., Al-Sibai M., Abdallah A., Zahra S., Obeissi M., Wolfer J., Huber M., Hennings V., Schelkes K. (2008a) A Decision Support System (DSS) for Water Resources Management, Design and Results from a Pilot Study in Syria. In: Zereini F. & Hötzl H. (Eds) *Climatic Changes and Water Resources in the Middle East and North Africa*. Springer, 199–225.

Droubi A., Al-Sibai M., Abdallah A., Wolfer J., Huber M., Hennings V., El Hajji K., Dechiech M. (2008b) Development and Application of a Decision Support System (DSS) for Water Resources Management in Zabadani Basin, Syria and Berrechid Basin, Morocco. *Technical Report from the Arab Center for the Studies of Arid Zones and Dry Lands (ACSAD)*, Damascus, Syria and the Federal Institute for Geosciences and Natural Resources (BGR) Hannover, Germany. http://www.bgr.bund.de/EN/Themen/Wasser/Projekte/laufend/TZ_Acsad/dss_fb_en.htm.

Fakir Y., Berjamy B., Tilborg H., Huber M., Wolfer J., Le Page M., Abourida A. (2010) Development of a decision support system for water management in the Haouz-Mejjate plain (Tensift basin, Morocco). *XXXVIII IAH Congress, Krakow 12–17 Sept. 2010*, Abstract Book: 281–284.

Harbaugh A.W., Banta E.R., Hill M.C., McDonald M.G. (2000) MODFLOW-2000, the U.S. Geological Survey modular ground-water model – User guide to modularization concepts and the ground-water flow process. USGS Open-File Report 00-92.

Koudstaal R., Rijsberman F.R., Savenije H. (1992) Water and sustainable development. *Natural Resources Forum* 16 (4): 277–290.

Kroes J.G., Van Dam J.C. (2003) Reference Manual SWAP Version 3.0.3 – Alterra-report 773. Alterra, Wageningen, Netherlands.

Kurbanov N., Zarjanov Y., Ponikarov V.P. (1968) The Geological Map of Syria, Explanatory Notes, scale 1:50000, Sheets Zabadani and Rayak.

Nouiri I., Maßmann J., Haddad R., Al-Mahamid J., Al–Sibai M., Tarhouni J. (2010) Optimization of Groundwater Resources Management by ALL_WATER_gw within the Framework of the WEAP-MODFLOW DSS. *Fifth Environmental Symposium of German-Arab-Scientific Form for Environmental Studies*, Byblos, Lebanon.

Pollok D.W. (1994) User's Guide for MODPATH/MODPATH-PLOT, Version 3: A particle tracking post-processing package for MODFLOW, the U.S. Geological Survey finite-difference ground-water flow model. USGS Open-File Report 94-464.

Sahli A., Jabloun M. (2005) MABIA-ETc: a tool to improve water use in field scale according to the FAO guidelines for computing water crop requirements. The WUEMED Workshop, Rome, Italy.

Shoemaker W.B., Kuniansky E.L., Birk S., Bauer S., Swain E.D. (2007) Documentation of a Conduit Flow Process (CFP) for MODFLOW-2005: U.S. Geological Survey Techniques and Methods, Book 6, Chapter A24.

Sieber J., Purkey D. (2011) WEAP Water Evaluation and Planning System User Guide. SEI Stockholm Environment Institute U.S. Center. http://www.weap21.org.

Sprague R.H., Carlson E.D. (1982) Building effective decision support systems. Prentice-Hall in Englewood Cliffs, N.J.

Yates D., Sieber J., Purkey D., Huber-Lee A. (2005) WEAP21 – A Demand-, Priority-, and Preference-Driven Water Planning Model. Part 1: *Model Characteristics. – IWRA, Water International*, 30(4): 487–500.

Zganjar C., Girvetz E., Raber George. (2011) ClimateWizard. The Nature Conservancy, the University of Washington, the University of Southern Mississippi. Online Resource: http://www.climatewizard.org.

Zheng C. (2009) MODFLOW-CFP: A New Conduit Flow Process for MODFLOW–2005. *Ground Water* 47(3):322–325.

Chapter 16

Groundwater recharge evaluation based on the infiltration method

Stanisław Staśko, Robert Tarka & Tomasz Olichwer
Institute of Geological Sciences, Wroclaw University, Wroclaw, Poland

ABSTRACT

The chapter presents a survey of methods of assessing groundwater recharge on local as well as regional scales. The effective infiltration rate was applied as a method of evaluating the level of recharge within a river catchment. This method, using the properties of the soil and/or rock occurring on the surface as well as atmospheric precipitation, allows for the calculation of groundwater recharge with high accuracy. The calculations were verified by means of separating river hydrograph from low flow and also by means of applying Wundt's method. Selected catchment areas in Lower Silesia demonstrate agreement between the figures based on the measurements and those calculated using Wundt's method. The effective recharge in these areas amounts on average to 18.5% of atmospheric precipitation. Dissimilar, significantly higher, values of recharge have been shown for mountainous regions, where the figures are in the range of 18–55%. The calculations were, in this case, based on the lysimeter infiltration monitoring, monitoring of groundwater drainage by gallery, river discharge hydrograph and groundwater table fluctuation.

16.1 INTRODUCTION

Groundwater recharge is an informative indicator of the water located beneath the ground surface. It has a direct impact on the size of renewable supplies of groundwater and, to a large extent, determines the degree of groundwater vulnerability to contamination. It is also of vital importance from the point of view of assessing the supplies, as well as protecting them against pollutants which penetrate an aquifer. An array of the methods and techniques used in evaluating groundwater recharge can be found in many studies, while a synthesis is presented in the works of de Vries & Simmers (2002), Scalon *et al.* (2002) and of Polish researchers (Pleczynski, 1981; Pazdro & Kozerski, 1990).

The assessment of groundwater recharge can be made following a number of approaches: water balance, lysimeter method, isotopic tracking, numerical method, heat transfer, groundwater table fluctuation or through defining river hydrograph separation method. The methods drawing on a water balance present considerable difficulties as they require, apart from defining atmospheric precipitation, an assessment of evapotranspiration. Lysimetric measurements are expensive and selective, albeit they allow for precise point calculations in shallow zones. Isotopic tracking, as well as heat transfer are viable on a local scale but do not always take into account regional variables. Even though groundwater table fluctuations are recorded on a regular long-term basis, the method has its drawbacks as it is constrained on the assumption that the parameters of an aquifer are constant. Based on the theory of aquifers, the numerical method is of significant importance in the context of a region, and so is the approach focusing on water-bearing systems, which shows how effective infiltration and recharge are in varying terrain (Brodie & Hostetler, 2005). Most of the above methods are adopted in Poland, the ones assessing water balance being the most popular (e.g. Pleczynski, 1981). The most frequent methods employed in recording and documenting groundwater resources are those monitoring water balance and effective infiltration, the latter being based mainly on the permeability of surface deposits (Paczynski, 1995). Cartographic studies rely mostly on the results of long-term experimental pumping and increasingly popular modelling studies.

Poland lies in the eastern part of the North European Plain. Lowlands constitute over 91.3% of the country area, while the uplands (highlands and mountains) situated in the south constitute only 8.7%. The average annual temperatures range from 6 to 8.8°C, except in mountainous areas, where the temperatures decrease with increase in altitude (up to 2499 above sea level in the Tatras). The annual rainfall varies depending on the altitude: the average for the whole country is around 600 mm. For the lowlands and uplands the figures range from 450 to 750 mm in the mountains, from 700 to 1200–1500 mm for the highest elevations in the Tatras and Sudetes. The rainfall peaks in the summer months. Most of the area of Poland is covered with Cenozoic deposits mainly dating back to the Quaternary period. Four glacials interspersed with interglacials have led to the accumulation of compact glacial drifts such as sand, gravel or clay, whose thickness ranges from 10 to 280 m. These deposits contain 75% of Poland's renewable groundwater resources. Groundwater trapped in Quaternary deposits occurs in several layers in which thickness increases from the south to the north of the country. The volume of renewable groundwater is equal to 18.2% of rainfall (Pazdro & Kozerski, 1990). There is a considerable spatial variation in the extent of recharge.

The assessment of groundwater recharge was carried out at the time when the groundwater vulnerability map of Poland was being prepared (Duda *et al.*, 2007). Two methods were tested in order to determine the amounts of water flowing to shallow water-bearing layers: the effective recharge method was chosen as the primary tool, and the results were verified by base flow analysis. In the case of the former method, groundwater recharge was defined on the basis of the storage capacity and soil permeability as specified in detailed soil maps.

16.2 INFILTRATION RATE METHOD

One of the fundamental ways of evaluating the renewable resources of groundwater is defining effective infiltration by means of the infiltration rate. This method assumes that a portion of rainfall reaches aquifers and then is discharged through subterranean outlets into springs, rivers, lakes or the sea. The volume of water seeping to water-bearing layers is defined by means of an effective infiltration rate showing the ratio of infiltrated water to the arithmetic mean of annual rainfall measured one several years. The infiltration within a given area can be calculated with the aid of geological or soil

Table 16.1 Soil protective capacities (Witczak *et al.*, 2003; Duda *et al.*, 2004a, b).

Soil protective capacity	*Soil category*	*Grain size group [acc. to soil classification]*	*Infiltration rate [%]*	*Field water volume* $[L^0]^{a)}$	t_{glm} *approximate time of water exchange in 1 m of soil profile*$^{b)}$ *[years]*
Very weak	Very light	Sand: fine, silt, loose, slightly clayey	30 (27*)	0.12	1.2
Weak	Light	Sand: very fine, light clayey, sandy silt	20	0.17	1.7
Medium	Medium	Loam: Light and powdery clay and silt loam	13 (20*)	0.24	2.4
Good	Heavy	Loam: medium and silt, heavy loam, clay loam	8	0.36	3.6

$^{a)}$average field water volume
$^{b)}$approximate time of water exchange was calculated for the average effective infiltration equal to 100 mm per year and based on the piston flow model
(27*) modified value

maps. In order to do this, the areas under study have to be marked on the map and assigned to separate infiltration classes. There are several types of classification which help to do the assigning, e.g. classifications of Pazdro & Kozerski (1990), Paczynski (1981), Schneider and Züschang (cf. Załuski, 1973), Singh (2003), Wright (1982), Daly (1994) and proposed by Hebrich *et al.* (2004).

In Pazdro's classification (Pazdro & Kozerski, 1990), the infiltration rate ranging from 0.05 to 0.3 was agreed on the strength of the infiltration capability of individual rocks. A similar approach was adopted in Wright *et al.* (1982), but here the rates are higher, e.g. the rate of 0.2 is used for less permeable clays, 0.5 for loamy sands and 0.8 for high-permeability sands and gravels. Daly (1994), on the other hand, proposes the infiltration rate of 0.3 for solid clays, 0.6 for thin-layered clays and 0.9 for thin layers of permeable soils covering karst limestone. The last two infiltration rates are applied in Ireland, where there is high average rainfall (from 750 mm in the driest areas to 1600 mm in the lowlands) and high evaporation (500–575 mm).

In Poland, the infiltration rate method was generally used in lowlands, which was in line with the approach adopted by Pazdro & Kozerski (1990) and Paczynski (1981). There were also attempts to use this method in upland areas (Tarka, 2001) or even in mountainous areas (Duda *et al.*, 2006). For instance, in the case of the model studies of the river Raba catchment area (the Outer Carpathians) infiltration rates ranging from 0.11 to 0.165 were selected for flysch sandstone, and of 0.085 for flysch slates (Duda *et al.*, 2006).

Soil maps are also suitable for determining the infiltration rate since the type and category of soil reflects the geological structure and weather conditions; hence a direct correlation between the lithological form and the grain size group of the co-occurring soil. This correlation was relied upon in the process of drawing up guidelines for data compilation for "Groundwater contamination vulnerability map" to a scale of 1:500,000 (Table 16.1).

Both soil and geological maps show a similar estimation of renewable groundwater supplies (Tarka, 2001). For instance, for the eastern and central part of the synclinorium of the northern part of the Sudetes, where the outcrops of Cretaceous formations are the main lithological configuration, the comparison of groundwater renewability was made by analysing a geological map to a scale of 1:50 000 (Szalamacha & Milewicz, 1988), and a soil map to a scale of 1:300,000 (Musierowicz, 1960) and 1:500,000 (Pawlak *et al.*, 1997). The values obtained for effective infiltration were comparable: 134.7 and 116.8 mm respectively, despite the deployment of different classifications defining close-to-surface layers (Tarka, 2010).

16.3 CALCULATION RESULTS AND DISCUSSION

The calculations have been made for selected catchment areas in Lower Silesia. Before assessing groundwater recharge by means of the infiltration rate method, an average annual rainfall is computed for the area under study. Next, the category of infiltration rate is selected depending on the type of rock occurring on the surface or the type of soil. Then the sectors assigned to the selected categories of infiltration rate are defined within the precipitation area. This forms a basis for the computation of the weighted mean of the infiltration rate for each precipitation area.

$$\alpha_r = \frac{\sum_{i=1}^{n} \alpha_i \cdot A_i}{\sum_{i=1}^{n} A_i} \tag{16.1}$$

where α_r is the average infiltration rate for precipitation area [effective fraction] r; α_i is the infiltration rate for the i-lithological configuration within precipitation area; A_i is the surface of the i-lithological configuration [L^2] within precipitation area r.

The aggregate recharge for a drainage basin is the sum of the recharge values for each precipitation area:

$$R = \frac{\sum_{r=1}^{m} \alpha_r \cdot P_r \cdot A_r}{A} \tag{16.2}$$

where P_r is the average annual rainfall in the precipitation area r [L]; A is the area under study [L^2]; m is the number of selected precipitation areas.

The digital soil map at a scale of 1:500 000 drawn by the Institute of Soil Science and Plant Cultivation in Puławy was referred to when determining the categories of infiltration rates for the map of groundwater recharge in Lower Silesia (Musierowicz *et al.*, 1960). The first step was to select infiltration rates for each soil formation, following Table 16.1, and the recommendations attached to the "Map of groundwater vulnerability to contamination" at a scale of 1:500 000 (Witczak *et al.*, 2003, Duda *et al.*, 2004a; 2004b). The calculations were made for fifteen catchment areas differing in size, altitude and geological structure (Fig. 16.1). The area under study was nearly 17 000 sq. km in size. The precipitation figures were quoted from the Climate Atlas

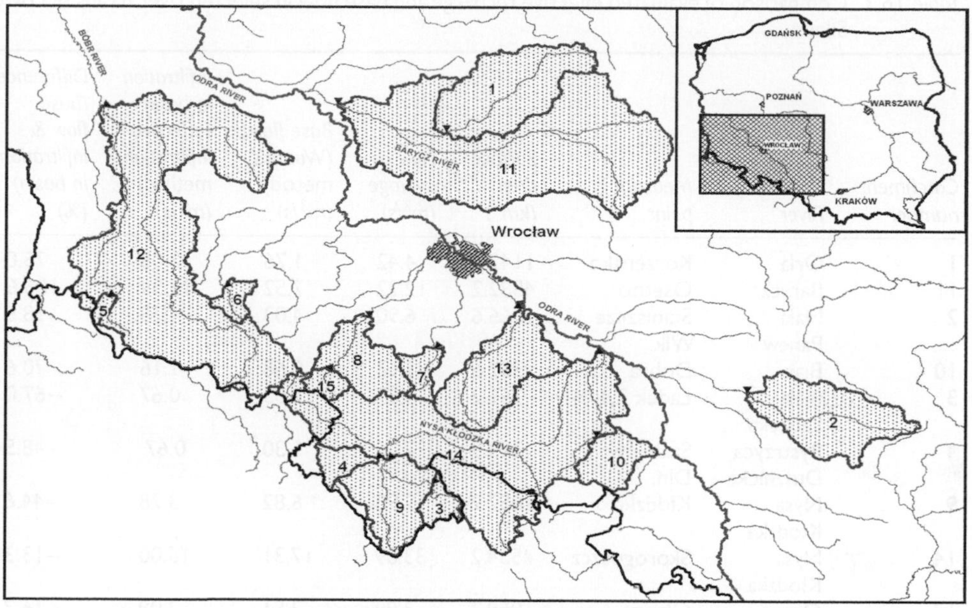

Figure 16.1 Location of catchment areas under study indicting gauging station in the region of Lower Silesia (number in conformity with Tables 16.2 and 16.3).

Table 16.2 Prevailing values for groundwater recharge calculation in selected catchments.

Catchment number	*Catchment area (km²)*	*Soil type*	*Rainfall amount (mm/y)*	*Infiltration rate (%)*	*Recharge (mm)*	*Recharge (m³/s)*
Orla – 1	1143.7	Medium	525	16.2	85	3.08
Mala Panew – 2	1066.6	Light	638	22.2	141	4.78
Bystrzyca – 8	676.1	Very light	577	27.0	165	3.78
Nysa Klodzka – 9	1057.2	Light	611	21.4	123	2.17
Bobr – 12	4192.3	Light	605	20.4	123	16.41
Oława – 13	959.2	Heavy	577	11.9	69	2.09
Barycz – 11	4582.2	Very light	537	26.5	143	15.54

of Poland (Lorenc, 2005). The estimated values for groundwater recharge were verified on the basis of groundwater drainage from the catchment areas under study. An example and applied values for calculation are illustrated in Table 16.2.

Groundwater drainage from the catchment area was determined with the Wundt method (Jokiel, 1994) and was based on the minimum monthly flow in the years 1976–2005. Smaller catchments were also analysed in a few selected cases. In the initial stages of computing, the results obtained were different from the volume of subterranean drainage occurring in the same areas. Therefore, in subsequent stages, infiltration rates were modified and the categories of soil formations readjusted in order to arrive at a probable volume of subterranean drainage. The analyses that followed indicated that the infiltration rate for "very light" soil had to be tuned from 30% down

Table 16.3 Comparison of evaluated effective recharge and base flow in selected river basins of Lower Silesia.

Catchment number	*River*	*Measuring point*	*Surface area (km²)*	*Flow as average (m³/s)*	*Base flow (Wundt's metod) (m³/s)*	*Infiltration in basin (effective infiltration method) (m³/s)*	*Difference (base flow & infiltration in basin) (%)*
1	Orla	Korzeńsko	1143.7	4.42	1.76	3.08	75.0
11	Barycz	Osetno	4582.2	12.22	7.52	15.54	106.6
2	Mała Panew	Staniszcze Wlk.	1066.6	6.90	4.03	4.78	18.6
10	Biała	Dobra	357.6	1.13	0.68	1.16	70.6
3	Biała Lądecka	Ladek Zdrój	161.2	3.37	2.03	0.67	−67.0
4	Bystrzyca Dusznicka	Szalejów Dln.	173.9	2.26	1.30	0.67	−48.5
9	Nysa Kłodzka	Kłodzko	1057.2	12.93	6.82	3.78	−44.6
14	Nysa Kłodzka	Skorogoszcz	4554.2	33.89	17.31	15.00	−13.3
13	Oława	Oława	959.2	3.88	2.51	2.09	−16.7
7	Ślęza	Białobrzezie	186.1	0.48	0.24	0.52	116.7
15	Bystrzyca	Jugowice	120.6	1.33	0.52	0.46	−11.5
8	Bystrzyca	Krasków	676.1	4.32	1.61	2.17	34.8
6	Kaczawa	Świerzawa	136.2	1.18	0.60	0.40	−33.3
5	Czarny Potok	Mirsk	51.1	0.90	0.35	0.24	−31.4
12	Bóbr	Żagań	4243.4	37.93	23.14	16.41	−29.1
						Sum	0.7

to 27% and the soil in the forests (Ls) needed a change in category: from medium to light, which lead to the change of the infiltration rate from 13% to 20%.

The results of the analysis are illustrated in Table 16.3. This provided the comparable values for nine catchment areas. The difference between the groundwater drainage and the recharge varies from 13% to 117%. The bigger discrepancies occur in a small mountainous catchment (Fig. 16.1). This leads to an underestimation of atmospheric precipitation which, in the highest part of the Sudetes is intensive. These differences occur in the catchment of the river Barycz (no. 11), which is attributed to a significant transformation of the area caused by man-made fish ponds. The other catchment areas (nos. 1, 7 & 10) contain errors due to the properties of soil and bedrock underneath the aquifer, as well as to the flow between the catchments. Nonetheless, the divergence between the recharge defined by means of infiltration rate and by the groundwater drainage is only 0.7% for the whole area under study.

The infiltration rate method yielded a mean groundwater recharge level of 109 mm against 587 mm of rainfall in all the catchment areas, while the subterranean drainage method yielded the figure of 108 mm. The recharge constitutes 18.5% of rainfall. The groundwater recharge map may be considered reliable on a regional scale.

The figures were compared with the results obtained using the numeric modelling technique, which is recommended for calculations on a regional scale. Model studies show weak recharge ranging between 52 and 84 mm per year for Quaternary formations in the river Odra Valley (Gurwin, 2000). In central Poland, where the precipitation is low, groundwater recharge fluctuates between 11 and 80 mm in the low-lying Tertiary aquifers (Dąbrowski *et al.*, 2007). The values for the mountainous regions differ markedly from those for the lowlands. The on-site measurements for the years 1998–2002 prove that the groundwater recharge in the mountains, as ascertained by various methods, is significantly higher at 244–680 mm (Staśko & Tarka, 2002).

Recent studies have been undertaken into recharge in small and medium mountainous hard rock catchments (7–160 km^2) in the Sudeten Mountains in SW Poland which are characterised by moderate to cold mountainous climate with the mean precipitation of 1360 mm/y showed interesting results. Four recharge assessment methods were applied: (1) lysimeter infiltration monitoring; (2) monitoring of groundwater drainage by gallery; (3) monitoring of river base flow; (4) monitoring of groundwater table fluctuation. The results showed that: (i) the recharge had an impulse character and was temporally variable; (ii) different methods of recharge assessment resulted in different recharge estimates mainly due to different spatial scales of assessment and different rainfall contributing areas; (iii) except for lysimeter method, the recharge-to-rainfall ratio was consistent and estimated at ~50% of precipitation; so a large recharge-to-rainfall ratio was attributable to high all-year-round soil moisture status and low thickness, 1–3 m, of weathered deposits implying relatively low soil retention capacity; (iv) the groundwater residence time was approximately 7–10 years as defined by tritium isotopic sampling and model simulation; (v) the aquifer, as with many other mountainous hard rock aquifers, receives a large quantity of recharge that results in significant groundwater flow; despite relatively low transmissivity, groundwater flow is efficiently transferred through the aquifer system to drainage lines (rivers and streams) and drainage points (springs) mainly thanks to large hydraulic gradients, typical for mountainous catchments (Stasko *et al.*, 2010).

ACKNOWLEDGEMENT

This study was partly supported by Wroclaw University Institute of Geological Science grants program 1017/S/ING/10. The authors also wish to express thanks to two anonymous reviews for their constructive comments and suggestions for improvement of the manuscript.

REFERENCES

Brodie R.S., Hostetler S. (2005) A review of techniques for analysing baseflow from stream hydrographs. *Bureau of Rural Sci*, Australia (http://www.connectedwater.gov.au/documents/IAH05_Baseflow.pdf).

Daly E.P. (1994) Groundwater resources of the Nore River basin. *Geological Survey of Ireland*, RS 94/1.

Dąbrowski S., Przybyłek J., Górski J. (2007) Region of the Warta river lowland. In: *Regional Hydrogeology of Poland*, pp. 369–407. Published by Polish Geological Institute (in Polish).

De Vries J.J., Simmers I. (2002) Groundwater recharge: an overview of processes and challenges. *Hydrogeol J* 10:5–17.

Duda R., Karlikowska J., Witczak S., Żurek A. (2004a) Methodology of elaboration Groundwater vulnerability map to contamination in 1:500,000 scale on selected test areas. Hydrogeology and Water Protection Dept, AGH Kraków (not published, in Polish).

Duda R., Karlikowska L., Witczak S., Żurek A. (2004b) Modification of the preparation information layers for the groundwater vulnerability map to contamination in 1:500,000 scale. Hydrogeology and Water Protection Dept AGH Kraków (not published, in Polish).

Duda R., Zdechlik R., Paszkiewicz M. (2006) Some remarks on numerical modeling of the Rava River basin. *Geologos* 10:47–57. Uniwersity of Adam Mickiewicz, Poznań, Poland.

Dubicki A. (ed) (2002) Water resources in upper and middle part of the Odra River basin in drought season. Institute of Meteorology and Water Management. *Atlases and Monographies* 88, Warszawa (in Polish).

Gurwin J. (2000) Groundwater flow model of the Odra ice-marginal valley aquifer system near Glogów, *Prace Geol.-Mineral.* LXX, Acta Universitatis Wratislavienisi 2215, Wrocław (in Polish, abstract in English).

Instruction of preparation hydrogeological map of Poland in 1:50,000 scale associated with information part 1, 2, 1999, Polish Geological Institute. Warszawa (in Polish).

Jokiel P. (1994) Resources, renewability and groundwater runoff in active zone of Poland. *Acta Geographica* 66–67, Łódź (in Polish).

Lorenc H. (ed) (2005) Climate atlas of Poland. *Inst Meteorology, Water Management.* Warszawa (in Polish)

Musierowicz A. (ed) (1960) Soil map of Poland In 1:300,000 scale. Geological Publ., Warszawa (in Polish).

Paczyński B. (1972) Methodology of groundwater resources evaluation in region scale. Instruction and method of geological survey. Book 17. Warszawa (in Polish).

Paczyński B. (ed) (1995) Hydrogeological atlas of Poland In 1:500,000 scale. Part II. Resources, quality and protection of fresh water. Polish Geological Institute. Warszawa (in Polish).

Pawlak J. (ed) (1997) Atlas of Lower and Opole Silesia. Wrocław University & Polish Academy Sci, Wroclaw.

Pazdro Z., Kozerski B. (1990) General hydrogeology. Geological Publisher. Warszawa (in Polish).

Pleczyński J. (1981) Groundwater resources renewability. Geological Pub, Warszawa (in Polish).

Scanlon R.B., Healy R.W., Cook P.G. (2002) Choosing appropriate techniques for quantifying groundwater recharge. *Hydrogeol J* 10(16.1):18–39.

Singh D.K. (2003) Assessment of groundwater potential. In *Proc: Winter School on Advanced Techniques and Their Applications in Water Management* (eds Singh A.K. & Khanna M.), Water Technology Centre, Indian Agricultural Research Institute, New Delhi, pp. 271–280.

Staśko S., Tarka R. (2002) Groundwater recharge and drainage processes in mountainous terrains based on research in the Śnieznik Massif (Sudetes, SW Poland). *Acta Uni Vratislaveinsis* 2528, Wroclaw (in Polish, abstract in English).

Staśko S., Michniewicz M. (2007) Sudeten region. In: *Reg Hydrogeol Poland*, pp. 306–327. Pub Polish Geolog Institute (in Polish)

Staśko S., Tarka R., Olichwer T., Lubczynski M. (2010) Groundwater recharge in mountainous terrains – case study from Sudeten Mountains in SW Poland. Chapter 28: *Global Groundwater Resources Management*. Scientific Publisher Jodpur, pp. 451–474

Szałamacha A.J., Milewicz J. (1988) Geological map of Poland, A – Map of surface sediments, Basic map in 1:50,000 scale, sheet Jelenia Góra. Polish Geological Institute, Warszawa (in Polish).

Tarka R. (2001) Discrepancies in groundwater renewable resources evaluation and permeability of surface rocks. *Współczesne Problemy Hydrogeologii X*, Vol. 1: 279–287, Sudetes Pub. Wroclaw (in Polish).
Witczak S., Duda R., Żurek A., Zieliński W. (2003) Concept of elaboration of the groundwater vulnerability map to contamination in 1:500,000 scale. Arcadis Ekokonrem, Kraków, Wrocław (not Published, in Polish).
Wright G.R., Aldwell C.R., Daly D., Daly E.P. (1982) Groundwater resources of the Republic of Ireland, Vol. 6. In: *European Community's atlas of groundwater resources*, SDG, Hanonover.
Zaluski M. (1973) Groundwater recharge base on selected elements and water balance evaluation. *On hydrogeological research in Poland*, Vol. III. Geological Institute, Bulletins 277:107–120.

Chapter 17

Seawater intrusion control by means of a reclaimed water injection barrier in the Llobregat delta, near Barcelona, Catalonia, Spain

Felip Ortuño[1], Emilio Custodio[2], Jorge Molinero[3], Iker Juarez[3], Teresa Garrido[1] & Josep Fraile[1]

[1]*Agencia Catalana de l'Aigua (ACA), Barcelona, Spain*
[2]*Departament of Geotechnical Engineering and Geosciences, Technical University of Catalonia (UPC), Barcelona, Spain*
[3]*Amphos XXI Consulting S.L., Barcelona, Spain*

ABSTRACT

The main aquifer of the Llobregat delta (Barcelona, Spain) has been affected by seawater intrusion since the 1960s. The Catalan Water Agency (ACA) has sponsored the construction of a positive hydraulic barrier in order to stop and redress seawater intrusion advance due to intensive aquifer development. The hydraulic barrier consists of 15 wells into which highly treated reclaimed water from the wastewater treatment plant of the Baix Llobregat is injected, with salinity reduction through reverse osmosis. A preliminary pilot phase of the project started in late 2007, with highly positive results. Hydrogeological and hydrochemical monitoring data indicate an efficient performance. Quantitative evaluation of such efficiency and operational costs has been analysed. The second phase started in mid 2010 and the full injection barrier is now in operation.

17.1 INTRODUCTION

17.1.1 Aquifers of the Lower valley and delta of the Llobregat

The geological make-up of the Llobregat delta has been well known since the 1960s (Marqués, 1984; Manzano, 1993; Simó *et al.*, 2005; Gámez, 2007). It is formed by a silt and clay wedge deposited in an estuarine environment that separates two sand and gravel aquifers: an upper one 15 m thick, below the current land surface, of Holocene age, and another aquifer 10 to 20 m thick, of late Pleistocene age, which is the main and most important aquifer (Fig. 17.1). This deep aquifer is confined and is highly transmissive (1000 to 5000 m^2/day). It is linked to the Lower Valley Aquifer. Abstracted water is primarily used for urban and industrial supply. It is a strategic water resource for supplying Barcelona and its metropolitan area in droughts and emergency situations (Niñerola *et al.*, 2009).

17.1.2 Seawater intrusion

Seawater intrusion processes have affected the main delta aquifer since the 1960s. The intensive exploitation of groundwater resources, along with the excavation of

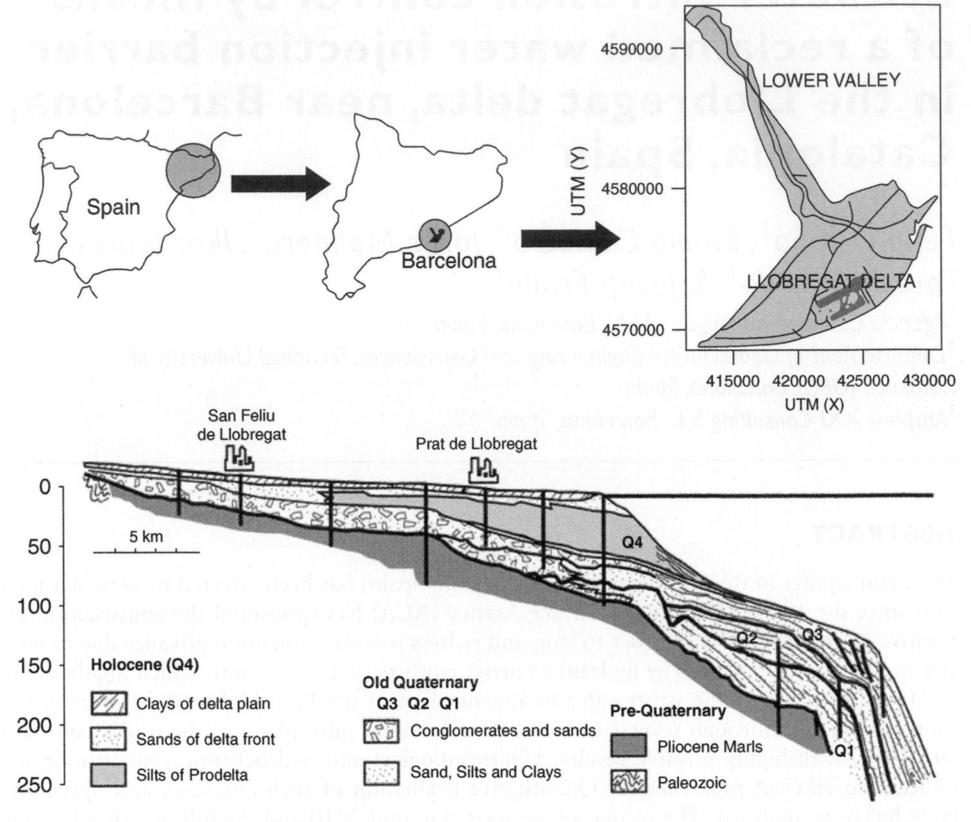

Figure 17.1 Location of the Llobregat delta and Lower valley (from Abarca *et al.*, 2006) and geological cross section perpendicular to the coast of the emerged and submerged Llobregat delta (from Simó *et al.*, 2005); the very thin layer below the silts off Q4 and the upper gravels of Q3 form the main aquifer.

part of the confining layer in the eastern coastal corner, has led to the progressive, serious deterioration of groundwater quality (Custodio, 1981, 2008, 2010; Iríbar, 1992; Iríbar *et al.*, 1997). Two main seawater intrusion fronts exist, one in the central coastal area and the other in the easternmost part. Currently the marine water affected area is about one third of the delta area (Fig. 17.2). Current groundwater abstraction is about 54 hm^3/year, but it exceeded 100 hm^3/year in the 1970s. By using numerical models (Vázquez-Suñé *et al.*, 2006), the sustainable exploitation rate to avoid further groundwater deterioration is calculated around 40 hm^3/year, for current well distribution.

17.1.3 Artificial recharge in the Llobregat aquifers

To mitigate water scarcity in the area and at the same time trying to recover the good status of the groundwater bodies, mandated by the European Water Framework Directive (Niñerola *et al.*, 2009), the Catalan Water Agency, along with the Metropolitan

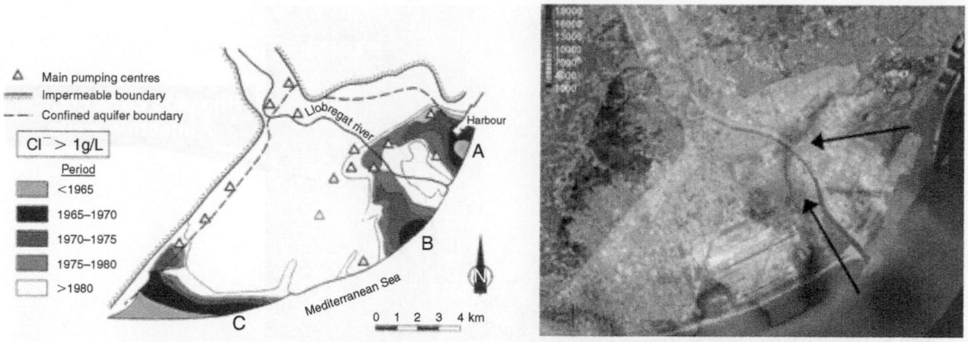

Figure 17.2 Progress of seawater intrusion (left) (Iríbar, 1992; Iríbar & Custodio, 1992), and chloride concentration (right) (Catalan Water Agency data) in 2007 in the main Llobregat delta aquifer; seawater intrusion currently extends over one third of the delta area, mostly following the preferential pathways shown by the arrows.

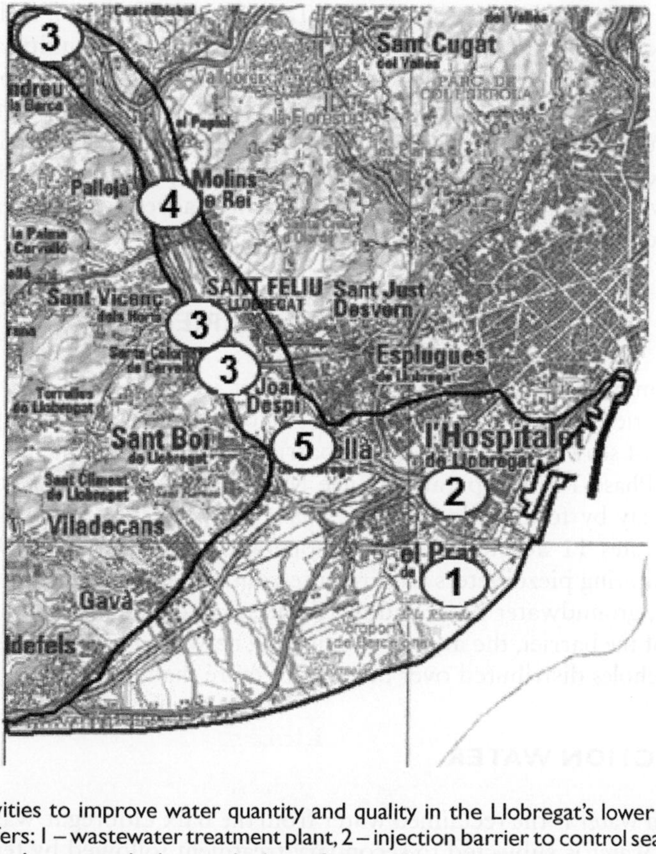

Figure 17.3 Activities to improve water quantity and quality in the Llobregat's lower valley and delta aquifers: 1 – wastewater treatment plant, 2 – injection barrier to control seawater intrusion, 3 – recharge ponds, 4 – riverbed scarification, 5 – well injection.

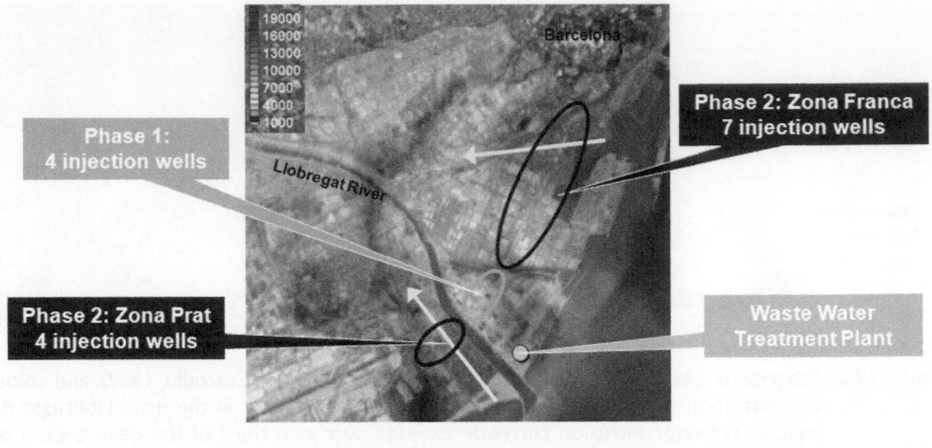

Figure 17.4 Hydraulic barrier configuration. The project has been carried out in two phases, with a total of 15 injection wells and a maximum injection rate of 15,000 m^3/day; currently one of the wells of Phase 1 is out of service and grouted.

Agency for Hydraulic Services and Waste Treatment (EMSHTR), Agbar as the main water abstractor for supply, and the Groundwater Users' Community, are carrying out various artificial recharge activities and defining a Groundwater Exploitation Strategy Plan (Ortuño *et al.*, 2009). In the lower valley of the Llobregat, recharge ponds are being constructed in three areas, which will provide a total additional recharge ranging from 6 to 10 hm^3/year, and Agbar is traditionally performing scarification of the Llobregat river bed to enhance river recharge, as well as recharge of treated river water through injection wells (Fig. 17.3).

17.2 THE LLOBREGAT HYDRAULIC BARRIER PROJECT

The key to improving the quality of the aquifer is the construction of the positive hydraulic barrier using reclaimed water (Ortuño *et al.*, 2008). The objective is to halt the advance of seawater intrusion. The barrier has been implemented in two phases (Fig. 17.4). Phase 1 is in operation since March 2007, with the total injection rate of 2400 m^3/day by four wells. Phase 2 has a total injection rate of 15 000 m^3/day and incorporates 11 additional wells in operation since April 2010. There are 17 specific monitoring piezometers with remote-control data gathering systems for water temperature, groundwater head and water electrical conductivity. In order to follow the impact of the barrier, the aquifer monitoring network also includes 13 wells and 7 existing boreholes distributed over the area of more than 30 km^2.

17.3 INJECTION WATER

The injection water is the reclaimed water from the Baix Llobregat Waste Water Treatment Plant. Water is subjected to secondary treatment followed by tertiary treatment

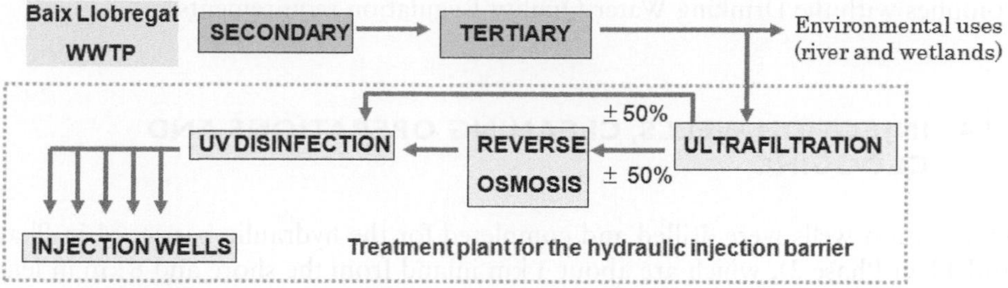

Figure 17.5 Water treatment stages prior to injection into the aquifer; the injection water is the reclaimed water from the tertiary effluent of the El Baix Llobregat wastewater treatment plant (WWTP) in Barcelona, after ultrafiltration, reverse osmosis and UV disinfection treatments.

that consists of ballasted coagulation-flocculation, lamellar decantation, filtration and disinfection. The Tertiary treated water is used for environmental purposes to increase the Llobregat river flow in the lower reaches, sustain wetlands (Cazurra, 2008), and feed the hydraulic barrier sub-treatment plant. At the Hydraulic Barrier Plant, prior to the distribution to the injection wells water suffers ultrafiltration, reverse osmosis and UV disinfection (Fig. 17.5). Water is not chlorinated before injection due to potentially harmful effects and possible appearance of trihalomethanes in the aquifer. Ultrafiltration of the injected water is performed for bacteriological purposes to avoid the introduction of nematode eggs and bacteria, prevent physical clogging in the injection wells, and extend the functional life of the reverse osmosis membranes. Reverse osmosis is necessary to reduce the salinity of 35% to 50% of the ultrafiltrated water. After tertiary treatment water usually contains more than 600 mg-Cl/l due to upstream saline water discharges to the Llobregat river basin, especially from current and past potash mines activities. Chloride content in the areas of the aquifer unafected by seawater intrusion is about 350 mg/l, which is the quality objective.

The water quality control is carried out in compliance with the Sanitation Authority requirements. Some injection water parameters (electrical conductivity, pH, temperature, ammonia and turbidity) are logged automatically, while bacteriological, physicochemical and chemical parameters (nematodes, E. Coli, P, N, Cl, NO_3, NH_4 and TOC) are monitored weekly or every two weeks. Major elements, metals, and volatile organohalogenated compounds are monitored on a bimonthly basis. A complete analysis of all the Drinking Water Quality Regulation parameters is performed once a year.

Injection at the Phase 1 began on 26 March 2007. By February 2011 around 3 500 000 m^3 of reclaimed water has been injected, with an average 347 mg-Cl/l, which is similar to that found in areas unaffected by seawater intrusion. The average values of injected water are 1849 μS/cm for electrical conductivity, 7.43 for pH, <0.09 NTU for turbidity and 14.7 mg/l for NO_3. Injected water temperature, measured in the injection wells, range between 14°C in winter and 29°C in summer. To date coliforms, *E. Coli*, clostridium perfringes and nematodes have not been detected in any of the water samples. No organochloride compounds have been detected. Injection water complies with the Drinking Water Quality Regulation requirements.

17.4 INJECTION WELLS, CLEANING OPERATIONS AND CLOGGING

15 injection wells were drilled and completed for the hydraulic barrier (4 in Phase 1 and 11 in Phase 2), which are about 1 km inland from the shore and 6 km in length. Wells are about 400 m apart (Fig. 17.6).

The wells are 70 m deep and penetrate the 6 to 10 m thickness of the main confined aquifer along the barrier. Wells drilling diameter is 610 mm, and they are cased with 350 mm diameter stainless steel or PVC tube. Wellheads are sealed. Wells are equipped with flowmeters, automated electrovalves to maintain constant flow, and a pressure sensor to control injection rate. The injection wells of Phase 2 have also temperature and electrical conductivity downhole sensors. All wells are controlled remotely from the barrier's water treatment plant, to which they are linked every 5 seconds. The injected volumes, water temperature and electrical conductivity and well hydraulic heads are recorded every hour.

Cleaning is necessary to avoid clogging of the well screen and surroundings. This is done through backflushing: 12 of the injection wells are equipped with submersible electropumps and 2 are equipped with compressed-air pumping devices. Cleaning is carried out periodically during a short period of time in which the water discharge is 3 times that of injection.

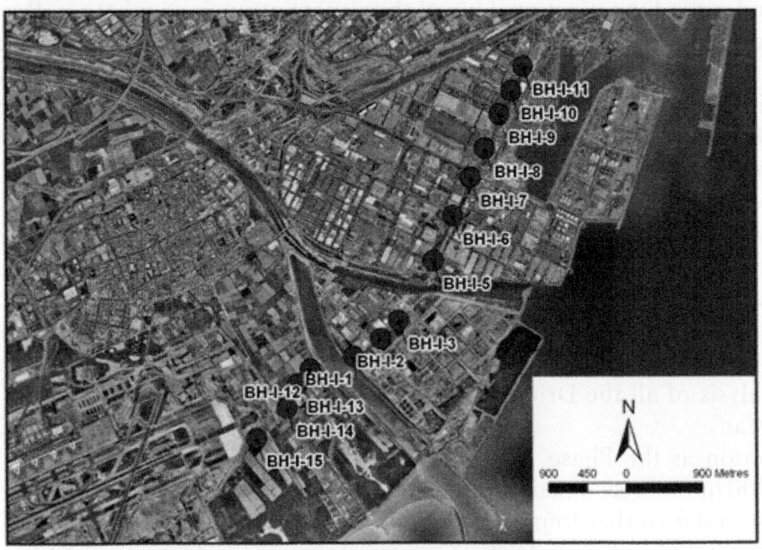

Figure 17.6 Injection wells of the Llobregat hydraulic barrier (BH-I). Wells 1 to 4 correspond to Phase 1, and wells 5 to 15 to Phase 2; well BH-I-4 is no longer in use due to grouting problems of the annular space, and is now discarded and backfilled.

The injection water is in equilibrium with silica, but oversaturated with respect to calcite. Different possible mixtures of the treated water with the osmotized water result in calcite saturated waters, which imply the possibility of carbonate precipitation and well clogging problems. Samples of the precipitates contained in backflushing water for injection wells BH-I-2 and BH-I-3 show the presence of calcite in XRD analyses.

An intensive cleaning programme of the injection wells has been established, based on these geochemical results. One hour cleaning is performed every two weeks in all wells. The cleaning programme is working well and no decrease of well efficiencies (flow/head increase) has been detected so far. This is attributed to the high quality of injected water as well as to the strict cleaning program. Reverse osmosis and ultrafiltration prevent physical clogging, while disinfection prevents bacteriological clogging.

17.5 THE AQUIFER MONITORING NETWORK

Water quality and quantity monitoring relies on a net consisting of existing boreholes and wells, as well as new boreholes that were constructed in the framework of the barrier project. During the first phase, 8 monitoring points were available, covering an area of 3 km^2. Currently there are 37 monitoring points (17 new piezometers and 20 existing points) (Fig. 17.7). All monitoring points are sampled every 2 months and chemical analyses are carried out to determine major elements, nitrogen compounds, metals and organochlorine compounds. Water temperature, electrical conductivity, Eh

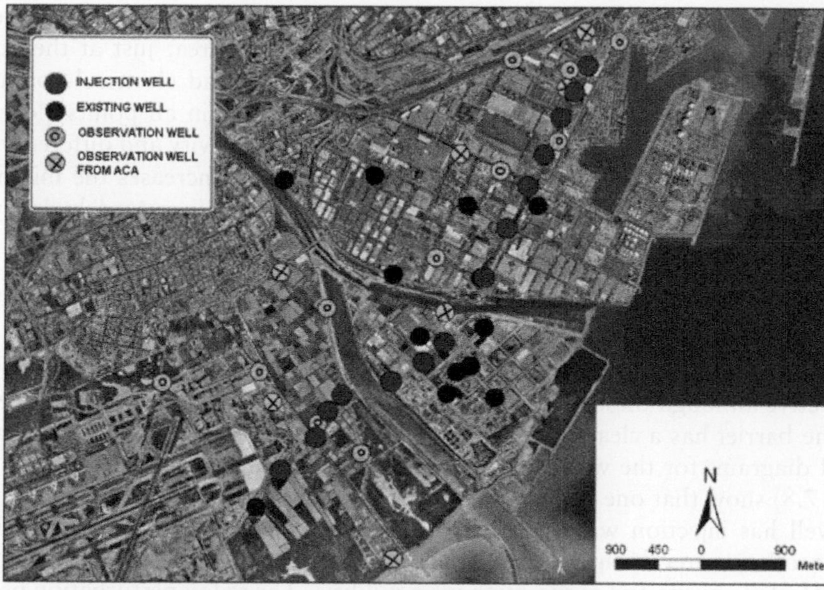

Figure 17.7 Aquifer monitoring network of the Llobregat hydraulic barrier; there are 17 specific monitoring boreholes fitted with remote-control data systems and 20 old wells and existing boreholes, covering more than 30 km^2.

and pH are measured in the field during sampling with down-hole sensors. The 17 new piezometres have also permanent sensors at the midpoint of the aquifer thickness that record hourly the water level, temperature and electrical conductivity. Down-hole temperature and electrical conductivity logs are also performed every 6 months in the 17 new boreholes, to provide three-dimensional information about the injected water movement through the aquifer.

17.6 AQUIFER IMPROVEMENT AND HYDROGEOCHEMICAL PROCESSES

Chemical data have been studied with the help of PHREEQC_i v.2.15.0 (Parkhurst & Appelo, 1999) for saturation indices and geochemical modelling. The analytical results and field monitoring of the aquifer during Phase 1 of the project show highly positive results (Ortuño *et al.*, 2009). Since 2007, in 8 monitoring points there is a progressive decrease of chloride, sodium, calcium, magnesium, iron and ammonium contents, and a slight nitrate increase since it is present in the injected water. The bicarbonate content is constant. The area influenced by recharged water is estimated to be between 1 and 2 km around the injection wells. This area is irregular with respect to two sides of the barrier, because groundwater flow from the sea side is halted or greatly reduced, while recharged water readily moves inland to the pumped areas. For Phase 1, recharged water has been found in 25% to 40% of the monitoring points in the sea-side strip, while inland from the barrier it has been found in 75% to 90% of them.

The initial spread of recharged water has been rapid because it moves preferentially through the more permeable layers of the aquifer and does not extend to the entire thickness until the injected water has not been spread. This was studied in a pilot reclaimed water injection well in the coastal Besós area, just at the northern side of Barcelona (Custodio *et al.*, 1976). Temperature and electrical conductivity logs carried out in the 17 new boreholes corroborate this in all points close to the injection wells, where the arrival of lower electrical conductivity and different temperature injected water is observed at irregular depths. This increases the mixing with existing aquifer water, which in practice shows up as an increased hydrodynamic dispersion.

The injected water interacts with the aquifer material, thus modifying ionic ratios. Some indices are introduced to monitor the presence and changes of the injected water. The Na/K, Ca/Mg, SO_4/Cl, (Na-Na_{sea})/Cl and (Ca-SO_4-HCO_3)/Cl ratios of injected water and water from the monitoring points and their evolution over time, appear to be effective although these indices may be modified with increasing experience.

The barrier has a clear oxidizing effect on the otherwise highly reducing aquifer. Eh/pH diagrams for the waters sampled from the monitoring network in May 2009 (Fig. 17.8) show that one observation point (BH-PP-1), the closest (4 m) to the injection well has injection water plot in the field of calcite – ferrihydrite equilibrium. All the other water samples are highly reducing and plot in the fields of calcite-siderite/calcite-pyrite and pyrite-methane equilibria. The redox perturbation produced by the oxidizing water injection is under continuous monitoring and modelling. The redox potential change in the aquifer could result in oxidation of reduced sulphur species and reduced iron in natural water and soil, and also could lead to oxidation of

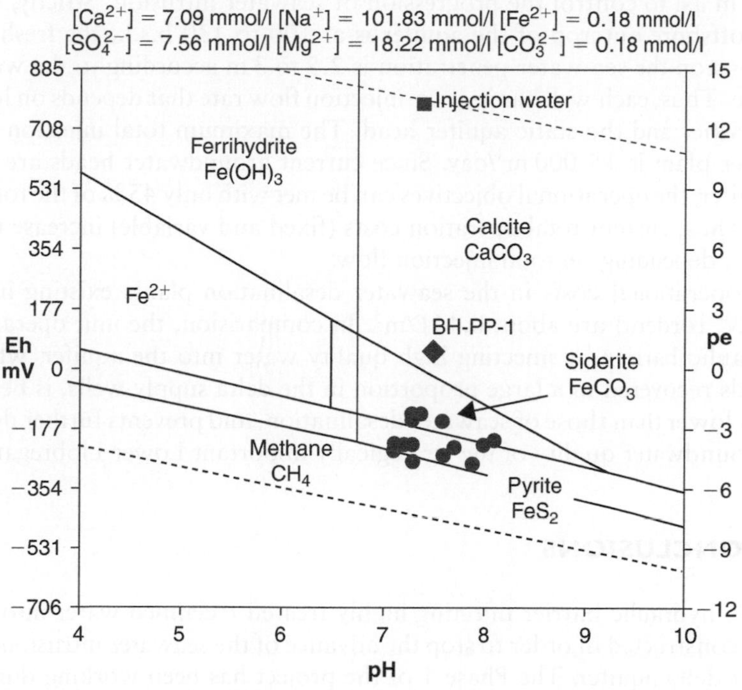

Figure 17.8 The Eh/pH diagram for the waters sampled at the aquifer monitoring network in May 2009; only the injection water and the closest-to-the-injection-well observation point plot in the field of calcite – ferrihydrite equilibrium.

the existing dissolved methane. The redox potential decrease in the mixing zone produces a reduction of nitrate in the injected water, probably to inert N_2. The behaviour of native NH_4^+ in the aquifer, is in part sorbed in the solids. So far, a clear redox perturbation is only observed in the close vicinity of injection wells. The other boreholes in the local monitoring network show what seems unaltered reducing conditions, up to present.

17.7 INVESTMENT AND EXPLOITATION COSTS OF THE BARRIER

The total investment costs of the construction of the Llobregat Hydraulic Barrier amount to €23 millions, contributed by the Catalan Water Agency, the Spanish Ministry of the Environment and Rural and Marine Affairs, and the European Commission. The total exploitation costs have been calculated from actual costs after one year of the whole hydraulic barrier operation. Fixed annual costs amount to 268 072€, and comprise the staff, treatment plant, wells maintenance and hydrogeological support. Thus, variable costs are of 0.04 €/m³ for chemicals and of 0.05 €/m³ for energy, with a total of 0.09 €/m³.

The hydraulic barrier is operated to maintain the aquifer dynamic head in the range of 1 to 3 m asl to control the progression of seawater intrusion. Strictly, considering that the offshore outcrop of the aquifer is at 100 to 120 b.s.l., the freshwater head needed to stop the sea water penetration is 2.5 to 3 m according to the water density difference. Thus, each well has its own-injection flow rate that depends on local aquifer characteristics and the static aquifer head. The maximum total injection capacity of the barrier plant is 15 000 m^3/day. Since current groundwater heads are high in the main aquifer, the operational objectives can be met with only 45% of the total injection volume. Thus, current total operation costs (fixed and variable) increase to 0.15 and 0.22 €/m^3, depending on total injection flow.

The operational costs in the seawater desalination plants existing in Catalonia (El Prat & Tordera) are about 0.44 €/m^3. In comparision, the unit operation cost of the hydraulic barrier by injecting high quality water into the aquifer, which will be afterwards recovered in a large proportion in the delta supply wells, is between 34% and 50% lower than those of seawater desalination, and prevents further deterioration of the groundwater quality of the strategically important Lower Llobregat aquifer.

17.8 CONCLUSIONS

A positive hydraulic barrier injecting highly treated reclaimed water into the aquifer has been constructed in order to stop the advance of the seawater intrusion in the main Llobregat delta aquifer. The Phase 1 of the project has been working during the last 3 years, showing highly positive results. No clogging has been observed in the injection wells, and this is attributed to the high water quality after ultrafiltration, reverse osmosis and UV disinfection processes, and also to the strict well cleaning program. The second phase of the project is now in operation. Substantial improvement of groundwater quality has been observed in wells around the injection points. This shows that the concept and technology of the hydraulic barrier are able to halt or slow the advance of the saline water fronts. Currently, hydrochemical studies and modelling are carried out to identify ion exchange and mixing processes in the aquifer and the spread and dispersion of the injected water. The aquifer monitoring network has been designed to monitor the influence and impact of the injected water. Hydrochemical studies, as well as those carried out to identify clogging processes, are important to understand the behaviour of the barrier and to implement appropriate management. Total operating costs vary between 0.15 and 0.22 €/m^3, depending on the actual injection flow rate, and a function of regional aquifer hydraulic head. Results from the two-phase development of this project have been extremely positive. This is a joint venture of the different teams, government agencies and entities involved, which is needed to guarantee the future success and good performance of the entire project.

ACKNOWLEDGEMENTS

Special thanks to Josep Maria Niñerola, Joan Jovés and Gabriel Borràs (Catalan Water Agency, ACA) for their support to the project. The authors acknowledge all the companies, universities, organizations and city councils involved in the hydraulic barrier project, special to AREMA and the Groundwater Users' Community of the Llobregat Delta (CUADLL). The financial support of the project has been provided by the Catalan Water Agency (ACA), the Spanish Ministry of the Environment and Rural and Marine Affairs, and the European Commission.

REFERENCES

Abarca E., Vázquez-Suñé E., Carrera J., Capino B., Ga'mez D., Batlle F. (2006) Optimal design of measures to correct seawater intrusion. *Water Resour. Res.* 42, W09415, doi:10.1029/2005WR004524.

Cazurra T. (2008) Water reuse of south Barcelona's wastewater reclamation plant. *Desalination* 218:43–51.

Custodio E. (1981) Sea water encroachment in the Llobregat and Besós areas, near Barcelona (Catalonia, Spain). In: *Intruded and Fosil Groundwater of Marine Origin*. Sveriges Geologiska Undersökning. Uppsala. Rapporter och Meddelanden 27: 120–152.

Custodio E. (2008) Acuíferos detríticos costeros del litoral mediterráneo penínsular: valle bajo y delta del Llobregat. *Monográfico: Las Aguas Subterráneas. Rev. Assoc. Española Enseñanza de las Ciencias de la Tierra* 15(3): 295–304, Madrid.

Custodio E. (2010) Coastal aquifers of Europe: an overview. *Hydrogeol* J 18: 269–280.

Custodio E., Suárez M., Galofré A. (1976) Ensayos para el análisis de la recarga de aguas residuales en el Delta del Besós. *Actas de la II Asamblea Nacional de Geodesia y Geofísica.* Instituto Geográfico y Catastral, Madrid, 1893–1936.

Gámez D. (2007) Sequence stratigraphy as a tool for water resource management in alluvial coastal aquifers: application to the Llobregat delta (Barcelona, Spain). Doctoral Thesis, Department of Geotechnical Engineering and Geo-Sciences (ETCG), Technical University of Catalonia (UPC), pp. 177 + An.

Iríbar V. (1992) Evolución hidroquímica e isotópica de los acuíferos del Baix Llobregat. Doctoral Thesis, Departamento de Geoquímica, Petrología y Prospección Geológica, University of Barcelona.

Iríbar V., Custodio E. (1992) Advancement of seawater intrusion in the Llobregat delta aquifer. In: *Study and Modelling of Salt Water Intrusion.* CIMNE–UPC. Barcelona, 35–50.

Iríbar V., Carrera J., Custodio E., Medina A. (1997) Inverse modelling of seawater intrusion in the Llobregat delta deep aquifer. *J Hydrol* 198 (1–4): 226–247.

Manzano M. (1993) Génesis del agua intersticial del acuitardo del Delta del Llobregat: origen de los solutos y transporte interactivo con el medio sólido. Tesis Doctoral, Escola Tècnica Superior d'Enginyers de Camins, Canals i Ports de Barcelona, Universitat Politècnica de Catalunya.

Marqués M.A. (1984) Las formaciones cuaternarias del delta del Llobregat. Doctoral Thesis, Facultad de Ciencias Geológicas, University of Barcelona: 1–280.

Niñerola J.M., Queralt E., Custodio E. (2009) Llobregat delta aquifer. In: Quevauviller P., Fouillac A.–M., Grath J. &. Ward R. (eds) *Groundwater Monitoring*, John Wiley & Sons: 289–301.

Ortuño F., Niñerola Pla J.M., Teijon G., Candela L. (2008) Desarrollo de la primera fase de la barrera hidráulica contra la intrusión marina en el acuífero principal del Delta del Llobregat. *IX Simposio de Hidrogeología*, Elche (Spain).

Ortuño F., Niñerola Pla J.M., Armenter J.L., Molinero Huguet J. (2009) La barrera hidráulica contra la intrusión marina y la recarga artificial en el acuífero del Llobregat (Barcelona, España). *Boletín Geológico y Minero (IGME)*, 120 (2): 235–250.

Parkhurst D.L., Appelo C.A.J. (1999) Users guide to PHREEQC (version 2) a computer program for speciation, batch-reaction, one-dimensional transport and inverse geochemical calculations. *Water Resour Invest Rep*, 99-4259.

Simó J.A., Gamez D., Salvany J.M., Vazquez-Suñé E., Carrera J., Barnolas A., Alcala F.J. (2005) Arquitectura de facies de los deltas cuaternarios del río Llobregat, Barcelona, España. *Geogaceta*, 38: 171–174.

Vázquez-Suñé E., Abarca E., Carrera J., Capino B., Gámez D., Pool M., Simó T., Batlle F., Niñerola J.M., Ibáñez X. (2006) Groundwater modelling as a tool for the European Water Framework Directive (WFD) application. *The Llobregat case. Physics and Chemistry of the Earth* 31 (17): 1015–1029.

Part 3

Groundwater – surface water interaction

Chapter 18

Sustainability of river bank filtration – examples from Germany

Thomas Grischek[1], Dagmar Schoenheinz[1], Paul Eckert[2] & Chittaranjan Ray[3]

[1]*Division of Water Sciences, Department of Civil Engineering & Architecture, University of Applied Sciences Dresden, Dresden, Germany*

[2]*Waterworks, Stadtwerke Duesseldorf AG, Duesseldorf, Germany*

[3]*Department of Civil and Environmental Engineering & Water Resources Research Center, University of Hawaii at Manoa, Honolulu, HI, USA*

ABSTRACT

River bank filtration (RBF) is a well established technique for water treatment in Germany and many other countries worldwide. Because RBF systems utilize natural filtration processes to treat water, this strategy can provide several advantages in terms of sustainability compared to conventional water treatment technologies. These advantages include lower energy and resource requirements, little or no generation of waste streams, reduced environmental impacts during construction and system operation, and greater adaptability to changing water supply conditions due to climate changes or changing water demands. Selected sustainability aspects are discussed based on two examples from RBF sites in Germany.

18.1 INTRODUCTION

River bank filtration (RBF) is a process during which surface water is subject to subsurface flow prior to extraction from vertical or horizontal wells. The raw water discharged from the production well consists of a mixture of infiltrated river water and groundwater recharged in the land catchment (Fig. 18.1). From a water resources perspective, RBF is normally characterised by improved water quality (Kuehn & Mueller, 2000). Therefore, RBF is a well-proven treatment step, which at numerous sites is part of a multi-barrier approach to drinking water supply. Grischek *et al.* (2002) report about the extensive application of RBF along the European rivers Danube, Rhine, and Elbe. In the United States, RBF is receiving increased attention especially with regard to the removal of parasites and precursors to disinfection by-products (Ray *et al.*, 2002; Tufenkji *et al.*, 2002).

From a sustainability point of view, RBF systems make more sense than full-scale treatment plants using surface water, since the energy and resource use in RBF is lower and little to no chemical residues are produced. RBF systems require less energy to operate and to deliver a unit amount of water than conventional surface water treatment systems. The ecological impact of surface water depletion using RBF is low compared to full-scale surface water treatment plants because a (typically substantial) portion of pumped water is groundwater.

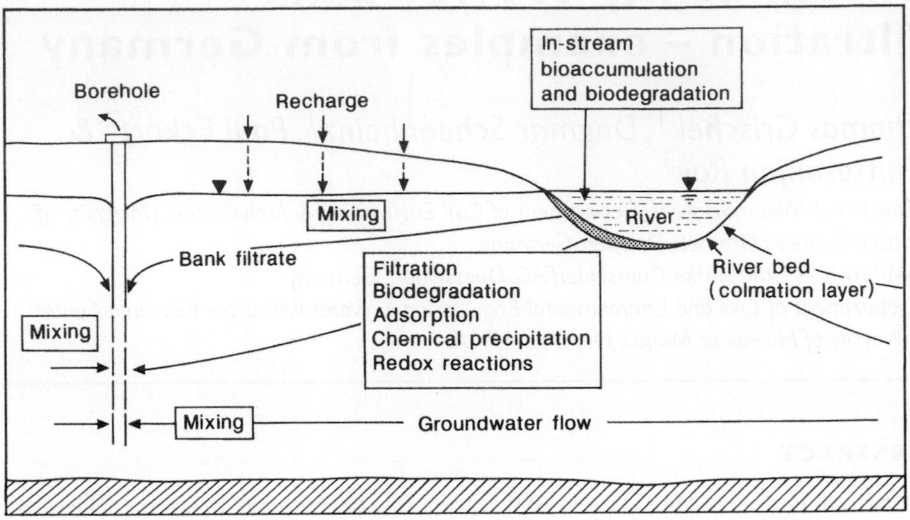

Figure 18.1 Schematic diagram of processes affecting water quality during bank filtration (reprinted from Hiscock & Grischek 2002, with permission from Elsevier).

Table 18.1 Characteristics of bank filtration sites in Germany (Kühn & Müller, 2005).

Condition	*Typical range*
Aquifer thickness	4 to 70 m
Hydraulic conductivity	0.0001 to 0.05 m/s
Distance bank – well	20 to 860 m
Well fields length along river	1 to 2 km
Travel times	3 days to 0.5 years

RBF in Germany provides about 8% of total drinking water supplies. The city of Duesseldorf, situated on the River Rhine, is entirely supplied with drinking water from RBF. In the Rhine basin, more than 20 million inhabitants receive drinking water which is directly or indirectly derived from river water, mostly via bank filtration. In the city of Dresden public water supply on average relies on up to approximately 32% bank filtrate and 66% surface water from reservoirs. Typical characteristics for RBF sites in Germany are given in Table 18.1.

18.2 HISTORY OF RIVER BANK FILTRATION IN DUESSELDORF AND DRESDEN

In the summer of 1866, there were 57 cases of cholera in the urban area of Duesseldorf. About half of those who contracted the disease died. This forced the town council to adopt a resolution to construct and operate waterworks. The English engineer William Lindley was called in to provide expert advice on the choice of location and planning

Figure 18.2 Photograph of vertical wells along the River Rhine at Duesseldorf-Flehe.

of the technical equipment. The first well field at Flehe of Duesseldorf, on the banks of the River Rhine, was put in operation for the first time on May 1, 1870 and has been continuously used since then. Up to that point in time, the population had obtained water from rainwater storage tanks, as well as from open and pumped wells. In the following years, the increasing water demands had to be met. Driven by the increasing population and the industrial water demands, the expansion of the water supply was the main task. In the period between 1948 and 1956, the water requirement almost doubled. While the increasing demand could be met by the continuous development of well fields, the simultaneous decrease of the river water quality posed an additional challenge (Eckert & Irmscher, 2006).

At present 600 000 inhabitants are supplied with treated bank filtrate by three waterworks. The water demands of about $55 \times 10^6\,m^3\,year^{-1}$ and up to $210 \times 10^3\,m^3 day^{-1}$ have to be met. The vertical wells and the horizontal collector wells are situated between 50 m and 300 m from the river bank. Figure 18.2 shows a line of vertical wells at Flehe waterworks that have been in operation since 1900.

The raw water is collected using a siphon system. Depending on the hydraulic situation, the residence time of the bank filtrate in the aquifer varies between one week and several months, determined from a monitoring cross-section as shown in Figure 18.3 (Schubert 2002).

In Dresden, three RBF waterworks exist. The first waterworks, the Dresden-Saloppe Waterworks was built between 1871 and 1875 on the bank of the River Elbe (Fig. 18.4). Drain pipes were installed near the river bank to abstract raw water. Due to geological boundary conditions, more than 90% of the abstracted water is bank filtrate. Today, the waterworks is still in operation and produces up to $12 \times 10^3\,m^3 day^{-1}$ for industrial water supply. Rising water demand at the end of the 1880s exceeded the capacity of the Dresden-Saloppe Waterworks. In 1891, the city council assigned the building officer, Bernhard Salbach, to write an expert report on the future water supply

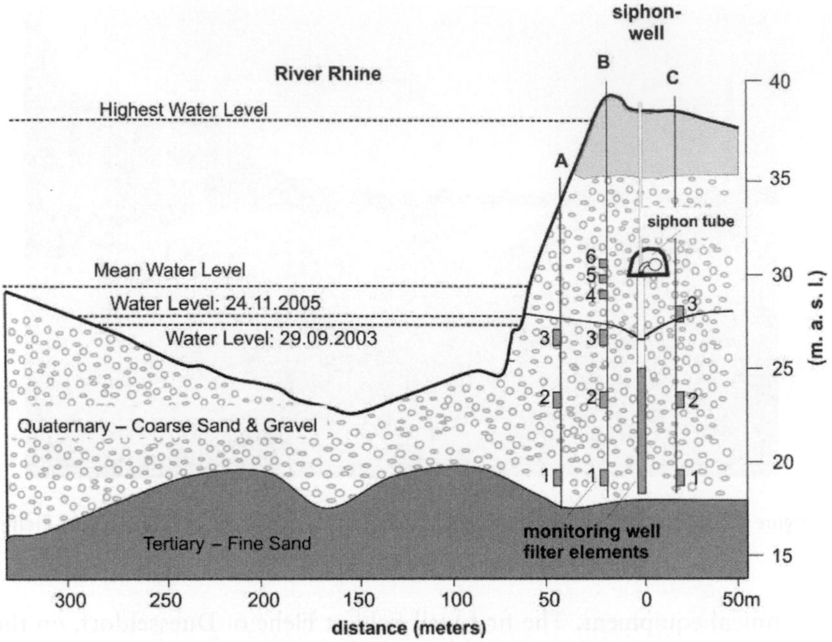

Figure 18.3 Monitoring cross-section at Duesseldorf-Flehe.

Figure 18.4 Photograph of Dresden-Saloppe Waterworks in 2009.

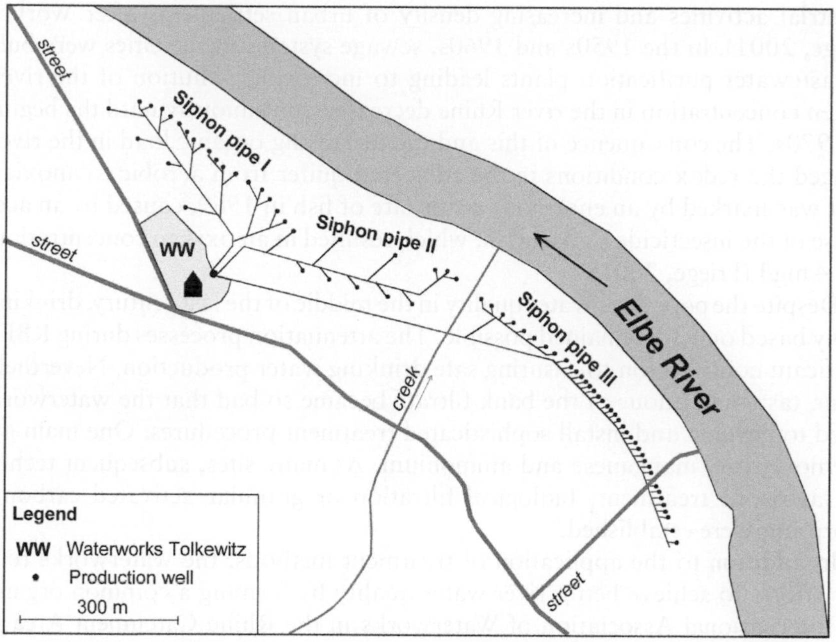

Figure 18.5 Location of the bank filtration scheme of Dresden-Tolkewitz Waterworks.

of the city. Salbach proposed a test well on the left bank of the river, which abstracted 4×10^3 $m^3 day^{-1}$ in 1891. Four more wells were completed in 1893 resulting in a total water abstraction from the left bank of 20×10^3 $m^3 day^{-1}$. Wells were connected using a siphon pipe and a collector well. Between 1896 and 1898, the second waterworks, the Dresden-Tolkewitz Waterworks, was constructed. A further rise in water demand resulted in the construction of four more wells and a second siphon pipe in 1901 to raise the capacity to 40×10^3 $m^3 day^{-1}$. In the 20th century, the number of wells was again increased and the water treatment facilities improved. Between 1919 and 1928 a third siphon pipe with 39 wells was built. Figure 18.5 shows the scheme with wells and siphon pipes. A significant decrease in the water demand after the reunification of Germany in 1989 allowed for the closure of the well fields in April 1992 in order to plan a general reconstruction to modernise the waterworks. After intensive construction, the Dresden-Tolkewitz Waterworks and the well fields were put into operation again in February 2000. The maximum capacity is now 35×10^3 $m^3 day^{-1}$.

18.3 RIVER WATER QUALITY

During the first 80 years (1870–1950), the quality of the River Rhine water in Germany permitted the production of drinking water without further treatment; the pumped bank filtrate had only to be disinfected. After 1950, the quality of the river water began to deteriorate. Increasing quantities and insufficient treatment of effluents from industry and domestic sources caused a noticeable drop in the oxygen concentration in the river water. In the Rhine valley, water pollution was caused by rapidly growing industrial activities and increasing density of urban settlements after World War II (Friege, 2001). In the 1950s and 1960s, sewage systems in the cities were built prior to wastewater purification plants leading to increasing pollution of the rivers. The oxygen concentration in the river Rhine decreased continuously until the beginning of the 1970s. The consequence of this and the increasing organic load in the river water changed the redox conditions in the adjacent aquifer from aerobic to anoxic. A low point was marked by an enormous death rate of fish in 1969, caused by an accidental release of the insecticide Endosulfan, which resulted in an oxygen concentration of less than 4 mg/l (Friege, 2001).

Despite the poor river water quality in the middle of the last century, drinking water supply based on RBF remained possible. The attenuation processes during RBF made a significant contribution to ensuring safe drinking water production. Nevertheless, the colour, taste and odour of the bank filtrate became so bad that the waterworks were forced to develop and install sophisticated treatment procedures. One main goal was to remove iron, manganese and ammonium. At many sites, subsequent technologies such as ozone treatment, biological filtration or granular activated carbon (GAC) adsorption were established.

In addition to the application of treatment methods, the waterworks reinforced their efforts to achieve better river water quality by forming a common organisation. The International Association of Waterworks in the Rhine Catchment Area (IAWR) was founded in 1970 in Duesseldorf. Its goal was to demand measures for water protection. In 1973, the IAWR published their first "Memorandum" on raw water quality that served as a "yardstick" for local government bodies and for public debate on Rhine water quality. Together with other stakeholders, such as environmental groups, the IAWR promoted a public discussion of water protection. At the end of this process the federal government issued its first programme for environmental protection which included measures to ensure that river water quality would reach a high standard within twenty years.

Furthermore, spectacular industrial spills underlined the need for remediation measures and pollution control. On November 1, 1986, a fire broke out in an agrochemical storage facility of a chemical plant in Basel, Switzerland. Insecticides, herbicides and fungicides were carried into the adjoining River Rhine with the fire-fighting water. The effects on the river were serious. On the stretch of the Rhine up to the Middle Rhine region, the entire stock of eel was destroyed. Other species of fish were also affected and damaging effects were detected on fish food organisms up to the mouth of the River Mosel. The question then arose, whether such a wave of poison could simultaneously contaminate the water source in the adjacent aquifer. This accident gave fresh impetus to the improvement of pollution control on the Rhine, and was the reason for projects aimed at understanding and managing the effects of accidental shock loads on RBF systems (Sontheimer, 1991).

The numerous measures taken to reduce nutrients and pollutants were consistent with the best available technology in wastewater treatment and production along the Rhine. Consequently, river water quality has improved significantly since the mid 1970's with the return of salmon to the river in 2000. The historical development of water pollution of the Rhine is illustrated in Figs. 18.6 and 18.7 by the concentration-time plots of oxygen and dissolved organic carbon (DOC).

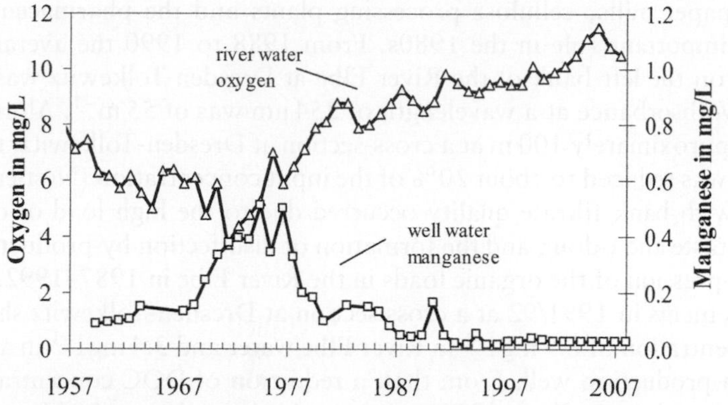

Figure 18.6 Oxygen concentration in the River Rhine water and manganese concentration in bank filtrate at Duesseldorf-Flehe.

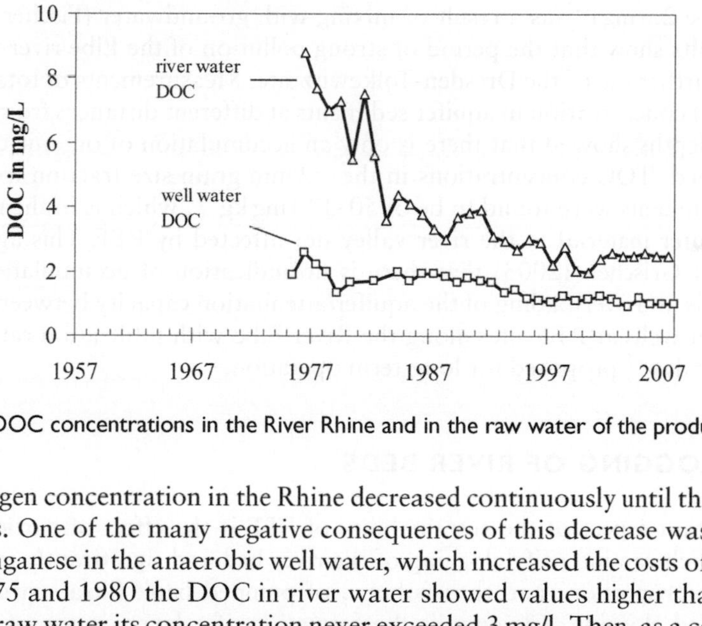

Figure 18.7 DOC concentrations in the River Rhine and in the raw water of the production wells.

The oxygen concentration in the Rhine decreased continuously until the beginning of the 1970s. One of the many negative consequences of this decrease was the occurrence of manganese in the anaerobic well water, which increased the costs of treatment. Between 1975 and 1980 the DOC in river water showed values higher than 7.5 mg/l, while in the raw water its concentration never exceeded 3 mg/l. Then, as a consequence of the restoration efforts, the DOC concentration decreased to a level between 2 and 4 mg/l and the oxygen concentrations returned to saturation level at the beginning of the 1990s. The higher oxidation capacity, combined with the lower oxygen demand of the infiltrating river water, led to more efficient natural attenuation processes within the aquifer. This, in turn, enabled the waterworks to reduce their treatment costs (Eckert & Irmscher, 2006). However, the occurrence of chemical pollutants in the river water, like pesticides and pharmaceuticals, remained an issue.

A similar situation has been reported for the River Elbe. The industries along the Upper Elbe River valley previously discharged a wide range of organic contaminants into the river. Hence, together with urban sewage, the DOC comprises a complex mixture of easily degradable and refractory substances. In addition to the industrial effluents, paper mills, cellulose processing plants and the pharmaceutical industry played an important role in the 1980s. From 1988 to 1990 the average DOC concentration on the left bank of the River Elbe at Dresden-Tolkewitz was of 24 mg l^{-1} and the UV-absorbance at a wavelength of 254 nm was of 55 m^{-1}. Along a flow path length of approximately 100 m at a cross-section at Dresden-Tolkewitz, the DOC concentration was reduced to about 20% of the input concentration (Nestler *et al.*, 1991). Problems with bank filtrate quality occurred due to the high load of organic pollutants, foul taste and odour, and the formation of disinfection by-products. Figure 18.8 gives an impression of the organic loads in the River Elbe in 1987–1992. Results from 17 measurements in 1991/92 at a cross section at Dresden-Tolkewitz showed a mean DOC concentration of 6.9 mg l^{-1} in River Elbe water and 3.4 mg l^{-1} in an observation well near a production well. From that, a reduction of DOC concentration of about 50% can be seen as an effect of RBF processes. Investigations in 2003 at the same cross section included 7 samples. In 2003 the mean DOC concentration in the River Elbe water was of 5.6 mg l^{-1} and of 3.2 mg l^{-1} in bank filtrate at the same observation well sampled in 1991/92. The mean DOC concentration in raw water from all wells was found to be 2.6 mg l^{-1} as a result of mixing with groundwater (Fischer *et al.*, 2006). These results show that the period of strong pollution of the Elbe river water did not limit the further use of the Dresden-Tolkewitz site. Measurements of total organic carbon (TOC) concentration in aquifer sediments at different distances from the river and different depths showed that there is only an accumulation of organic compounds in the river bed. TOC concentrations in the <2 mm grain size fraction from Tolkewitz aquifer sediments were found to be of 50–120 mg kg^{-1}, which is in the range of TOC in the aquifer material in the river valley not affected by RBF. This agrees with the findings of Grischek (2003) that there is no indication of accumulation of organic compounds and overloading of the aquifer/attenuation capacity between the river and abstraction wells at RBF sites along the River Elbe with infiltration rates lower than 0.2 $m^3\,m^{-2}day^{-1}$ proposed for long-term operation.

18.4 CLOGGING OF RIVER BEDS

An important aspect of the sustainability of RBF is the effect of particulate organic matter which can intensify clogging of the river bed and significantly reduce the well yield. The proportion, and thus volume, of pumped bank filtrate strongly depends on river bed clogging. Clogging is the formation of a layer on top of or within the river bed which has a lower hydraulic conductivity and reduces the flow rate of the filtrate through the river bed. It is the result of the infiltration and accumulation of both organic and inorganic suspended solids, precipitation of carbonates, iron- and manganese-(hydro) oxides and from biological processes. Erosive conditions in the river and floods limit the formation of a clogging layer by disturbing the river bed via increased flow velocity and shear stress. The permeability of clogged areas varies with the flow dynamics of the river. There are not only variations in the pressure head between the river and the aquifer but also variations in the concentration of suspended solids in the river water. The concentration of suspended solids in the River Rhine varies from 10 to more than 400 mg l^{-1} with an average concentration of less

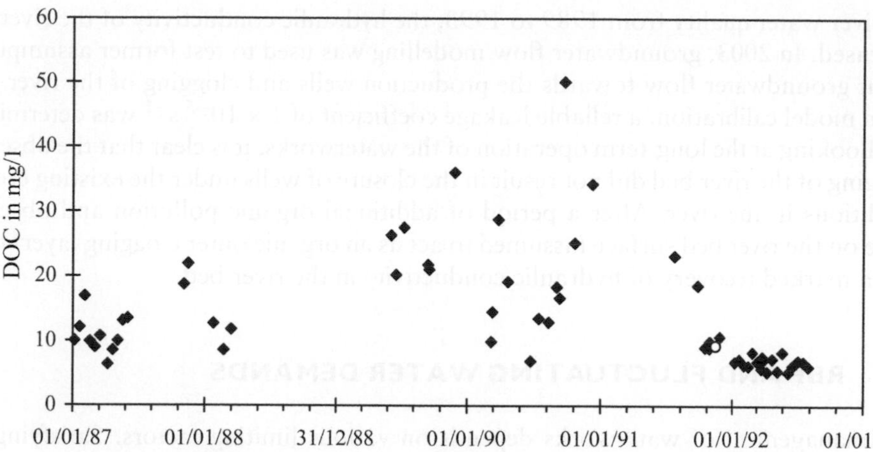

Figure 18.8 DOC concentration in the River Elbe water 1987–1992 (Grischek, 2003).

than 40 mg l^{-1}. Highest values appear in periods of rising water levels following storm events. Due to difficulties in determining the thickness of the clogging layer, the term leakage coefficient is introduced, which is defined as hydraulic conductivity of the clogging layer in metres per second divided by the thickness of the clogging layer in metres. Under specific conditions, the leakage coefficient can be calculated for RBF sites using water levels in the river and two observation wells positioned between the river and the production borehole using an analytical solution. Otherwise it has to be determined by calibration procedures in groundwater flow modelling. Based on water levels and known pumping rates, the leakage coefficients of the River Rhine in Duesseldorf and the River Elbe in Dresden were previously determined for different river stages and measuring campaigns and compared with historical data.

An early field study of the river bed adjacent to the Flehe Waterworks was done in 1953 and 1954 with a diving cabin. In 1987, a second study of the river bed at the Flehe Waterworks was carried out. This investigation revealed a zone of almost 80 m which had a fixed surface and was entirely clogged by suspended sediments (Schubert, 2002). The expansion of the clogged area is limited especially by bed load transport in the river. In regions with sufficient shear force, the deposits are removed. The zones at the Flehe site are characterised by different permeabilities. The infiltration occurs mainly in the middle of the river.

At Dresden-Tolkewitz, a significant decrease in groundwater levels was observed between 1914 and 1930 and attributed to river bed clogging by suspended materials caused by increased infiltration rates since 1901. In the 1980s severe river water pollution caused by organic compounds from pulp and paper mills in conjunction with high water abstraction led to unsaturated conditions beneath the river bed, especially at the Dresden-Tolkewitz Waterworks. However, investigations of river beds using a dive-chamber showed that the material responsible for the pore clogging in the gravel bed consisted of up to 90% inorganic materials (Heeger, 1987). Heeger (1987) calculated a leakage coefficient of about 1×10^{-4} s^{-1} for the river bed without bank filtration and a mean value of 5×10^{-7} s^{-1} at RBF sites in and around Dresden. After improvement of the river water quality from 1989 to 1993, the hydraulic conductivity of the river bed increased. In 2003, groundwater flow modelling was used to test former assumptions about groundwater flow towards the production wells and clogging of the river bed. From model calibration, a reliable leakage coefficient of 1×10^{-5} s^{-1} was determined.

Looking at the long-term operation of the waterworks, it is clear that the observed clogging of the river bed did not result in the closure of wells under the existing erosive conditions in the river. After a period of additional organic pollution and observed slime on the river bed surface (assumed to act as an organic outer clogging layer) there was a marked recovery of hydraulic conductivity in the river bed.

18.5 RBF AND FLUCTUATING WATER DEMANDS

The management of waterworks depends on various limiting factors. Satisfying the fluctuating water demands is of principal importance, although this is often opposed to the preferred continuous operation of wells that ensures stable flow conditions and abstraction rates. Another important factor is the mixing of river bank filtrate and groundwater to obtain an optimum quality with regard to raw water treatment. At most sites, the main aims of water quality management include achieving the maximum attenuation of organic compounds during aquifer passage and low concentrations of DOC, dissolved iron and nitrate in raw water. At all sites with long flow paths, mixing ratios of bank filtrate and groundwater were found to be of primary importance for the concentrations of DOC, nitrate, sulphate, dissolved iron and manganese in the abstracted raw water (Grischek *et al.*, 2010a).

Before 1990, the planning and construction of waterworks and well galleries in Germany was based on predicted water demands of up to 200 litres per capita per day. Since 1990 many waterworks, especially in East Germany, have been facing drastic reductions in water consumption. Mean water abstraction rates for public water supply decreased due to political changes, the "water price shock" after the re-unification of Germany, demographic changes, and changes in consumption patterns. In some regions, water use decreased by more than 50% within 10 years. Due to the expected demographic development, a further decline in water consumption and thus water production is expected. At many bank filtration sites, reduced water abstraction results in a lower portion of bank filtrate in the abstracted raw water. Thus, the quality of the groundwater becomes more important for the subsequent water treatment. To mitigate this effect, a reduction in the number of wells seems reasonable. However, since peak demands still have to be met, this option is not practical everywhere. In the short term, optimisation of production well operation is the most promising and probably the most economic way to handle higher portions of groundwater and changes in raw water quality due to changing mixing ratios (Grischek *et al.*, 2010a). This entails controlling the flow path length and travel time as well as mixing by selection of the most suitable wells from a well field. For that, a detailed investigation of groundwater flow conditions and portions of bank filtrate in the raw water is important for decisions concerning the most effective water quality management measures. During periods of low water demands, wells at Dresden-Tolkewitz are operated in a way that ensures a certain portion of bank filtrate (having low nitrate and sulphate concentration) in the pumped water to prevent high concentrations brought by the groundwater. Here, siphon pipe systems have the disadvantage of limited flexibility in well operation to achieve specific mixing ratios. But in Dresden-Tolkewitz three siphon pipes can be operated independently and single wells can be disconnected from the siphon pipes. Furthermore, this disadvantage concerning flexible well operation is compensated by significantly lower energy consumption of the whole system compared to a large number of vertical wells each equipped with a submersible pump.

18.6 RBF AND CLIMATE CHANGE

The future use of RBF waterworks requires an integrated assessment of the sustainability of bank filtration under changing boundary conditions, e.g. caused by potential climate change. Boundary conditions of bank filtration influenced by climate change are mainly the frequency, duration and peak behaviour of floods and droughts affecting the available water quantity, and the river water temperature resulting in changing biomass production and biological activity and thus influencing the water quality. The expected effects with respect to water quantity are: increasing drawdown or lower portions of bank filtrate abstracted due to the formation of clogging layers during low flow periods, contamination of production wells and damage of power supply systems during flooding (Schoenheinz & Grischek, 2010). Anticipated effects with respect to water quality of bank filtrate are:

- higher global radiation and higher temperatures resulting in an increase of algae growth, lower dissolved oxygen concentration and additional depletion due to degradation of biomass in the surface water,
- increased algae growth resulting in higher concentrations of TOC and DOC in the surface water,
- greater release of adsorbed, poorly degradable DOC components from the sediment due to higher water temperature,
- decrease of the dilution potential during low flow periods,
- longer retention times during low flow periods resulting in longer contact times for degradation of organic compounds,
- decreased attenuation potential of soil passage with respect to microbial and organic loads due to reduced travel times during flood periods.

To our present knowledge, the potential climate changes do not jeopardise the bank filtration effectiveness, although adaptation strategies have to be developed to account for an increase in extreme events (Grischek *et al.*, 2010b). To cover low flow periods, integrated water resources management becomes important. In the United States, horizontal collector wells have been installed with laterals directly beneath the river bed. Thus, there is sufficient water abstraction during low flow periods, but the pre-treatment effect is limited due to short retention times (several hours to a few days). Furthermore, high abstraction rates cause clogging of the river bed.

Low flow conditions are critical for RBF operations if the extraction rates per unit area of the river bed and river bank are high. From long-term experiences, an average infiltration rate of less than $0.2\,m^3\,m^{-2}\,day^{-1}$ over the river bed ensures limited clogging and stable infiltration conditions (Grischek *et al.*, 2007). Furthermore, the amount or percentage of induced river water infiltration during low flow conditions is an important requirement for obtaining operational permits for bank filtration sites. As an example, the maximum pumping rate of the wells for waterworks employing RBF along the River Elbe in the city of Dresden is of 97×10^3 m^3 day^{-1}. At low flow conditions in summer, the river discharge is about 8600×10^3 m^3 day^{-1}, thus the induced infiltration reduces the river discharge by 1.1% in the worst case. A few kilometres downstream from the RBF site, treated effluent is discharged back to the River Elbe.

Mixing ratios of river bank filtrate and groundwater as well as travel times can be determined using tracers such as chloride, boron and oxygen-18. Measurements of oxygen-18 were used successfully to determine the portion of bank filtrate at RBF sites at an island in the River Danube (Stichler *et al.*, 1986), and to calculate travel times of bank filtrate at Torgau at the River Elbe (Trettin *et al.*, 1999).

Eckert *et al.* (2006) studied the potential impact of droughts in the River Rhine on the operation of the siphon system at Duesseldorf-Flehe. Due to the foresighted construction of the siphon system and the depth of the siphon pipe of 9 m b.g.l., the wells can be operated even at very low water levels in the river. Nevertheless, effects of long-term erosion of river beds and increased drawdown caused by well clogging have to be carefully checked for RBF sites to cope with future droughts.

Sprenger *et al.* (2010) evaluated the vulnerability of bank filtration systems to climate change. They state that bank filtration is vulnerable to climate change, but less vulnerable than surface water or groundwater abstraction alone, as RBF uses water from two sources. Furthermore, they conclude that only bank filtration systems comprising an oxic to anoxic redox sequence ensure maximum removal efficiency due to the redox-dependent degradation rate of many contaminants. Schoenheinz & Grischek (2011) discussed the effect of climate change on the removal of DOC during bank filtration and gave an overview of possible changes in boundary conditions for bank filtration sites as a consequence of the anticipated climate change and their effects on the quality of bank filtrate and optional adaptive measures.

18.7 IMPLICATIONS FOR SUSTAINABILITY

Three major advantages the RBF systems possess compared to traditional water treatment plants are: (a) minimization of energy and other resource use and waste generation, (b) lower environmental impact, and (c) operational flexibility. Klein *et al.* (2005) reported that 20% of a community's total energy use is associated with the treatment, conveyance, and delivery of water. Greenhouse gas emission (GHG) can be reduced when an elaborate treatment unit is eliminated in favour of RBF. Energy intensive coagulation, flocculation, sedimentation, and membrane filtration processes are typically avoided with RBF. Elimination of the coagulation and flocculation process reduces the need for sludge disposal.

In the United States, in many RBF systems water is simply pumped and disinfected prior to distribution. The footprint of RBF systems is typically smaller than that of full-scale treatment plants. In the case of ecologically sensitive river systems, RBF wells draw a portion of groundwater, and this reduces the stream flow to less than that for a surface water treatment plant of an equal size.

RBF systems in the case of rivers with highly variable flow can use infiltration ponds or inflatable dams to produce water. Systems that are fully dependent on surface water need a storage reservoir and elaborate intake structures for steady flow to the treatment plant. These reservoirs and intakes are not only expensive but also environmentally unpopular. Aquifer storage and recovery (ASR) can be used to store excess water in aquifers during periods of high river flows. When the river is dry or near base flow, the ASR wells can pump the stored water from aquifers to supply the population.

18.8 CONCLUSIONS

Two examples from Germany – the Lower Rhine region and the Upper Elbe River – have been presented, where RBF has been employed for more than 140 years. During this time the RBF systems were able to overcome extreme conditions with respect to poor river water quality, and to withstand contaminant spills in the rivers. Drain pipes at the Dresden-Saloppe Waterworks have been in operation for more than 140 years whilst four production wells at the Dresden-Tolkewitz Waterworks had to be replaced after 60 years. In Dresden, severe clogging of the river bed occurred in the 1980s mainly due to high loads of organic compounds from pulp and paper mills upstream. Following improvement of the river water quality in the 1990s, no problems with river bed clogging or foul taste and odour have been encountered.

Field studies are part of ongoing efforts to evaluate the risks of RBF and to obtain knowledge of the best practice for sustainable operation of bank filtration plants. Raw water quality and treatment are optimised by managing specific mixing ratios of bank filtrate and groundwater. Pumping rates can be reduced to get longer retention times in the aquifer and higher attenuation rates of organic compounds. No indication of a decrease in attenuation capacity of the aquifer over time was observed. Long-term experiences and results of the evaluation of historic and recent data and of investigations using modelling tools strongly indicate that RBF is a sustainable option for water supply in Germany.

There is an enormous potential for wider use of RBF worldwide, especially given that the removal of microbial pathogens from surface water through RBF would be a crucial factor (Ray, 2008; Sandhu *et al.*, 2010). Thus, it could serve as a preferable alternative to direct river water intake. At a minimum, bank filtration acts as a pre-treatment step in drinking water production. In some instances, it can serve as the final treatment just before disinfection. High-quality drinking water is not the only long-term benefit of RBF, as it also leads to reduced medical costs and improved productivity for the consumer. Bank filtration also serves as an asset to water suppliers by way of capital cost reduction and lower maintenance, improved reliability of source water and enhanced community supply by reduction of total dissolved solids concentrations. Nevertheless, the application and adaptation of RBF is very much site-specific and demands careful investigation on hydrological, hydrogeological, hydrochemical and hydrobiological conditions, especially clogging of river or lake beds and redox reactions in the aquifer.

Bank filtration is an important technology for producing drinking water and will continue to be so in the foreseeable future due to favourable regulatory provisions (Ray, 2008). It can serve as a pre-treatment or full treatment step for waterworks.

The efficiency of RBF is very much site-specific and depends on design and operation features. Experiences from existing sites could be used as a benchmark for planning new sites as is being done in India and Russia, but have to be adapted to the local conditions, e.g. climate, river morphology and water quality. Further international scientific collaboration is needed to provide adaptable guidelines for design and operation of RBF sites.

ACKNOWLEDGEMENTS

The authors are grateful to the German Federal Ministry for Education and Research (BMBF, IND 08/156) and the Ministry for Science and Art, Saxony (grant no. 475316002512007) for funding RBF research projects.

REFERENCES

Eckert P., Lamberts R., Irmscher R. (2006) Einfluss extremer Niedrigwasserereignisse im Rhein auf den Betrieb einer Heberbrunnenanlage [Impact of extreme droughts in the River Rhine on the operation of a siphon system]. *DVGW Energie-Wasser Praxis* 12:20–23.

Eckert P., Irmscher R. (2006) Over 130 years of experience with river bank filtration in Duesseldorf, Germany. *J Water SRT – Aqua* 55(4):283–291.

Fischer T., Day K., Grischek T. (2006) Sustainability of riverbank filtration in Dresden, Germany. In: UNESCO IHP-VI Series on Groundwater No. 13, Recharge systems for protecting and enhancing groundwater resources. *Proc. Int Sym Management of Artificial Recharge*, 11–16 June 2005, Berlin, 23–28.

Friege H. (2001) Incentives for the improvement of the quality of river water. *Proc. Int. RBF Conf.*, Duesseldorf, Germany, 2–4 Nov. 2000, IAWR-Rheinthemen 4:13–29.

Grischek T., Schoenheinz D., Syhre C., Saupe K. (2010a) Impact of decreasing water demand on bank filtration in Saxony, Germany. *Drink Water Eng Sc* 3:11–20.

Grischek T., Schoenheinz D., Herlitzius J., Kulakov V. (2010b) River bank filtration in towns and communities of the XXI century: Design and operation. Proc. *IWA Specialist Conference on Water and Wastewater Treatment Plants in Towns and Communities of the XXI Century: Technologies, Design and Operation*, 2–4 June 2010, Moscow.

Grischek T., Schubert J., Jasperse J.L., Stowe S.M., Collins M.R. (2007) What is the appropriate site for RBF? In: Fox P (ed.) Management of aquifer recharge for sustainability. *Proc. ISMAR 6, Acacia, Phoenix*, 466–474.

Grischek T. (2003) Zur Bewirtschaftung von Uferfiltratfassungen an der Elbe [Management of bank filtration sites along the River Elbe]. *PhD thesis, Department of Forestry, Geo- and Hydrosciences*, Dresden Univ Technology.

Grischek T., Schoenheinz D., Worch E., Hiscock K.M. (2002) Bank filtration in Europe – An overview of aquifer conditions and hydraulic controls. In: Dillon P (ed.) *Management of aquifer recharge for sustainability. Swets and Zeitlinger*, Balkema, Lisse, 485–488.

Heeger D. (1987) Untersuchungen zur Kolmationsentwicklung in Fließgewässern [Investigations on clogging of river beds]. *PhD thesis*, Bergakademie Freiberg.

Hiscock K., Grischek T. (2002) Attenuation of groundwater pollution by bank filtration. *J. Hydrol.* 266:139–144.

Klein G., Krebs M., Hall V., O'Brien T., Blevins B.B. (2005) California's water–energy relationship. Final staff report. California Energy Commission. http://www.energy.ca.gov/2005 publications/ CEC-700-2005-011/CEC-700-2005-011-SF.PDF.

Kuehn W., Mueller U. (2000) River bank filtration – an overview. *J AWWA* 92(12):60–69.

Kühn W., Müller U. (2005) Exportorientierte Forschung und Entwicklung auf dem Gebiet der Wasserver- und –entsorgung, Teil 1 Trinkwasser [Export oriented research and development in the field of water supply and treatment, Part 1 Drinking water]. DVGW/BMBF report, Germany.

Nestler W., Socher M., Grischek T., Schwan M. (1991) River bank infiltration in the Upper Elbe River Valley – hydrochemical aspects. IAHS Publ 202:247–356.

Ray C. (2008) World wide potential of river bank filtration. *Clean Techn. Environ. Policy* 10(3):223–225.

Ray C., Grischek T., Schubert J., Wang Z., Speth T.F. (2002) A perspective of river bank filtration. *J AWWA* 94(4):149–160.

Sandhu C., Grischek T., Kumar P., Ray C. (2010) *Potential for Riverbank filtration in India. Clean Techn Environ Policy*, DOI: 10.1007/s10098-010-0298-0 (online first).

Schoenheinz D., Grischek T. (2010) River bank filtration under extreme climate conditions. In: Ray C, Shamrukh M (eds.) Riverbank filtration for water security in desert countries. *Springer Science+Business Media B.V.*, 51–67.

Schubert J. (2002) Hydraulic aspects of river bank filtration–field studies. *J Hydrol* 266: 145–161.

Sontheimer H. (1991) Trinkwasser aus dem Rhein? [Drinking water from the River Rhine?]. Academia, Sankt Augustin (in German).

Sprenger C., Lorenzen G., Hülshoff I., Grützmacher G., Ronghang M., Pekdeger A. (2011) Vulnerability of bank filtration systems to climate change. *Sci Total Environ* 409:655–663.

Stichler W., Maloszewski P., Moser H. (1986) Modelling of river water infiltration using oxygen-18 data. *J Hydrol* 83:355–365.

Trettin R., Grischek T., Strauch G., Mallen G., Nestler W. (1999) The suitability and usage of ^{18}O and chloride as natural tracers for bank filtrate at the Middle River Elbe. *Isotopes Environ Health Stud* 35:331–350.

Tufenkji N., Ryan J.N., Elimelech M. (2002) The promise of bank filtration. *Environ Sc Technol* 36:422A–428A.

Chapter 19

Comparison of established methods to determine infiltration rates in river banks and lake shore sediments

Marcus Soares[1], *Günter Gunkel*[1] *& Thomas Grischek*[2]

[1] *Berlin University of Technology, Department of Water Quality Control, Berlin, Germany*

[2] *Division of Water Sciences, University of Applied Sciences Dresden, Dresden, Germany*

ABSTRACT

Lakes and rivers have been and continue to be major sources of water supply. In many countries of the world, alluvial aquifers hydraulically connected to water courses are preferred sites for drinking water production. Lake shore and river bank filtration has been used for drinking water supply e.g. in Germany, Finland, the Netherlands, Hungary, India and Brazil. The determination of infiltration rates is necessary for optimal design and operation of bank filtration sites. Various methods have been used during the last few years to answer questions about infiltration regimes during bank filtration. This study provides a comparison of established methods to determine infiltration rates in littoral sediments.

19.1 INTRODUCTION

For more than 130 years, rivers and lakes have been used as a source for bank filtration in Europe. Many developing countries such as Brazil, India and Egypt have also used this system for drinking water supply in the last few years due to its proven efficiency in removal of pathogens and persistent contaminants. The amount of river bank filtered water is dependent on a variety of factors, including stream bed permeability, stream bed thickness, river elevation, water viscosity and pumping rate. Fischer *et al.* (2005) reported that proportion (i.e. land side versus stream side) and volume of bank filtrate strongly depend on riverbed clogging. Deposition of fine-grained sediments on the stream bed normally happens during periods of low stream velocities. In addition, algal growth in higher temperatures can result in deposition of organic matter at the bottom of the stream. This increases streambed thickness and reduces bed permeability. During lower temperatures water viscosity increases, thus reducing water flow through the sediments. Increased flow velocity during floods causes cleaning of the river bed and re-suspension of sediments, thereby decreasing the thickness of the stream bed and raising bed permeability.

Interest in analysing infiltration rates has increased due to the need for water management at sites with high abstraction rates as well as sites with decreasing abstraction rates. At the same time, a variety of techniques and methods have been developed to examine and monitor infiltration of surface water. An improved understanding of the connection between surface and groundwater is viewed as an important prerequisite to effectively managing these resources (Sophocleos, 2002).

This chapter provides a comparison of common established techniques to determine infiltration rates. Methods using Rn-222, fluorescent tracers such as melamine resin particles and air-dried FPOM (Fine Particulate Organic Matter) from alder leaves, heating, the use of monitoring wells, and seepage meters will be discussed and evaluated.

19.2 ESTABLISHED METHODS

19.2.1 Temperature gradient

The continuous exchange of water between surface and groundwater provides an opportunity to use heat as a natural tracer to estimate infiltration rates indirectly. Many authors have used this method to trace water exchange between surface and groundwater in recent years (e.g. Stonestrom & Constantz, 2004), to delineate flows in the hyporheic zone (Constantz *et al.*, 2003a; Constantz *et al.*, 2003b; Conant, 2004) and also in bank filtration studies (Doussan *et al.*, 1994; Wang, 2002; Mutiti & Levy, 2010). Other uses are related to recover water content information (Behaegel *et al.*, 2007) and estimate groundwater advection rates in coastal estuaries (Land & Paull, 2001).

Use of heat as a tracer relies on measurement of temperature gradients. Water and sediment temperature can be measured either directly by using a probe driven into the ground or indirectly by measuring water temperature at different depths in an observation well (Fig. 19.1). In both cases it is necessary to use a probe connected to a data logger that will take readings at least every hour. The measurements are quick,

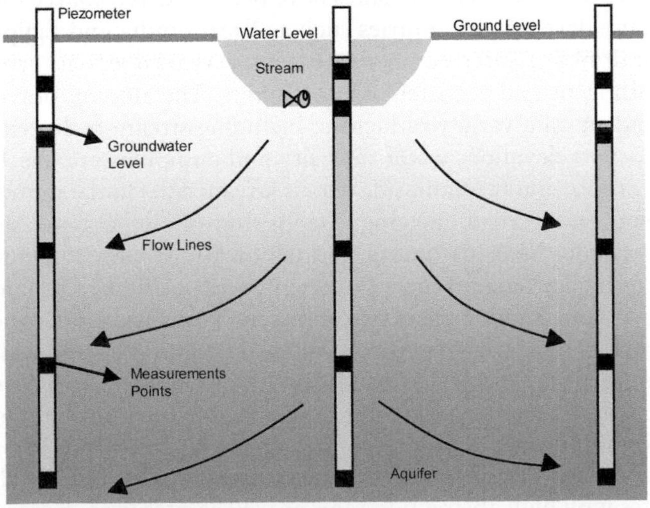

Figure 19.1 Temperature measurements at different depths in a losing stream (adapted from Stonestrom & Constantz 2004).

low-cost and easy to perform, making this method very attractive. Stonestrom & Blasch (2003), for instance, described some types and characteristics of temperature sensors that could be used to assess temperature gradients in groundwater-surface water studies. Conant (2004) identified different types of water flow within the stream bed by measuring water temperature and subsurface water temperature using only data loggers, and developed a conceptual model of water movement within the stream bed. The effects of temperature changes on hydraulic conductivity are cited in literature extensively (Constantz, 1998; Stonestrom & Constantz, 2004; Constantz *et al.*, 2006; Mutiti & Levy, 2010). Constantz (1998) reported a change of 11% at a gaining stream site and 30% at a losing stream site in diurnal stream flow variations in a single month; fluctuation in infiltration rates in different seasons must be considered not only because of the gradient difference between the river water and a measurement point in the river bed, but also because water density (ρ_w) and viscosity (μ_w) depend on temperature. During infiltration of surface water into groundwater and discharge of groundwater into surface water temperature changes occur due to advective heat transport (heat transport due to the fluid's movement in a particular direction) and conductive heat transport (heat transport by heat conduction through the solid and fluid phase of the sediment) (Kipp, 1987; Conant, 2004; Anderson, 2005). In a stream bed, the vertical and horizontal distribution of the temperature is a function of both processes, the conductive and advective heat transport.

According to Domenico & Schwartz (1998), the 3-D heat transport equation is based on variables such as temperature (T in K), time (t), porosity (n), fluid density (ρ in kg/m^3), specific heat of the fluid (c in J/(g · K)), and effective thermal conductivity of the fluid (k_e in W/(m · K)). The term represents a 3-D vector and ρ', c' is the heat removed or gained from the unit volume when temperature changes by one degree (eq. 19.1).

$$k_e \nabla^2 T - n\rho_{w'} c_{w'} v \cdot \nabla T = \rho' c' \frac{\partial T}{\partial t} \tag{19.1}$$

It assumes steady groundwater flow and that the variables ρ, c, k_e and n are constant. Water temperature and solid temperature are also assumed to be equal. In the absence of fluid movement the term thermal dispersivity (λ) appears as

$$\lambda = \frac{k_e}{\rho' c'} \tag{19.2}$$

where λ is thermal dispersivity (m^2/s); k_e is effective thermal conductivity (W/(m · K)); ρ is fluid density (g/m^3); c is specific heat (J/(g · K)); ($\rho' c'$) is the effective heat capacity of the unit volume.

Equation 19.2 is based on the Fourier's Law, which describes the conduction of heat from areas of higher temperature to lower temperature, associated with the forced convective transport (based on Darcy fluid motion). Detailed information about equation 19.1, as well as the assumptions, which must be taken into account are given by Kipp (1987). The first term of equation 19.1, where effective thermal conductivity appears, describes heat transport for conduction and dispersion. This term includes parameters such as effective thermal conductivity, soil porosity, thermal conductivity of the fluid and thermal conductivity of the soil grains, because k_e includes effects of conduction and thermal dispersion (Anderson, 2005). The second term describes heat transport through advection or convection in groundwater, which is divided into forced convection, represented by a dimensionless Nuselt number, or the commonly used Peclet number, and free convection that is apparently much less influential than the forced convection in sedimentary aquifers (Conant, 2004; Anderson, 2005).

Many models of heat and groundwater transport have been developed and applied in recent years, such as: FEFLOW (Kolditz *et al.*, 1998; Trefry & Muffels, 2007; Diersch & Kolditz, 2008), ROCKFLOW (Kolditz *et al.*, 1998), HST3D (Kipp, 1987), SUTRA (Voss, 1984), STRIVE (Anibas *et al.*, 2011), and VS2DH (Constantz, 1998; Rosenberry & LaBaugh, 2008). A more detailed example of estimating infiltration rates is a computer modelling program from the U.S. Geological Survey called VS2DH, used extensively nowadays, in which stream bed temperature profiles are defined as input by the user. This two-dimensional simulation code is based on inverse modelling to match simulated temperatures against measured temperatures to estimate heat and water fluxes into or out of the stream bed. This model has been used to simulate diurnal variations in stream bed temperature, moisture content and infiltration in the beds of ephemeral streams. It can also be used for any kind of streambed covering (Ronan *et al.*, 1998), although Rosenberry & LaBaugh (2008) do not recommend temperature measurements in places with very low hydraulic gradients and/or extensive clay-textured streambeds, due to interference with gradient measurements.

19.2.2 Seepage meters

This method is one of the most commonly used for measuring infiltration rates directly at the sediment-water interface. Installation of a seepage meter is normally simple. First, a plastic bag is filled with water and its weight obtained using a balance. In the field, the seepage meter is installed manually in the top few centimetres of the soil (Fig. 19.2). After installation, the bag is attached to the half-cylinder under water, and at the end of the experiment the bag is reweighed. The volumetric flow rate through the infiltration chamber is obtained by observing the weight change (change in volume) during the experiment and dividing it by the area of the chamber mouth to yield the infiltration rate. High groundwater level together with low water level in the river can

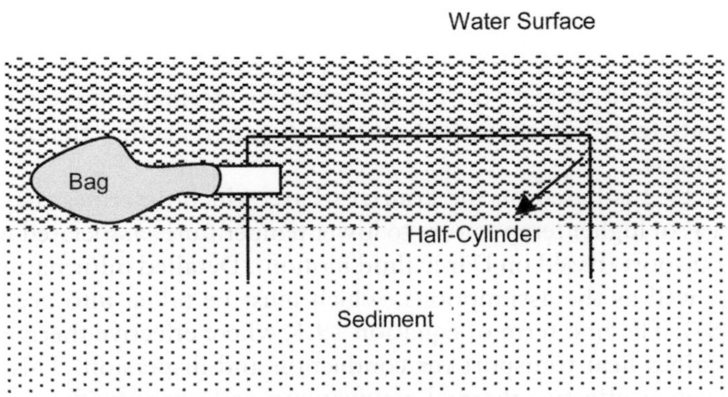

Figure 19.2 A typical seepage meter used to estimate infiltration rate (adapted from Rosenberry & LaBaugh, 2008).

lead to exfiltration conditions being registered by seepage meters, if sufficiently flexible plastic bags are used. According to some authors, seepage meters can measure natural water fluxes as low as 0.01 cm/d (Cable *et al.*, 1997) and as high as 12 cm/d (Woessner & Sullivan, 1984). High infiltration rates of up to 43 cm/d were found in lake shores with bank filtration facilities (Gunkel *et al.*, 2009). Some problems in using seepage meters are due to placement of the chamber in the sediment (e.g. Lee, 1977; Kalbus *et al.*, 2006). Gas emission, mainly methane, can lead to a malfunction, but using acrylic glass as the chamber material with an additional valve on top enables removal of gas bubbles.

A rapid attachment of the meter in the sediment can form blowouts of gases and for a short time excessive pressure. After installing the seepage meter, it is recommended to wait some time before the first measurement because the sediment and the flow need to equilibrate before the bag is attached. Landon *et al.* (2001) suggested waiting between 10 and 15 minutes for sandy sediments before making the first measurement, and Cey *et al.* (1998) left the seepage meter overnight in order to reach equilibrium. An occasional cleaning of the inflow/outflow tube to avoid algae and shell growth can be necessary as well. Additionally, Rosenberry & Labaugh (2008) recommended care to ensure a good seal between the chamber and the sediment, especially when inserting the meter into irregular sediments that are rocky and woody. Lee & Cherry (1978) state that 8 centimetres is normally enough to achieve a good seal if the sediment is not too soft or irregular. The bag properties are also of significance for a good working seepage meter. Cable *et al.* (1997) and Shaw & Prepas (1989) have found anomalies in measurements during the first few minutes when using empty bags; pre-filling of the bag with water can solve not only this problem but also problems related to flexing/folding the bag which might cause positive/negative flow (Cable *et al.*, 1997; Shinn *et al.*, 2002). Murdoch & Kelly (2003) estimated an error factor that should be used for different collection bags, which is related to a mathematical equation to estimate performance of seepage meters. In contrast, Isiorho & Meyer (1999) have not found any influence of bag type on measurements.

Sediments in lakes and rivers are very irregular and small-scale variability in physical and chemical parameters is common. Thus, the clogging layer thickness can vary significantly in a small area (Hubbs, 2004), which means that seepage meter diameter might have an influence on measurements. Shin *et al.* (2002) indicate that the smaller the diameter of the seepage chamber the higher the infiltration variance, and Rosenberry & LaBaugh (2008) pointed out that a chamber with larger diameter can more precisely measure variance of small flows, although it is difficult to ensure a good seal between the sediment and the meter. Libelo & MacIntyre (1994) evaluated the effects of surface water velocity on seepage meter measurements, and found that changes in pressure caused by flow of surface water in streams can occur, consequently also within a seepage meter. The lowering of the water pressure is transmitted to the bag and to the seepage meter, and this pressure decrease can induce flow through the sediment-water interface. In this way, it might affect seepage-flux measurements by 50% or more, depending on water velocity or currents. Covering the collection bag in order to protect from flow-induced pressure gradients can improve the measurements significantly. Seepage meters have been modified through the years in order to adapt them to different environments and measurement requirements, as shown in Table 19.1.

Table 19.1 Modified seepage meters for application in different environments.

References	*Device type/Characteristics*
Cherkauer & Mcbride, 1988	Seepage meter able for use in large lakes with strong hydraulic currents
Taniguchi & Fukuo, 1993	Automatic seepage meter using a heat pulse method able to measure continuously the water flow
Paulsen *et al.*, 2001	Ultrasonic flow meter to be applied in submarine environment
Sholkovitz *et al.*, 2003	Automated dye-dilution-based seepage meter to quantify groundwater flow and seawater infiltration
Rosenberry & Morin, 2004	Electromagnetic seepage meter that provides continuous measurements
Rosenberry, 2008	Modified seepage meter for use in flowing water
Gunkel *et al.*, 2009	Small plexiglas infiltration chamber for use in lakes and deep water by divers

19.2.3 Piezometers

A common method to estimate flow of groundwater through lake/stream sediments at a bank filtration sites is the use of monitoring wells or piezometers, or the segment approach method (Rosenberry & LaBaugh, 2008), in order to determine infiltration/exfiltration flows. Piezometers are simple and provide local-scale results of water levels to determine surface water exchange into or out of streams and lakes. This method also offers a good idea of the dynamic nature of in- and exfiltration flow in the field, as shown by Oxtobee & Novakowski (2002).

Darcy's equation (eq. 19.3) is commonly used to calculate groundwater flow:

$$Q = KA\frac{dh}{dl} \tag{19.3}$$

where Q is the hydraulic flow or groundwater flux (L^3/T), K is the horizontal hydraulic conductivity (L/T), A is the area through which occurs the flux (L^2), dh/dl is the hydraulic gradient (−).

Small differences in hydraulic head relative to surface water can be read after installing a piezometer. This difference can be above or below the surface water depending on the water flow direction. This differential head Δh is used to calculate the vertical hydraulic gradient $\Delta h/\Delta l$, where Δl is the distance of the piezometer screen beneath the sediment/water line. In order to obtain the flow rate, it is necessary to measure not only the head gradient but also permeability. To observe temporal variations in hydraulic head, data loggers or pressure transducers might be used (Kalbus *et al.*, 2006). Piezometers are normally installed in boreholes drilled by different means: rotary drill, wash boring, auger on hammering (Lee & Cherry, 1978; Baxter *et al.*, 2003) (see Fig. 19.3), but a good embedding must be ensured. Different shapes are described e.g. by Hoffmann (2008), who used an expanded piezometer to determine water drawdown, or even more sophisticated piezometers which use an attached manometer (Winter *et al.*, 1988). According to Obbink (1969), the method of installation varies with the piezometer depth, diameter of the pipe

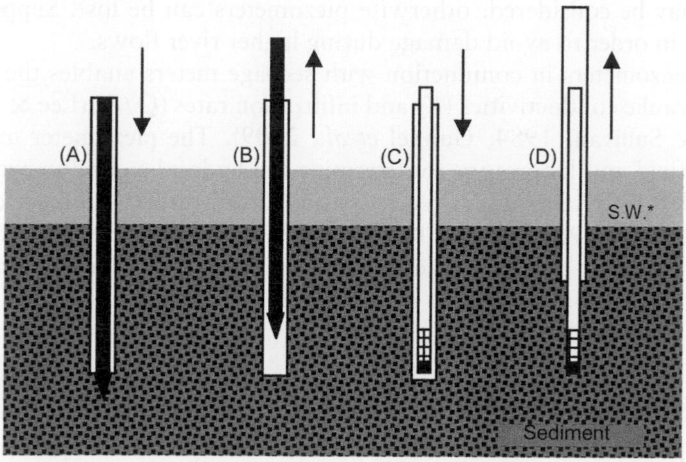

Figure 19.3 Mini-piezometer installation procedure proposed by Baxter *et al.* (2003). From left to right: (A) insertion of a capsule with the driver rod by hammering; (B) driver removal; (C) piezometer insertion into the capsule, and (D) capsule removal from the streambed (adapted from Baxter *et al.*, 2003); *S.W. – surface water.

and soil type. Depth and pipe diameter can vary considerably from a few centimetres to many metres (Lee & Cherry, 1978; Belanger & Mikutel, 1985; Morice *et al.*, 1997; Gunkel *et al.*, 2009; Grischek *et al.*, 2010). Small piezometers or the so-called mini-piezometers are normally made of plastic or metal tubes with a perforated tip rolled with a nylon mesh netting or fibreglass cloth (Lee & Cherry, 1978; Cey *et al.*, 1998; Baxter *et al.*, 2003). Actually, the material depends on the sediment characteristic, as noted by Baxter *et al.*, 2003. The same technique has been used in the past for bigger piezometers (Obbink, 1969). Today, however, bigger piezometers normally require different techniques not only during drilling but also during installation, as stated by LWBC (2003) which offers some guidelines for piezometer construction. The difference between both piezometers is that mini-piezometers are normally designed to monitor shallow groundwater conditions, as opposed to deeper aquifers measured by the bigger ones.

Some factors, such as heterogeneity, positioning of monitoring wells and accuracy of hydraulic gradient readings can influence measurements (Cey *et al.*, 1998; Baxter *et al.*, 2003). The waves caused by the wind in lakes or by strong currents in rivers can disturb hydraulic gradient readings from the stream surface. When the difference between the stream surface and the drawdown are on the order of just a few centimetres, even a tiny wave can disturb the measurement. To reduce this error and get accurate and precise readings Baxter *et al.* (2003) developed a "stilling tube" attached to the piezometer. Problems associated with piezometer installation and maintenance must also be considered. Geist *et al.* (1998) found that installation by driving or vibrating causes less sediment disturbance and the seal between the mesh netting improves considerably. But nevertheless sealing decreases with smaller diameter of the piezometer and coarser sediment structure. In addition, Geist *et al.* (1998) warn that in rivers strong water level variations and boat traffic can interrupt measurements. An extension pipe may be considered; otherwise piezometers can be lost. Support pipes also can be used in order to avoid damage during higher river flows.

Using piezometers in conjunction with seepage meters enables the estimation of aquifer hydraulic conductivities (K) and infiltration rates (Q/A) (Lee & Cherry, 1978; Woessner & Sullivan, 1984; Gunkel *et al.*, 2009). The piezometer method is used also in the field and laboratory to determine the hydraulic conductivity in sediment cores sampled from lakes and rivers, especially undisturbed sediment cores taken in deep lakes (DEV, 1998; Lang, 2008). Piezometers and mini-piezometers can be used for additional purposes such as thermal data collection and water chemistry analysis (Doussan *et al.*, 1994; Grischek *et al.*, 1997; Trettin *et al.*, 1999).

19.2.4 Radon-222 isotope

Many isotopes have been used to assess exfiltration of groundwater into surface water and vice-versa. For example, ^{14}C has been used to measure the flow of groundwater on the order of years because of its half-life of approximately 5730 years. ^{3}H has a half-life of approximately 12.3 years and can be used to study groundwater in the span of months. Radon (^{222}Rn) is an inert natural gas found in different concentrations in groundwater and surface water and has a half-life of approximately 3.82 days. It is the decomposition product of the isotope ^{226}Ra, which also occurs naturally in soil. In groundwater it is in equilibrium with ^{226}Ra, but in surface water its high volatility limits its presence. As a consequence, it can be used as a natural tracer to estimate infiltration rates indirectly and/or to identify areas of significant groundwater discharge, based on its equilibrium concentration and retention time. Up to now some investigations have demonstrated the potential of this method concerning analytical determination (Gundersen, 1992; Burnett *et al.*, 2001; Lee *et al.*, 2010), applications (Bertin & Bourg, 1994; Macheleidt *et al.*, 2006; Kluge *et al.*, 2007) and assessing residence times in the hyporheic zone (Hoehn & Cirpka, 2006) (see Fig. 19.4).

For application of this method, homogeneity of ^{222}Rn emanation in soil is assumed. The emanation power is the fraction of radon generated in the materials that can be released in soil pore volume (Macheleidt *et al.*, 2006). During infiltration from surface water into the aquifer, radon concentrations from grains are released into the water until an equilibrium concentration is reached. Based on the increase of radon concentration, water retention time can be obtained from (Macheleidt *et al.*, 2006):

$$t = \lambda^{-1} \cdot \ln \frac{A_e - A_0}{A_e - A_t} \tag{19.4}$$

where A_t is radon concentration at the time t (Bq L^{-1}); A_e is radon equilibrium concentration (Bq L^{-1}); A_0 is radon concentration in the surface water (Bq L^{-1}); t is retention time (s); λ is decay rate ($\lambda_{Rn} = 0.18$ d^{-1}).

Through a derivation of Darcy's equation and assuming 1-D flow, it is possible to estimate the velocity between the surface water and sampling point by eq. 19.5:

$$v = \frac{z}{t} \tag{19.5}$$

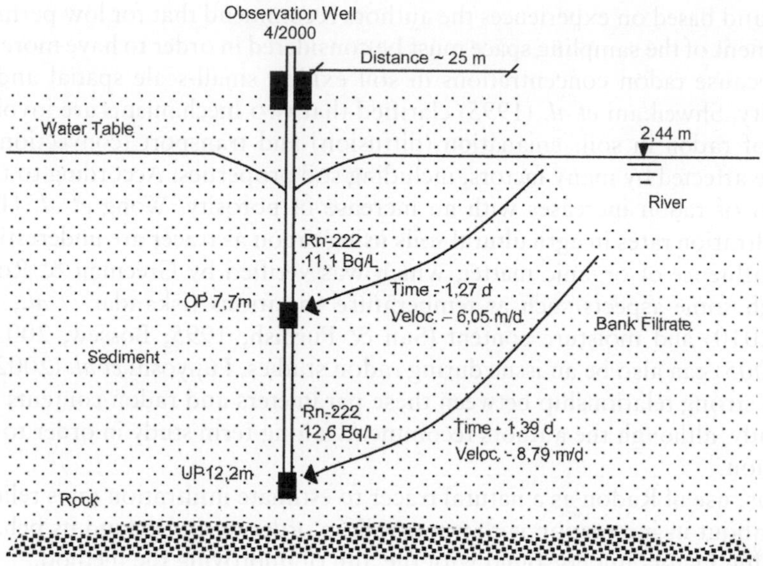

Figure 19.4 Results of ^{222}Rn field measurements in Tolkewitz, Dresden to estimate infiltration rates.

where v is water flow velocity (m/d); z is sampling point depth (m); t is retention time (d).

Many authors have used radon to estimate groundwater influx into superficial water, such as: lakes (Schmidt *et al.*, 2009), ponds (Corbett *et al.*, 1997), groundwater in saline regions (Hussain *et al.*, 1999; Burnett & Dulaiova, 2003), rivers/streams (Yoneda *et al.*, 1991; Cook *et al.*, 2003) and aquifers (Low, 1996). The limitation of this method is the local sedimentary stratification (geologic structure), which can influence equilibrium concentration of ^{222}Rn. In addition, gas in the saturated upper zone of the aquifer can influence the concentration of ^{222}Rn and its measurements (Macheleidt *et al.*, 2006). Pereira *et al.* (2010) related that radon concentration can be variable on regional and local scales depending on mineralogy, lithology and faulting. Uranium content of bedrock is expected to correlate with soil-gas radon concentration as a source of radon gas. In groundwater the concentration of radon depends on uranium enriched faults, which can give different results depending on percolation. Lee *et al.* (2010) explain that the groundwater radon concentration depends on the radon emanation coefficient of the aquifer rock; the radon emanation coefficient (or emanating power, emanating fraction or release ratio) is only a fraction of the total amount produced by radium decay that escapes from the rocks and attaches to the groundwater. Baixeras *et al.* (2001) added that the emanation of radon depends on many factors such as permeability, organic content, grain size, porosity, moisture content, internal structure of the materials and temperature. Additionally, the authors showed that the smaller the soil grains the higher radon emanation can be. Soils with clay (grain size <0.01 mm) showed a radon concentration approximately fifty times higher than in sand with grain size between 0.1 and 2.0 mm. Matolin *et al.* (2000, cited in Fujiyoshi *et al.*, 2002) related that radon concentration increases with sampling depths, and based on experiences the authors recommend that for low permeable soils enlargement of the sampling space must be considered in order to have more consistent data, because radon concentrations in soil exhibit small-scale spatial and temporal variability. Shweikani *et al.* (1995) clarified that two mechanisms are involved in the release of radon in soil: emanation (diffusion) and transport (convection). Both of them are affected by many factors, including soil properties. According to the authors diffusion of radon increases with an increase of porosity. Wong *et al.* (1992) state that infiltration rates in agricultural soils using radon as tracer are underestimated due to the influence of organic matter, which was verified by Greeman & Rose (1996). Although some aspects such as temperature variations (Iskandar *et al.*, 2004; Lee *et al.*, 2010) and moisture content (Sun & Furbish, 1995; Bossew, 2003; Barillon *et al.*, 2005) can also be an issue during radon studies, Fujiyoshi *et al.* (2002) have not found a strong relationship between these parameters and radon contents in a short-term study, although the authors recommend a long-term study in order to detect any correlation.

Thus, use of Radon as a natural tracer to estimate infiltration rates relies on more studies about its interaction with water/soil, in order to understand its behaviour and controlling factors underground with the aim of improving the method.

19.2.5 Fluorescence labelled tracers

Coloured tracers are commonly used to estimate travel times and consequently infiltration velocities; used with sediment data they can give also information about hydraulic conductivity, porosity, dispersivity, particle migration and other parameters concerning surface/groundwater interaction. At bank filtration sites, where water infiltration velocity is normally higher compared to undisturbed aquifers, control of residence times of the bank filtrate is extremely useful because metabolism and mineralisation processes of particulate and dissolved organic matter occur along the flow path. Up to now many tracers have been studied and applied to trace water movement, such as: solid material in suspension, dissolved salts or ionized substances, stable and radioactive isotopes, dyes, and gases (Skibitzke, 1958; Davis *et al.*, 1980; Evans, 1983; Flury & Flühler, 1994; Schudel *et al.*, 2003; Anderson, 2005; Macheleidt *et al.*, 2006). Even pharmaceutical residues have been used to determine residence times at bank filtration sites (Massmann *et al.*, 2008). However, dyes are most successfully applied as tracers in hydrological studies. Schudel *et al.* (2003) mentioned that an ideal tracer should imitate water movement without any adsorption or complexation from active uptake by organisms; they should be mobile, visible, easy to detect, nontoxic, chemically stable, show low tendency of sorption, good solubility and very small or no concentration in natural waters (Davis *et al.*, 1980; Flury & Flühler, 1995; Schudel *et al.*, 2003). Moreover, some dyes can also be affected by light, pH variations and temperature changes which lead to their decomposition (Davis *et al.*, 1980). Most inorganic dyes tend to sorb onto soil particles and Davis *et al.* (1980) recommended using only organic dyes in very permeable aquifers to investigate predominantly short travel distances. Many dyes are available and have been tested and compared in order to evaluate surface/groundwater interactions at different sites. Table 19.2 shows characteristics of commonly used dyes,

Table 19.2 Main characteristics of the most commonly used dyes.

Dye	*Characteristics*	*References*
Rhodamine WT	contains toxic properties*, inclination to adsorption, good detection limit, fluorescence intensity is a function of temperature and low pH	Trudgill *et al.*, 1983; Bencala *et al.*, 1983; Sabatine & Al Austin, 1991; Field *et al.*, 1995; Ptak & Schmid, 1996; Allaire-Leung *et al.*, 2000; Schudel *et al.*, 2003
Naphthionate	low sorption properties, sample impurity might cause optical interference, good to use at pH 4–9, short distance tests are preferable, invisible at concentrations used	Schudel *et al.*, 2003; Ammann *et al.*, 2003; Winter, 2006
Uranine	nontoxic at low concentrations*, low adsorption properties, high detection sensitivity, high light sensitivity, better to use at pH > 5.5, oxidants destroy the Uranine (e.g. chlorine dioxide), considered a quasi-ideal tracer	Ptak & Teutsch, 1994; Field *et al.*, 1995; Ptak & Schmid, 1996; Niehren & Kinzelbach, 1998; Corbett *et al.*, 2000; Ammann *et al.*, 2003; Winter, 2006; Chua *et al.*, 2007
Brilliant Blue	nontoxic, slowly degrading, can be neutral or ionic, depending on pH, low adsorption properties, good visibility, good for detecting flow patterns as affected by clogging, not ideal for tracing travel times	Flury & Flühler, 1995; Kildsgaard & Engesgaard, 2001
Eosin	low sensitivity to pH changes and light, low toxicity or non toxic*, good tracer but with more sorption properties than Uranine, not good for low hydraulic conductivity or long distances	Field *et al.*, 1995; Schudel *et al.*, 2003

* Toxicity not conclusively proven by Field *et al.* (1995).

such as Rhodamine WT, Naphtionate, Uranine/Fluorescein, Eosin and Brilliant Blue. An ideal tracer does not exist, however, and each one has different characteristics that best suits a particular purpose (Davis *et al.*, 1980; Schudel *et al.*, 2003).

The use of fluorescent dyes for colouring particles (resin particles, bacteria, yeast cells, organic matter) is an innovative technique and might be used to estimate infiltration rates and related particle transport (Hall *et al.*, 1996; Wanner & Pusch, 2000; Paul & Hall, 2002). The fluorescent tracers 7-amino-4-methylcoumarin (AMC) and fluorescein-5-isothiocyanate (FITC) were used for the quantification of particle transport in sediment interstices and retention of small particles by straining, a physical-chemical retention process (Gunkel *et al.*, 2009; Gunkel & Hoffmann, 2009). In this procedure, melanine resin particles (produced by microParticles, Berlin, Germany) are labelled with 7-amino-4-methylcoumarin (AMC). The fluorescing MF-AMC particles have a blue colour (λ_{ex} 360 nm/λ_{em} 429 nm), a density of 1.5 g cm^{-3} and a positively or negatively charged surface, particles are available with 1–10 μm diameter. In addition, fluorescein-5-isothiocyanate (FITC) fluorescent labelling (λ_{ex} 560 nm/ λ_{em} 529 nm) is also a method that allows the study of vertical particle transfer in sediments. This fluorescence dye is used for labelling of organic matter such as yeast or algae cells and leaves (microParticles, Berlin, Germany). Therefore, natural organic substrates from the sampling site can be marked with FITC and used for analysis of the biological crushing, decomposition and vertical transport. This method has been applied in the interstices to quantify vertical fluxes of POC in-situ in plastic tubes (Gunkel *et al.*, 2009).

19.3 CONCLUSIONS

Interest in determining infiltration rates has increased significantly in recent years, in order to assess or quantify surface/groundwater interaction, especially at bank filtration sites. For bank filtration, aspects such as clogging, infiltration/exfiltration processes, transport of contaminants, implementation of bank filtration in tropical and semi-arid countries are important issues. At the same time, techniques have been created and developed in order to examine and monitor such features.

The clogging effect, its consequences and importance of monitoring are extensively discussed in the literature. This phenomenon is known to alter the exchange between surface water and groundwater. It is identified by the formation of a thin layer on the stream/lake bed, which is caused by deposition of (eroded) particles, chemical reactions, gas formation, biological activities with formation of extracellular polymeric substances etc. Nowadays infiltration is known as a complex process with a high variability in time and space. Filtration variance in response to flow dynamics of the river on clogged areas at bank filtration sites was demonstrated by Schubert (2004). Changes in riverbed hydraulic conductivity, as well as specific capacity of a radial production well due to clogging, were also demonstrated by Hubbs (2004), and clogging coverage effects on infiltration rates were discussed by Macheleidt *et al.* (2006). Biological clogging has also been recognized as important and it needs to be monitored. A filtration effect of organic particles was analysed by Gunkel *et al.* (2009) in order to understand transport of fine particulate organic matter into littoral interstices. In addition, an extensive literature review was made by Baveye *et al.* (1998) with the purpose of comprehending the bio-clogging properties and aspects, not only on hydraulic characteristics but also on water quality.

Regular methods used to investigate infiltration rates are already established and utilised in scientific studies. They became largely applicable in surface/groundwater interaction studies and issues involving the clogging. Innovative methods, such as ^{222}Radon and fluorescence labelled tracers (AMC and FITC), have been used in infiltration studies and have proved to be very useful, although there are still some points to be discussed and studied to improve these tools.

Selection of the method to be used depends on many factors. Availability of material, field site conditions, stream bed grain size distribution and clogging layer thickness are some aspects that should be considered. In general, all field measurements must be conducted at many locations in the river/lake bed to obtain reliable data for infiltration rates. Table 19.3 summarizes some positive and negative aspects of the methods described in this chapter.

Table 19.3 Brief comparison between common and new methods.

Method	*Positive Aspects*	*Negative Aspects*
Seepage meters	• direct measurement of infiltration rates • possibility of having a good idea of the real distribution of the flux • well suited for calm and shallow water settings • relatively easy to handle • possibility to measure low and high infiltration rates • low construction costs	• not recommended for surface water with strong hydraulic movement • not applicable to coarse sediments (gravel, pebbles) • dense vegetation interferes with the measurements • small diameters may not give real site infiltration rates
Piezometers	• to quantify flow between groundwater and surface water • interesting for homogeneous aquifers • to get a general idea of the aquifer characteristics • small ones are normally low in cost	• no information about infiltration rates at a specific area • large numbers of monitoring wells might be needed • waves might cause reading errors
Temperature	• heat is a natural tracer free from issues of contamination associated with the use of chemical tracers • well-suited for investigations of stream/groundwater exchanges • temperature data are immediately available as opposed to most chemical tracers • quick and low cost measurements	• very low hydraulic gradients and hydraulic conductivities can affect temperature measurements • assumptions for models can lead to results differing from reality
Radon-222	• natural tracer found at different concentrations in groundwater and surface water • possibility to get a good idea of infiltration rate regimes	• high volatility of ^{222}Rn when in contact with atmosphere • demands specific sampling • expensive Rn analyser needed • sediment stratification can influence the ^{222}Rn concentration • temperature, material structure, organic matter content and soil porosity influences must be considered
Fluorescent labelled tracers	• possibility of measuring infiltration rates by labelling natural organic substances from the sampling site	• high costs of fluorescent particles • severe clogging can limit the resin passage through the sediment • ingestion of tracers by invertebrates • particle agglomeration • sorption on soil not studied

REFERENCES

Allaire-Leung S.E., Gupta S.C., Moncrief J.F. (2000) Water and solute movement in soil as influenced by macropore characteristics. 1. Macropore continuity. *J Con Hydrol* 41:283–301.

Ammann A.A., Hoehn E., Koch S. (2003) Ground water pollution by roof runoff infiltration evidenced with multi-tracer experiments. *Water Research* 37:1143–1153.

Anderson M.P. (2005) Heat as a groundwater tracer. *Ground Water* 43(6):951–968.

Anibas C., Buis K., Verhoeven R., Meire P., Batelaan O. (2011) A simple thermal mapping method for seasonal spatial patterns of groundwater-surface water interaction. *J Hydrol* 397:93–104.

Baixeras C., Erlandsson B., Font L., Jönsson G. (2001) Radon emanation from soil samples. *Radiation Measurements* 34: 441–443.

Barillon R., Özgümüs A., Chambaudet A. (2005) Direct recoil radon emanation from crystalline phases. Influence of moisture content. *Geochimica et Cosmochimica Acta* 69(11): 2735–2744.

Baxter C., Hauer F.R., Worsner W.W. (2003) Measuring groundwater-stream water exchange: New techniques for installing mini-piezometers and estimating hydraulic conductivity. *Transactions of the American Fisheries Society* 132:493–502.

Baveye P., Vandevivere P., Hoyle B.L., DeLeo P.C., de Lozada D.S. (1998) Environmental impact and mechanisms of the biological clogging of saturated soils and aquifer materials. *Crit Rev Envir Sci Tech* 28(2): 123–191.

Behaegel M., Sailhac P., Marquis G. (2007) On the use of surface and ground temperature data to recover soil water content information. *J Applied Geophysics* 62:234–243.

Belanger T.V., Mikutel D.F. (1985) On the use of seepage meters to estimate groundwater nutrient loading to lakes. *Water Resour Bul Amer Water Resour Ass* 21(2):265–272.

Bencala K.E., Rathbun R.E., Jackman A.P., Kennedy V.C., Zellweger G.W., Avanzino R.J. (1983) Rhodamine WT dye losses in a mountain stream environment. *Water Resour Bul. Amer Water Resour Ass* 19(6):943–950.

Bertin C., Bourg A.C.M. (1994) Radon-222 and chloride as natural tracers of the infiltration of river water into a alluvial aquifer in which there is significant river/groundwater mixing. *Environ. Sci. Technol* 28:794–798.

Bossew P. (2003) The radon emanation power of building materials, soil and rocks. *App Rad Isot* 59:389–392.

Burnett W.C., Dulaiova H. (2003) Estimating the dynamics of groundwater input into the coastal zone via continuous radon-222 measurements. *J Environ Radioactivity* 69:21–35.

Burnett W.C., Kim G., Lane-Smith D. (2001) A continuous monitor for assessment of 222Rn in the coastal ocean. *J Radioanal Nuclear Chemistry* 249(1):167–172.

Cable J.E., Burnett W.C., Chanton J.P., Corbett D.R., Cable P.H. (1997) Field evaluation of seepage meters in the coastal marine environment. *Estuarine, Coastal and Shelf Science* 45:367–375.

Cey E.E., Rudolph D.L., Parkin G.W., Aravena R. (1998) Quantifying groundwater discharge to a small perennial stream in southern Ontario, Canada. *J Hydrol* 210:21–37.

Cherkauer D.A., Mcbride J.M. (1988) A remotely operated seepage meter for use in large lakes and rivers. *Ground Water* 26(2):165–171.

Chua L.H.C., Robertson A.P., Yee W.K., Shuy E.B., Lo E.Y.M., Lim T.T., Tan S.K. (2007) Use of fluorescein as a ground water tracer in brackish water aquifers. *Ground Water* 45(1): 85–88.

Conant Jr. B. (2004) Delineating and quantifying ground water discharge zones using streambed temperatures. *Ground Water* 42(2):243–257.

Constantz J., Jasperse J., Seymour D., Su G.W. (2003a) Heat tracing in the streambed along the Russian River o northern California. In: Stonestrom D.A. & Constantz J. (eds.) *Heat as a tool for studying the movement of ground water near streams, USGS Circ. 1260*. Reston, Virginia, pp. 17–20.

Constantz J., Cox M.H., Sarma L., Mendez G. (2003b) The Santa Clara River – the last natural river in Los Angeles. In: Stonestrom D.A. & Constantz J. (eds.) *Heat as a tool for studying the movement of ground water near streams, USGS Circ. 1260*. Reston, Virginia, pp. 21–27.

Constantz J., Su G.W., Hatch C. (2006) Heat as a ground-water tracer at the Russian River RBF facility, Sonoma County, California. In: Hubbs S.A. (ed.) *Riverbank filtration hydrology*. Louisville, U.S.A., pp. 243–258.

Constantz J. (1998) Interaction between stream temperature, streamflow and groundwater exchanges in alpine streams. *Water Resour Res* 34(7):1609–1615.

Cook P.G., Favreau G., Dighton J.C., Tickell S. (2003) Determining natural groundwater influx to a tropical river using radon, chlorofluorocarbons and ionic environmental tracers. *J Hydrol* 277:74–88.

Corbett D.R., Burnett W.C., Cable P.H., Clark S.B. (1997) Radon tracing of groundwater input into Par Pond, Savannah River Site. *J Hydrol* 203:209–227.

Corbett D.R., Dillon K., Burnett W. (2000) Tracing groundwater flow on a barrier island in the north-east Gulf of Mexico. *Estuarine, Coastal and Shelf Science* 51:227–242.

Davis S.N., Thompson G.M., Bentley H.W., Stiles G. (1980) Ground-water tracers – A short review. *Ground Water* 18(1):14–23.

DEV. (1998) DIN 18130-1, Bestimmung des Wasserdurchlässigkeitsbeiwerts. Teil 1: Laborversuche. Deutsche Einheitsverfahren zur Wasser-, Abwasser- und Schlammuntersuchung; Schlamm und Sedimente.

Diersch H.J.G., Kolditz O. (2008) Coupled groundwater flow and transport: 2. Thermohaline and 3D convection systems. *Adv Water Resour* 21:401–425.

Domenico P.A., Schwartz F.W. (1998) *Physical and Chemical Hydrogeology*. 2nd Ed, John Wiley & Sons, Inc.

Doussan C., Toma A., Paris B., Poitevin G., Ledoux E., Detay M. (1994) Coupled use of thermal and hydraulic head data to characterize river – groundwater exchanges. *J Hydrol* 153:215–229.

Evans G.V. (1983) Tracer techniques in hydrology. *J Appl Radiat Isot* 34(1):451–475.

Field M.S., Wilhelm R.G., Quinlan J.F., Aley T.J. (1995) An assessment of the potential adverse properties of fluorescent tracer dyes used for groundwater tracing. *Environ Monitor Assess* 38:75–96.

Fischer T., Day K., Grischek T. (2005) Sustainability of riverbank filtration in Dresden, Germany. Recharge systems for protecting and enhancing groundwater resources. In: *Proceedings of 5th Int Sym Management of Aquifer Recharge*. UNESCO Publishing, Paris, pp. 23–28.

Flury M., Flühler H. (1994) Brilliant Blue FCF as a dye tracer for solute transport studies – A toxicological overview. *J Environ Quality* 23(4):1108–1112.

Flury M., Flühler H. (1995) Tracer characteristics of Brilliant Blue FCF. *Soil Sci Soc Amer J* 59(1):22–27.

Fujiyoshi R., Morimoto H., Sawamura S. (2002) Investigation of soil radon variation during the winter months in Sapporo, Japan. *Chemosphere* 47:369–373.

Geist D.R., Joy M.C., Lee D.R., Gonser T. (1998) A method for installing piezometers in large cobble bed rivers. *Ground Water Monitoring and Remediation* 28:78–82.

Greeman D.J., Rose A.W. (1996) Factors controlling the emanation of radon and thoron in soils of the eastern U.S.A. *Chem Geol* 129:1–14.

Grischek T., Hiscock K.M., Metschies T., Dennis P.F., Nestler W. (1997) Factors affecting denitrification during infiltration of river water into a sand and gravel aquifer in Saxony, Germany. *Water Res* 32(2):450–460.

Grischek T., Schoenheinz D., Syhre C., Saupe K. (2010) Impact of decreasing water demand on bank filtration in Saxony, Germany. *Drink. Water Eng. Sci.* 3:11–20.

Gundersen L.C.S. (1992) The effect of rock type, grain size, sorting, permeability and moisture on measurements of radon in soil gas. A comparison of two measurements techniques. *J Radio Nucl Chem* 161(2):325–337.

Gunkel G., Beulker C., Hoffmann A., Kosmol J. (2009) Fine particulate organic matter (FPOM) transport and processing in littoral interstices – use of fluorescent markers. *Limnologica* 39(3): 185–199.

Gunkel G., Hoffmann A. (2009) Bank filtration of rivers and lakes to improve the raw water quality for drinking water supply. In: *Water Purification*. Gertsen N. & Sonderby L. (eds). Nova Sci Pub, Chap. 3, pp. 137–169.

Hall S.H., Luttrell S.P., Cronin W.E. (1996) A method to estimate effective porosity and ground-water velocity. *Ground Water* 29(2):171–174.

Hoehn E., Cirpka O.A. (2006) Assessing residence times of hyporheic ground water in two alluvial flood plains of Southern Alps using water temperature and tracers. *Hydrol Earth System Sci* 10:553–563.

Hoffmann A. (2008) Biologische Funktionsfähigkeit und Dynamik des sandigen Interstitials unter dem Einfluss induzierter Uferfiltration am Tegeler See (Berlin). (Biological functions and dynamics of the sandy interstitial under the influence of bank filtration in Lake Tegel (Berlin), Dis, Berlin Technical Uni, Fac III.

Hubbs S.A. (2004) Changes in riverbed hydraulic conductivity and specific capacity at Louisville. In *River Bank Filtration Hydrology. Impacts on System Capacity and Water Quality.* S.A. Hubbs (ed.). NATO Science Series. IV. Earth and Environmental Sciences. Springer, Dordrecht, 60:199–220.

Hussain N., Church T.M., Kim G. (1999) Use of 222Rn and 226Ra to trace groundwater discharge into the Chesapeake Bay. *Marine Chem* 65:127–134.

Isiorho S.A., Meyer J.H. (1999) The effects of bag type and meter size on seepage meter measurements. *Ground Water* 37(2):411–413.

Iskandar D., Yamazawa H., Iida T. (2004) Quantification of the dependency of radon emanation power on soil temperature. *App Radiat Isot* 60:971–973.

Kalbus E., Reinstorf F., Schirmer M. (2006) Measuring methods for groundwater, surface water and their interactions: a review. *Hydrology and Earth Systems Sciences Discussions* 3:1809–1850.

Kildsgaard J., Engesgaard P. (2001) Numerical analysis of biological clogging in two-dimensional sand box experiments. *J Cont Hydrol* 50:261–285.

Kipp K.L. (1987) HST3D: A computer code for simulation of heat and solute transport in three-dimensional ground-water flow systems. U.S. Geol Sur. *Water-Resources Invest Rep* 86-4095, Denver, Colorado.

Kluge T., Ilmberger J., von Rohden C., Aeschbach-Hertig W. (2007) Tracing and quantifying groundwater inflow into lakes using radon-222. *Hydrology and Earth System Sciences* 4:1519–1548.

Kolditz O., Ratke R., Diersch H.J.G, Zielke W. (1998) Coupled groundwater flow and transport: 1. Verification of variable density flow and transport models. *Adv Water Resour* 21(1): 27–46.

Land L.A., Paull C.K. (2001) Thermal gradients as a tool for estimating groundwater advective rates in a coastal estuary: White Oak River, North Carolina, USA. *J Hydrol* 248:198–215.

Landon M.K., Rus D.L., Harvey F.E. (2001) Comparison of instream methods for measuring hydraulic conductivity in sandy streambeds. *Ground Water* 39(6):870–885.

Lang D. (2008) Durchlässigkeit von Böden – Übliche Verfahren und deren Überprüfung im Labor, (Hydraulic conductivity of soils – Current methods and their validation in the laboratory) Diplomica Verlag Igel.

Lee D.R. (1977) A device for measuring seepage flux in lakes and estuaries. *Lim Ocean* 22(1):140–147.

Lee D.R., Cherry J.A. (1978) A field exercise on groundwater flow using seepage meters and mini-piezometers. *J Geol Educ* 27:6–10.

Lee K.Y., Yoon Y.Y., Ko K.S. (2010) Determination of the emanation coefficient and the Henry's law constant for the groundwater radon. *J Radioanal Nucl. Chem.* DOI 10.1007/s10967-010-0730-2.

Libelo E.L., MacIntyre W.G. (1994) Effects of surface-water movement on seepage-meter measurements of flow through the sediment-water interface. *App Hydrogeology* 4:49–54.

Low R. (1996) Radon as a natural groundwater tracer in the Chalk Aquifer, UK. *Environ Int* 22(1):S333–S338.

LWBC. (2003) Land and Water Biodiversity Committee. Minimum construction requirements for water bores in Australia, 2nd ed. Land and Water Biodiversity Committee.

Macheleidt W., Grischek T., Nestler W. (2006) New approaches for estimating streambed infiltration rates. In: S. Hubbs (ed.) *Riverbank Filtration Hydrology.* NATO Sci Ser IV, Earth and Environ Sci 60:73–91.

Massmann G., Sültenfuß J., Dünnbier U., Knappe A., Taute T., Pekdeger A. (2008) Investigation of groundwater residence times during bank filtration in Berlin: A multi-tracer approach. *Hydrol Proc* 22:788–801.

Matolin M., Neznal J.Z., Neznal M. (2000) Geometry of soil gas sampling, soil permeability and radon activity concentration. In: *Proceedings of the 5th Int Workshop on the Geol Aspects of Radon Risk Map*, pp. 27–29.

Morice J.A., Valett H.M., Dahm C.N., Campana M.E. (1997) Alluvial characteristics, groundwater-surface water exchange and hydrological retention in headwater streams. *Hydrol Proc* 11:253–267.

Murdoch L.C., Kelly S.E. (2003) Factors affecting the performance of conventional seepage meters. *Water Resour Res* 39(6): 1163.

Mutiti S., Levy J. (2010) Using temperature modelling to investigate the temporal variability of the riverbed hydraulic conductivity during storms events. *J Hydrol* 388:321–334.

Niehren S., Kinzelbach W. (1998) Artificial colloid tracer tests: development of a compact on-line microsphere counter and application to soil column experiments. *J Cont Hydrol* 35:249–259.

Obbink J.G. (1969) Constructions of piezometers and method of installation for ground water observation in aquifers. *J Hydrol* 7:434–443.

Oxtobee J.P.A., Novakowski K. (2002) A field investigation of groundwater/surface water interaction in a fractured bedrock environment. *J Hydrol* 269:169–193.

Paul M.J., Hall Jr. R.O. (2002) Particle transport and transient storage along a stream-size gradient in the hubbard brook experimental forest. *J. N. Am. Benthol. Soc.* 21:195–205.

Paulsen R.J., Smith C.F., O'Rourke D., Wong T.F. (2001) Development and evaluation of an ultrasonic ground water seepage meter. *Ground Water* 39(6):904–911.

Pereira A.J.S.C., Godinho M.M., Neves L.J.P.F. (2010) On the influence of faulting on small-scale soil-gas radon variability: A case study in the Iberian Province. *J Environ Radioactivity* 101:875–882.

Ptak T., Schmid G. (1996) Dual-tracer transport experiments in a physically and chemically heterogeneous porous aquifer: affective transport parameters and spatial variability. *J Hydrol* 183:117–138.

Ptak T., Teutsch G. (1994) Forced and natural gradient tracer tests in a highly heterogeneous porous aquifer: instrumentation and measurements. *J Hydrol* 159:79–104.

Ronan A.D., Prudic D.E., Thodal C.E., Constantz J. (1998) Field study and simulation of diurnal temperature effects on infiltration and variably saturated flow beneath an ephemeral stream. *Water Resour Res* 34(9):2137–2153.

Rosenberry D.O. (2008) A seepage meter designed for use in flowing water. *J Hydrol* 359:118–130.

Rosenberry D.O., LaBaugh J.W. (2008) Field techniques for estimating water fluxes between surface water and ground water. *US Geol Sur Techniques and Methods 4-D2*, p. 128.

Rosenberry D.O., Morin R.H. (2004) Use of an electromagnetic seepage meter to investigate temporal variability in lake seepage. *Ground Water* 42(1):68–77.

Sabatine D.A., Al Austin T. (1991) Characteristics of Rhodamine WT and Fluorescein as adsorbing ground-water tracers. *Ground Water* 29(2):341–349.

Schubert J. (2004) Experience with riverbed clogging along the Rhine River. In: *River Bank Filtration Hydrol. Impacts on System Capacity Water Quality*. Hubbs S. (ed). NATO Sci Ser IV. *Earth Env Sci* 60:221–242.

Schmidt A., Stringer C.E., Haferkorn U., Schubert M. (2009) Quantification of groundwater discharge into lakes using radon-222 as naturally occurring tracer. *Environ. Geol.* 56:855–863.

Schudel B., Biaggi D., Dervey T., Kozel R., Müller I., Ross J.H., Schindler U. (2003) Application of artificial tracers in hydrogeology – Guideline. Lang P. (ed) no. 20, p. 99.

Shaw R.D., Prepas E.E. (1989) Anomalous short-term influx of water into seepage meters. *Lim Ocean.* 34:1343–1351.

Shinn E.A., Reich C.D., Hickey T.D. (2002) Seepage meters and Bernoulli's revenge. *Estuaries* 25(1):126–132.

Sholkovitz E., Herbold C., Charette M. (2003) An automated dye-dilution based seepage meter for the time series measurement of submarine groundwater discharge. *Lim Ocean Methods* 1:16–28.

Shweikani R., Giaddui T.G., Durrani S.A. (1995) The effect of soil parameters on the radon concentration values in the environment. *Radiation Measurements* 25(1–4):581–584.

Skibitzke H.E. (1958) The use of radioactive tracers in hydrologic field studies of ground-water motion. *US Geol Sur. Ground Water Branch*, pp. 243–252.

Sophocleos M. (2002) Interactions between groundwater and surface water—The state of the science: *Hydrogeol J* 10:52–67.

Stonestrom D.A., Blasch K.W. (2003) Determining temperature and thermal properties for heat based studies of surface-water ground-water interactions. In: Stonestrom D.A. & Constantz J. (eds) *Heat as a tool for studying the movement of ground water near streams*, USGS Circ. 1260. Reston, Virginia, App A, pp. 73–80.

Stonestrom D.A., Constantz J. (2004) Using temperature to study stream-ground water exchange. *US Geol Sur on-line fact sheet* 2004–3010.

Sun H., Furbish D.J. (1995) Moisture content effect on radon emanation in porous media. *J Cont Hydrol* 18:239–255.

Taniguchi M., Fukuo Y. (1993) Continuous measurement of ground-water seepage using an automatic seepage meter. *Ground Water* 31(3):675–679.

Trefry M.G., Muffels C. (2007) FEFLOW: A finite-element ground water flow and transport modelling tool. *Ground Water*, 525–528.

Trettin R., Grischek T., Strauch G., Mallén G., Nestler W. (1999) The suitability and usage of 18O and chloride as natural tracer for bank filtrate at the middle River Elbe. *Isotop Env Health Stud* 35:331–350.

Trudgill S.T., Pickles A.M., Smettem K.R.J., Crabtree R.W. (1983) Soil-water residence time and solute uptake. 1. Dye tracing and rainfall events. *J Hydrol* 60:257–279.

Voss C.I. (1984) A finite-element simulation model for saturated-unsaturated, fluid-density-dependent ground-water flow with energy transport or chemically-reactive single-species solute transport. *US Geol Sur*, p. 49.

Wang J. (2002) Riverbank filtration case study at Louisville, Kentucky. In: Ray C., Merlin G. & Linsky R.B. (eds) *Riverbank filtration: Improving source-water quality*. Kluwer Academic Publishers, pp. 117–145.

Wanner S.C., Pusch M. (2000) Use of fluorescently labeled Lycopodium spores as a tracer for suspended particles in a lowland river. *J N Am. Benthol Soc* 19:648–658.

Winter F. (2006) Using tracer techniques to investigate groundwater recharge in the Mount Carmel Aquifer, Israel. Diploma thesis, Institute of Hydrology, University Freiburg, p. 128.

Winter T.C., LaBaugh J.W., Rosenberry D.O. (1988) The design and use of a hydraulic potentiomanometer for direct measurement of differences in hydraulic head between groundwater and surface water. *Lim. Ocean* 33(4):1209–1214.

Woessner W.W., Sullivan K.E. (1984) Results of seepage meter and mini-piezometer study, Lake Mead, Nevada. *Ground Water* 22(4):561–568.

Wong C.S., Chin Y., Gschwend P.M. (1992) Sorption of radon-222 to natural sediments. *Geochimica et Cosmochimica Acta* 56:3923–3932.

Yoneda M., Inoue Y., Takine N. (1991) Location of groundwater seepage points into a river by measurement of 222Rn concentration in water using activated charcoal passive collectors. *J Hydrol* 124:307–316.

Chapter 20

Estimation of ratio of interception by shelterbelts from the saturated zone to evaporative loss

Andrzej Kędziora[1] *& Dariusz Kayzer*[2]
[1]*Institute for Agricultural and Forest Environment, Polish Academy of Sciences, Poznan, Poland*
[2]*Department of Mathematical and Statistical Methods, Poznan University of Life Sciences, Poznan, Poland*

ABSTRACT

Intensification of agricultural activity, in order to achieve higher production, brings many threats to the environment and ecosystem services. One of the most important services for sustainable development of rural areas is providing sufficient water for human consumption. Water pollution by chemical compounds originated from fertilisers and pesticides used by farmers is now impacting groundwater resources. Long term studies carried out by the Institute for Agricultural and Forest Environment showed that one of the best ways to counteract these negative impacts is increasing the complexity of the landscape structure by introducing of shelterbelt of selected vegetation which can control and limit the spreading of diffuse pollutants. Efficiency of the reduction of the concentration of chemical compounds in groundwater depends on the quantity of water taken by the shelterbelts from the saturated zone of the soil.

20.1 INTRODUCTION

Capitalising on the knowledge that ecosystems can change the water chemistry, modify microclimatic conditions and help sustain biodiversity, the concept of ecosystem services was developed. Using the classification of ecosystem services developed by the world Millennium Ecosystem Assessment (MEA, 2005), four groups of services have been distinguished: basic, provisioning regulatory and cultural (social).

The most important for human well-being are regulatory services connected with water cycling in the agricultural landscape and services that ensure clean water supplies. Many past practices in agriculture impacted landscape functions with the loss of ecosystem services (Kedziora and Ryszkowski, 2004). Conversion to more stable ecosystems, like woodland, meadow and wetland into less stable arable land causes increased threats for the fundamental processes such as energy flow and chemical cycling in environment. Simplification of the landscape structure, principally deforestation, and reduced surface and soil water retention are the most damaging changes threatening the sustainable development of the countryside (Ryszkowski *et al.*, 2002). One of the most important outcomes for sustainable development of the countryside are reducing storage of water and deteriorating water quality. Intensification of agricultural activities (crop and animal production) required the intensive application of fertilizers and pesticides, and this has impacted the groundwater resources. The concentration of NO_3 was variously higher than $50\,mg\,l^{-1}$.

One of the most active remedial plant cover elements in the agricultural landscape, which control the water regime, are shelterbelts and meadows (Kedziora and Olejnik, 2002). Ryszkowski and Bartoszewicz (1989) and Ryszkowski *et al.* (1997; 1999) presented evidence that shelterbelts located in upland parts of a watershed have a cleansing effect if groundwater is within direct or indirect (capillary rise) reach of the root system. The amount of nitrogen concentration in water outputs from agricultural landscape depend on a share of the biogeochemical barriers in the total landscape. The nitrogen concentration decreases exponentially with increase of landscape complexity (Fig. 20.1).

Long term studies carried out by Institute for Agricultural and Forest Environment show that one of the best tools for counteracting the negative effects of intensive agriculture is increasing landscape structure mainly by introducing shelterbelt areas which effectively control and limit spreading of non-point pollution, especially chemical compounds of fertilizers and means of plant protection. The efficiency of the shelterbelt in reducing chemical concentrations in groundwater depends on the quantity of groundwater taken up by the shelterbelts from the saturated zone of soil. Not all transpired water used for evapotranspiration is taken from groundwater; part is taken also from the unsaturated zone. The aim of the experimental research was to develop a model that provides an estimate of the share of groundwater used by the plants for evapotranspiration on the basis of simple agrometeorological indicators.

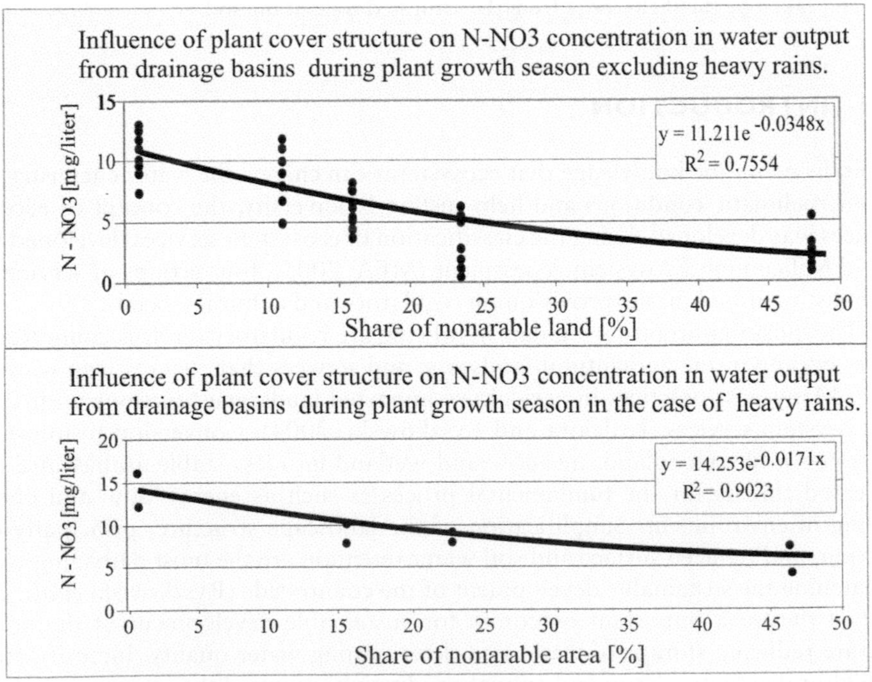

Figure 20.1 Impact of saturation of agricultural landscape on the concentration of N-NO_3 in water output from catchment. Wielkopolska, Turew.

20.2 CHARACTERISTICS OF THE TUREW LANDSCAPE

The studies were carried out in a mosaic of diverse agriculture 182 km^2 in area and located in Wielkopolska region. The landscape reflects the Pleistocene glaciation. The geographic location of the area is between 16°45′ to 17°05′ E and 51°55′ to 50°05′ N and it is situated about 50 km from the large town of Poznan. Although the differences in altitude are small across the area (from 75 m a.s.l. to 90 m a.s.l.) the area consists of a rolling plain made up of slightly undulating ground moraine with many drainage valleys. Light textured soils (Hapludalfs, Glossudalfs and less frequently met Udipsamments) occur with favourable water infiltration conditions particularly in the uplands. The deeper strata are poorly permeable. Endoaquolls, which are poorly drained and store water are found in depressions.

The natural forests were replaced by woodlands and cultivated fields. The landscape has many shelterbelts and there are many small patches of forests. Cultivated fields comprise 70% of the total area, shelterbelts, and small forests 16% and grasslands 9%. The other 5% consists of villages, roads, and marginal areas.

The climate of the region is shaped by the conflicting air masses from the Atlantic, Eastern Europe and Asia, which are modified by strong Arctic and Mediterranean influences (Wos, 1994). It results in changeable weather conditions and the predominance of western winds brings strong oceanic influence with milder winters and cooler summers than occur in the centre and east of Poland. The annual air temperature is 8.0–9.0°C (range from 6.9 to 10.0°C). The mean plant growth season, with temperatures above 5°C, lasts from late March until late October.

Average of annual evapotranspiration amounts to 500 mm, and surface runoff is equal to 95 mm. However, this is now changing (Fig. 20.2).

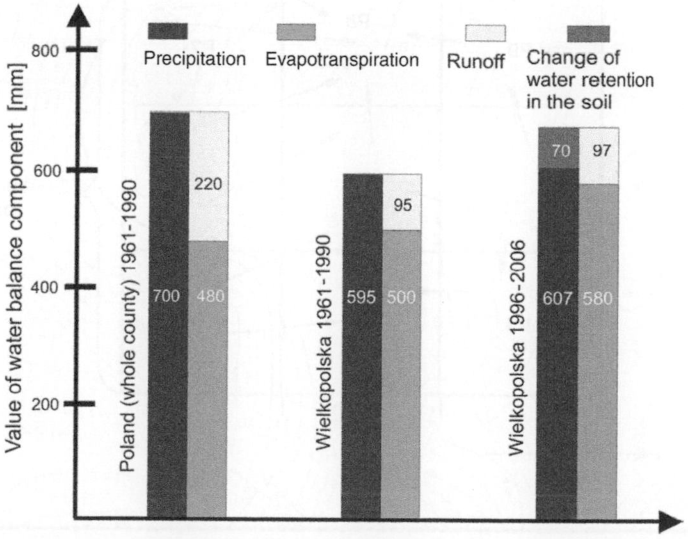

Figure 20.2 Structure of water balance in the periods of 1961–1991 and 1996–2006 in Poland (whole country and Wielkopolska region).

20.3 METHODOLOGY

20.3.1 Description of field experiment

Eighteen plots were arranged as shown in Fig. 20.3. They consists of three plots of shelterbelts (P10 to P12), one plot with a pond (P8), and the other plots straddle the shelterbelt. Plots located in the shelterbelt and plots P7 to P9 had a size of 50 by 40 m, the other ones had a size of 50 by 50 m. The model was based on the data obtained for 4 plots located inside plots P5, P8, P11, and P14. The plots surrounding the inside plots provide the boundary conditions.

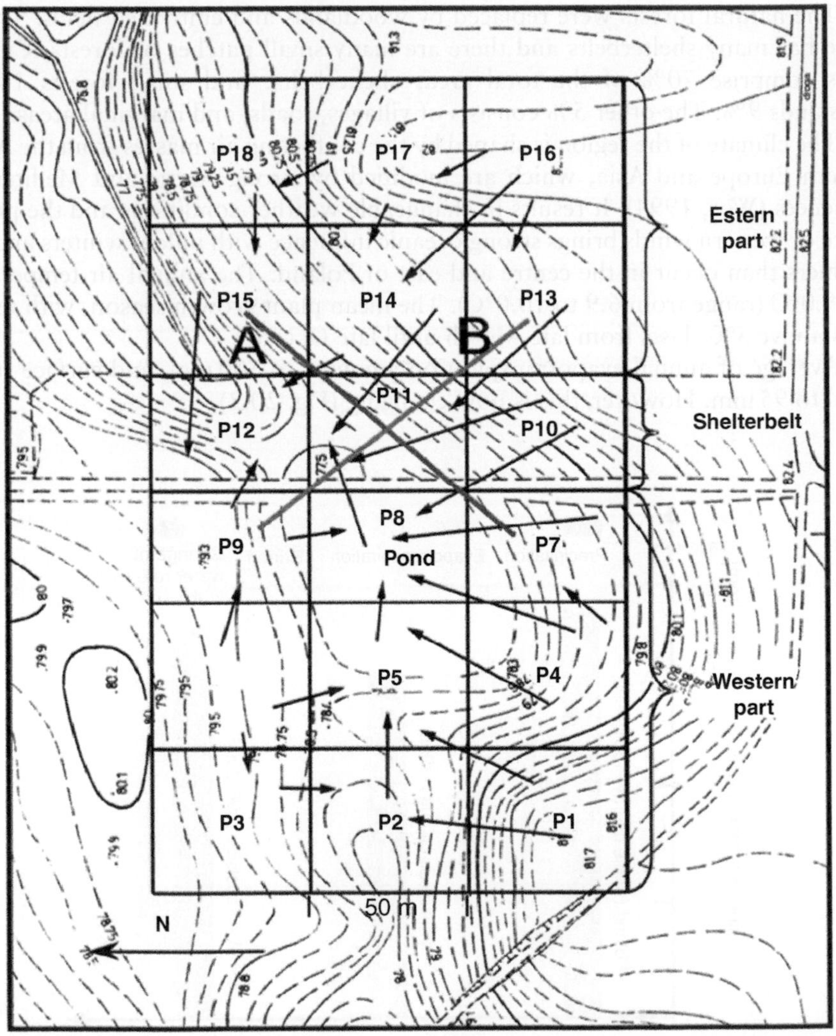

Figure 20.3 Scheme of experimental field. Arrows indicate the main directions of ground water flow. A and B – cross-section for which the sketches are presented at the Figure 20.7.

20.3.2 Methods of meteorological and soil measurements

Meteorological data were taken from a meteorological station located about 1 km from the experiment site. An automatic rainfall gauge was installed in experimental site. The boreholes for measuring depth to groundwater were installed in each plot. Soil pits were dug in each plot. The following parameters were determined for the soil profile at 25 mm intervals: distribution of soil particles by the use of Cassagrande method after Prószyński modification (Mocek *et al.*, 1997), soil bulk density ${}^{b}\rho_s$ [$kg \cdot m^{-3}$], soil particle density ρ_s [$kg \cdot m^{-3}$] and coefficient of total porosity No [$m^3 \cdot m^{-3}$] (Mocek et al., 1997). In addition hydraulic conductivity K [$m \cdot 24$ $hour^{-1}$] (Kessler & Oosterbaan 1980), basic infiltration Ib [$m \cdot h^{-1}$] (van der Meer & Messemaeckers van de Graaff 1979), field capacity FC [$m^3 \cdot m^{-3}$] and wilting point WP [$m^3 \cdot m^{-3}$] were determined for each plot (Kędziora 2008).

Four series were prepared for measuring soil moisture for each 25 cm layer above the water table in each plot. Soil samples for moisture determination were collected using a drill with a diameter of 110 mm. Groundwater depth measurements were made at weekly intervals in 2001 and ten day intervals in 2002.

20.3.3 Model description

The model is designed to estimate:

1. Real evapotranspiration for the different ecosystems on the basis of the calculation of all the components of the water balance and any vertical and horizontal water fluxes.
2. Ratio of water uptake by plants from the saturated zone to total water uptake of the plants.

The time resolution is one day calculation step. The data obtained during 2001 have been used for parametrisation and optimisation of the model, while the data from 2002 served for verification of the model.

20.3.4 Structure of the model

The model is composed of three blocks (Fig. 20.4): 1. Input data, 2. Core of the model, 3. Output data.

D20100874 The functions that are applied in the model are created from the laws of physics and the field measurements, and can be modified in the future on the basis of further research.

20.3.4.1 *Input data*

Meteorological data (24 hour values): extraterrestrial radiation $R_0(\tau)$ [$W\,m^{-2}$], surface albedo α [dimensionless], relative sunshine $u(\tau)$ [dimensionless], air temperature $t(\tau)$ [°C], saturation water vapour pressure deficit $d(\tau)$ [hPa], water vapour pressure $e(\tau)$ [hPa], wind speed $v(\tau)$ [$m\,s^{-1}$], precipitation $Pr(\tau)$ [mm] and duration of precipitation $tp(\tau)$ [h], (τ is a symbol of day).

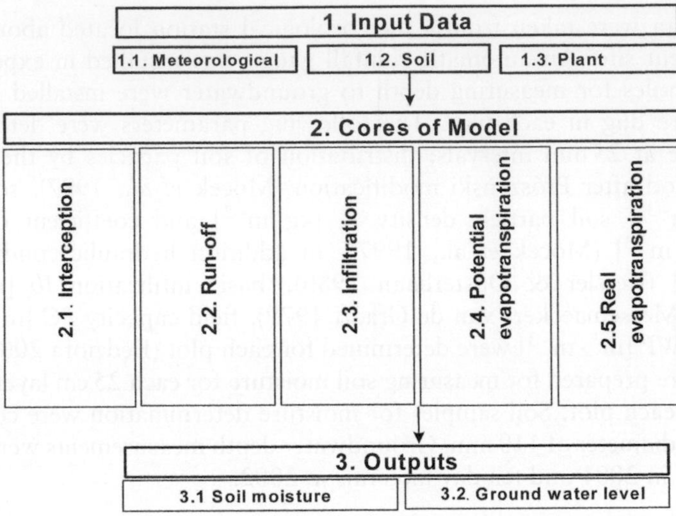

Figure 20.4 Structure of the model.

Soil data: for each plot (u): width of plot (a) [m], length of plot (b) [m], basic infiltration Ib [mm h^{-1}], groundwater level asl. $GWL(\tau_0)$ [m], for each layer: thickness m_i [m], hydraulic conductivity (K) [m 24hour^{-1}], total porosity No_i [dimensionless], temporary water content $WA_i(\tau_0)$ [m$^3 \cdot$ m^{-3}], water content at field capacity FC_i [m$^3 \cdot$ m^{-3}], wilting point water content WP_i [m$^3 \cdot$ m^{-3}], (i – layer number, layer "1" denotes the first upper layer of soil profile). The groundwater level above sea level (asl.) and the temporary water content are initial values at the first day τ_0.

Plant data: plant development stage f_u [dimensionless] (ranges from 0 to 1), range of root zone Dk_u [m] for each day, and maximal root range Mr_u [m] for each plot.

20.3.4.2 *Calculation of model cores*

(All values for the τ day and for the u plot are expressed in mm/24 hours)

Interception

Interception $It_u(\tau)$ is determined by formula:

$$It_u(\tau) = \min(\Pr_u(\tau); f_u(\tau)Ic), \tag{20.1}$$

where Ic stands for capacity of interception for established vegetation, and changes from 1 mm for crop to 2 mm for mixed shelterbelts.

Surface run-off

Surface run-off $Rs_u(\tau)$ from any plot u is calculated as follows:

$$Rs_u(\tau) = Rc_u(\tau)(Op_u^{\delta}(\tau) - It_u(\tau) - tp(\tau)Ib_u), \tag{20.2}$$

where the reduction coefficient $Rc_u(\tau)$ depending on the slope Sl_u and plant development stage $f_u(\tau)$ is calculated as follows:

$$Rc_u(\tau) = (0.1 - 0.05f_u(\tau))Sl_u. \tag{20.3}$$

Sl_u [degree] is determined from formula:

$$Sl_u = \frac{180}{\pi}\text{arc tg}\left(\max_{u^\delta \in \Omega_u}\left(\frac{Rt_u - Rt_{u^\delta}}{c_u}\right)\right), \tag{20.4}$$

where:
Ω_u – is a set of indices of plots u^δ neighbouring with plot u; c_u – distance between centre of plot u and centre of plot u^δ; Rt_u – ordinate of terrain; $Pr_u^\delta(\tau)$ – sum of precipitation and surface inflow from neighbouring plots obtained by recurrent procedure, starting with plot having the highest value of Rt. When for each $u^\delta \in \Omega_u$ Rt_u is smaller than Rt_{u^δ} then $Sl_u = 0$.

Surface run-off from a given plot is split into different directions according to weights
ω_u determined by the formula:

$$\omega_u = \frac{Sl_{u^\delta}}{\sum Sl_{u^\delta}}, \tag{20.5}$$

where summing pass through the all plots u^δ neighbouring with plot u, and having Rt_{u^δ} smaller then Rt_u.

Infiltration

The amount of water infiltrates into whole soil profile $If_u(\tau)$ is calculated as follows:

$$If_u(\tau) = Pr_u^\delta(\tau) - Rs_u(\tau). \tag{20.6}$$

But the amount of water infiltrating into the unsaturated soil layer i is given by formula:

$$If_{u,i}(\tau) = \min\left[1000(FC_{u,i} - WA_{u,i}(\tau - 1))m_{u,i} + ETR_{u,i}(\tau);\ If_u(\tau) - \sum_{j=1}^{i-1} If_{u,j}(\tau)\right] \tag{20.7}$$

And the amount of water infiltrating into the saturated soil layer z is given by formula:

$$If_{u,z}(\tau) = If_u(\tau) - \sum_{i=1}^{z-1} If_{u,i}(\tau). \tag{20.8}$$

Potential evapotranspiration

Potential evapotranspiration is calculated according to the Penman formula (Penman, 1949):

$$LE = \frac{\vartheta(Rn + G) + Ea}{1 + \vartheta}, \tag{20.9}$$

where:
LE – density of latent heat flux [W m^{-2}],
Rn – net radiation [W m^{-2}]
G – density of soil heat flux [W m^{-2}],
Ea – atmospheric water vapour demand [W m^{-2}],
ϑ – coefficient determining share of energetic factor ($Rn + G$) and atmospheric factor (Ea) in formation of potential evapotranspiration. It can be calculated from the formula:

$$\vartheta = 0.688 \exp(0.05662t). \tag{20.10}$$

Net radiation is calculated by the formula:

$$Rn = (1 - \alpha)(0.22 + 0.54u)R_0 - 5.68 \cdot 10^{-8}(t + 273)^4(0.56 - 0.08e^{0.5})(0.1 + 0.9u), \tag{20.11}$$

– density of soil heat flux by the formula:

$$G = -0.2Rn(1 - 0.75f)(\sin\frac{\pi}{6}(l - 1)), \tag{20.12}$$

– and atmospheric water vapour demand by the formula:

$$Ea = 7.44(1 + 0.54v)d. \tag{20.13}$$

l – ordinal number of the month. Other symbols as in Input Data.

Finally potential evapotranspiration [mm/24 hours] are calculated as follows:

$$ETP(\tau) = \frac{LE}{28.34}. \tag{20.14}$$

Actual evapotranspiration

Actual evapotranspiration is calculated in 5 few steps.
Step 1. The ratio of available water $x_{u,i}(\tau)$ for each layer i in any plot u is determined according to the formula:

$$x_{u,i}(\tau) = \frac{WA_{u,i}(\tau) - WP_{u,i}}{FC_{u,i} - WP_{u,i}}, \tag{20.15}$$

and soil-plant coefficient $k_{u,i}(\tau)$:

$$k_{u,i}(\tau) = 0.2(1 + 2x_{u,i}(\tau))(1 + f_u(\tau)). \tag{20.16}$$

Step 2. Ratio of available water $x_u(\tau)$ for the whole plot is determined as follows:

$$x_u(\tau) = \frac{\sum_i m_{u,i} x_{u,i}(\tau)}{\sum_i m_{u,i}}, \tag{20.17}$$

where the water passes through all layers and meet the condition:

$$\sum_{j=1}^{i-1} m_{u,j} < Dk_u(\tau) \tag{20.18}$$

and $Dk_u(\tau) = f_u(\tau) Mr_u$; only for crop fields.

Step3. Soil-plant coefficient $k_u(\tau)$ for each plot with plant cover is calculated according to the formula:

$$k_u(\tau) = 0.2(1 + 2x_u(\tau))(1 + f_u(\tau)). \tag{20.19}$$

In the case of bare soil the formula is as follows:

$$k_u(\tau) = 0.2(1 + 2x_{u,1}(\tau)). \tag{20.20}$$

For water reservoir $k_u(\tau) = 1$.

Step 4. Actual evapotranspiration $ETR_u(\tau)$ for the whole plot u located in arable land is calculated according to the formula:

$$ETR_u(\tau) = k_u(\tau) ETP_u(\tau). \tag{20.21}$$

But for shelterbelts and tree stands according to the formula:

$$ETR_u(\tau) = 1.2 k_u(\tau) ETP_u(\tau). \tag{20.22}$$

Step 5. Split of actual evapotranspiration into different soil layer is calculated according to the formula:

$$ETR_{u,i}(\tau) = \frac{d_{u,i}(\tau) ETR_u(\tau)}{\sum_j d_{u,j}(\tau)}, \tag{20.23}$$

where j through all layers in the unsaturated zone and water table.

Coefficient $d_{u,j}$ for the unsaturated zone is calculated as follows:

$$d_{u,i}(\tau) = k_{u,i}(\tau) \left(\frac{DGWL_u(\tau - 1)}{g_{u,i}} \right)^{1/3} \tag{20.24}$$

when the deph of the centre of layer i $g_{u,i}$ is smaller than the groundwater level depth $DGWL_u(\tau)$,

where:
$DGWL_u(\tau) = Rt_u - GWL_u(\tau)$,
$g_{u,i} = \sum_{j=1}^{i-1} m_{u,j} + 0.5m_{u,i}$ is depth of centre of layer i.

In the other case:

$$d_{u,i}(\tau) = \max\left(0; (1 - g_{u,i} + Dk_u(\tau))k_{u,i}(\tau)\left(\frac{DGWL_u(\tau-1)}{g_{u,i}}\right)^{1/3}\right), \tag{20.25}$$

For the saturated zone the formulae are as follows:

- for cultivated fields:

$$d_{u,z}(\tau) = k_{u,z}(\tau)f_u(\tau)\min(4.5; 9 - 3DGWL_u(\tau-1)), \tag{20.26}$$

- for shelterbelts:

$$d_{u,z}(\tau) = 5.2k_{u,z}(\tau)f_u(\tau)\exp(-0.05DGWL_u(\tau-1)), \tag{20.27}$$

Actual evapotranspiration from layer i of the bare soil is calculated according to the formula:

$$ETR_{u,i}(\tau) = \left[ETR_u(\tau) - \sum_{j=1}^{i-1} ETR_{u,j}(\tau)\right](x_i(\tau)). \tag{20.28}$$

Groundwater inflow and outflow

Inflow or outflow (direction is determined by the slope ofthe groundwater table $Gf_u(\tau)$ for each plot u is calculated according to the formula:

$$Gf_u(\tau) = \sum_{u^\delta \in \Omega_u} \frac{10^3 K_u(GWL_{u^\delta}(\tau-1) - GWL_u(\tau-1))}{c_{u^\delta}} \frac{A_{u^\delta}}{B_u}, \tag{20.29}$$

where: Ω_u – is a set of indices of plots u^δ neighbouring with plot u; c_{u^δ} – distance between centre of plot u and centre of plot u^δ; B_u – area of plot u; A_{u^δ} – area of cross section of the layer conducting groundwater. It can be calculated as follows:

$$A_{u^\delta} = 0.5\left([GWL_u(\tau-1) - GWL_{u^\delta}(\tau-1)] - \left[\sum_{i=1}^{s_u} m_{u,i} - \sum_{i=1}^{s_{u^\delta}} m_{u^\delta,i}\right]\right) a_u, \tag{20.30}$$

where a_u – length of boundary between plot u and plot u^δ, s_u – number of layers form the surface to the impermeable layer.

20.3.4.3 Outputs

Soil moisture

Because horizontal water flow in the unsaturated zone is small in comparison with vertical transport it is neglected in the daily soil moisture calculation. Thus the soil moisture of a given layer j of plot u at given day $WA_{u,i}(\tau)$ is equal to the soil moisture in the previous day $WA_{u,i}(\tau-1)$ minus evapotranspiration $ETR_{u,i}(\tau)$ and plus water that has percolated into the layer $If_{u,i}(\tau)$. It is calculated as follows:

$$WA_{u,i}(\tau) = WA_{u,i}(\tau-1) + \frac{If_{u,i}(\tau) - ETR_{u,i}(\tau)}{10^3 m_{u,i}}. \tag{20.31}$$

Groundwater level

The groundwater level asl. for the centre of plot u in a given day is calculated according to the formula:

$$GWL_u(\tau) = GWL_u(\tau-1) + \frac{If_{u,z}(\tau) - ETR_{u,z}(\tau) + Gf_u(\tau)}{10^3 Ne_{u,z}}, \tag{20.32}$$

where: $Ne_{u,z}$ – effective porosity of soil layer in saturated zone.

Error of model

The root mean square error (RMSE) and normalised root mean square error (NRMSE) were calculated to assess the usefulness of the model to estimate the share of groundwater used by the plants for evapotranspiration. The following equations were used:

$$RMSE = \sqrt{\frac{\sum_1^n (y-\hat{y})^2}{n}} \tag{20.33}$$

and

$$NRMSE = \frac{RMSE}{y_{\max} - y_{\min}}, \tag{20.34}$$

where: y – measured values, $\hat{y}$ – values calculated by model.

Calculations were carried out for model outputs for soil moisture and groundwater level.

20.4 RESULTS AND DISCUSSION

The model was used to calculate components of the water balance in order estimate the ratio of water taken by shelterbelts from groundwater to total evapotranspiration and to determine the groundwater level. Shelterbelts take in the excess chemical compounds from groundwater and reduce their concentration. The effectiveness of this depends on the amount of water taken from groundwater by the vegetation. In turn, knowledge of the share of groundwater in the total evapotranspiration allow the calculation of

Table 20.1 Water balance of plot P11 in individual half-month period in 2001 and 2002 years [mm].

Period	Precipitation	ETR	Run-off	Underground outflow	Change of soil retention: Saturation zone	Change of soil retention: Aeration zone	Balance
Year 2001							
16–30 Jun.	46	−75	0	20	14	−6	−1
1–15 Jul.	27	−92	0	31	12	18	−4
16–31 Jul.	101	−87	0	17	12	−40	2
1–15 Aug.	21	−77	0	12	20	23	−2
16–31 Aug.	27	−61	0	9	13	10	−2
1–15 Sep.	34	−36	0	15	−7	−8	−1
16–30 Sep.	49	−20	0	30	−37	−16	5
16 Jun.–30 Sep.	**305**	**−448**	**0**	**134**	**27**	**−19**	**−3**
Year 2002							
26 Mar.–15 Apr.	4	−19	0	2	2	9	−1
16–30 Apr.	6	−30	0	10	4	8	−1
1–15 May	14	−59	0	−6	43	9	1
16–31 May	23	−63	0	2	32	5	−1
1–15 Jun.	8	−59	0	2	26	21	−1
16–30 Jun.	11	−73	0	3	33	26	0
1–15 Jul.	40	−68	0	10	19	−2	−2
16–31 Jul.	6	−62	0	3	24	30	0
1–15 Aug.	45	−50	0	10	8	−15	−2
16–31 Aug.	0	−61	0	11	14	35	−1
1–15 Sep.	29	−30	0	6	1	−6	−1
16–30 Sep.	9	−14	0	2	0	2	0
16 Jun.–30 Sep.	**140**	**−358**	**0**	**45**	**99**	**70**	**−6**
26 Mar.–30 Sep.	**195**	**−588**	**0**	**55**	**206**	**122**	**−9**

the width of the shelterbelt to accomplish the desired effect in controlling the spread of diffuse pollution from agriculture.

The analysis of the data obtained during the two year experiment is focused on the four internal plots to determine the following issues: 1. Water balance, 2. Changes of groundwater level, 3. Share of groundwater in total evapotranspiration of the shelterbelt.

Water balance

The measurements in 2001 were carried out between 16 June and the end of September while in 2002 from the end of March to the end of September. The values given in the last row in Table 20.1 are representative of the whole vegetation season. Plot P11 is located in flat land, and is the lowest place in shelterbelt, so no runoff was observed; all precipitation penetrated into soil. Groundwater flowing into the experimental area from adjoining plots are oriented towards plot P11 (Fig. 20.3). Evapotranspiration was higher in the wetter year 2001 (precipitation was 305 mm) reaching 448 mm than in the dryer 2002 year (precipitation was only 140 mm) when it reached 358 mm. As a result of such meteorological conditions groundwater flowing into plot P11 was three times higher in 2001(134 mm) than in 2002 (45 mm). However, extraction of water

Table 20.2 Groundwater level at the beginning and at the end of the measuring period [m asl].

	Plot number			
Date	*P5*	*P8*	*P11*	*P14*
2001 year				
June 14	76.88	76.94	76.68	78.64
October 1	77.25	76.86	76.54	78.91
Difference [cm]	37	−8	−14	27
2002 year				
March 26	77.76	77.20	76.97	79.10
June 14	76.90	76.85	76.52	78.85
October 1	77.02	76.63	75.92	78.60
Diference [cm] June–October	17	−22	−60	−25
Difference [cm] March–October	−74	−57	−105	−50

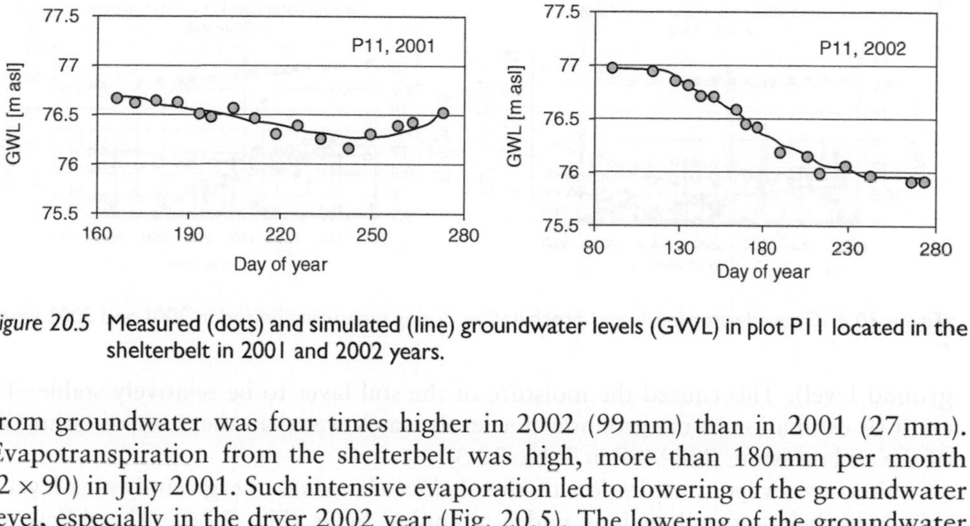

Figure 20.5 Measured (dots) and simulated (line) groundwater levels (GWL) in plot P11 located in the shelterbelt in 2001 and 2002 years.

from groundwater was four times higher in 2002 (99 mm) than in 2001 (27 mm). Evapotranspiration from the shelterbelt was high, more than 180 mm per month (2 × 90) in July 2001. Such intensive evaporation led to lowering of the groundwater level, especially in the dryer 2002 year (Fig. 20.5). The lowering of the groundwater level under the shelterbelt (Plot 11) is much higher than under other areas (Table 20.2). Plots P5 and P14 are crop fields and Plot P8 is the pond. Under the shelterbelt the groundwater level lowering was as much as 105 cm in 2002. Taken into consideration that effective porosity for the loamy soil under the shelterbelt is 0.20, the amount of water taken from the saturated zone was 210 mm (105 cm × 0.20 = 210 mm). This agrees with the value of water taken from the saturated zone estimated by the water balance (Tab. 20.1). The change in water level in the in 2002 was 206 mm.

The error in estimating the soil moisture by the model changed from 0.015 $m^3\ m^{-3}$ at plot 5 (rape seed field) in 2002 to 0.072 $m^3\ m^{-3}$ at plot 11 (shelterbelt) in 2001 (RMSE Table 20.3). The error in estimating the groundwater levels changed from 36 mm at plot 5 in 2001 year to 135 mm at the same plot in 2002.

The normalized errors (NRMSE) fluctuated between 12% and 28% for soil moisture, and between 7% and 19% for groundwater levels. Higher errors for soil moisture estimation occurred in 2001 as a result of a high groundwater level (0.5 m beneath

Table 20.3 Error of model. Soil moisture in $m^3 m^{-3}$ (RMSE) and in % (NRMSE), Groundwater level in m asl (RMSE) and in % (NRMSE).

	Plot					
	P5		*P11*		*P14*	
Year	*RMSE*	*NRMSE*	*RMSE*	*NRMSE*	*RMSE*	*NRMSE*
Soil moisture						
2001	0.035	27.7	0.072	31.5	0.031	16.7
2002	0.015	20.8	0.037	12.6	0.023	23.8
Groundwater level [m asl.]						
2001	0.036	6.9	0.065	12.4	0.070	18.9
2002	0.135	13.9	0.081	7.7	0.051	10.2

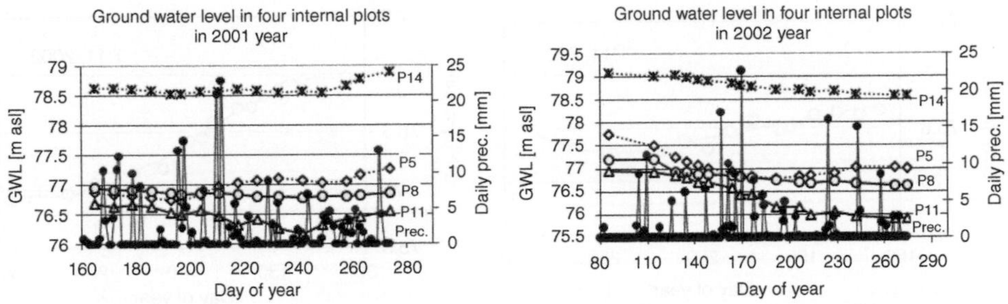

Figure 20.6 Groundwater levels and precipitation during measuring period in 2001 and 2002 years.

ground level). This caused the moisture of the soil layer to be relatively stable. The error of estimation of the groundwater levels was lower than the error of estimation of the soil moisture (NRMSE in Table 20.3).

Intensive evapotranspiration of the shelterbelt makes the groundwater level in the shelterbelt lower than it is under adjoining fields (Fig. 20.6). The differences between evapotranspiration in the cropped field (plot P5) and the shelterbelt (plot P11) are apparent in 2002 (Fig. 20.6). Relatively high precipitation between 155 and 180 mm day^{-1} is not reflected in the groundwater levels under the shelterbelt. The lowering trend was still observed, while in plot P5 an increase in groundwater level occurred. The effect of plants and weather conditions on groundwater levels is evident from comparison of precipitation and evapotranspiration over the two years (Table 20.4).

In the wetter 2001 year the ratio E/P was 1.45 for the shelterbelt, while for the cropped field it was 0.84. This means that even in wet years the shelterbelt used more water than it received from precipitation whereas the cropped field used only 84% of the precipitation. In the dryer year 2002, the shelterbelt used 256% of the precipitation while the cropped field used only 142%. this explains the impact of the shelterbelt on the groundwater levels, and that it can change the local direction of groundwater flow (Fig. 20.7). Lowering of the groundwater level under the shelterbelt occurred in both

Table 20.4 Precipitation and evapotranspiration of shelterbelts and winter wheat field.

	Year 2001			*Year 2002*		
Ecosystem	*Precipitation mm/day*	*Evapotranspiration mm/day*	*E/P*	*Precipitation mm/day*	*Evapotranspiration mm/day*	*E/P*
Shelterbelt	2.80	4.06	1.45	1.28	3.28	2.56
Winter wheat	2.80	2.41	0.86	1.28	1.82	1.42

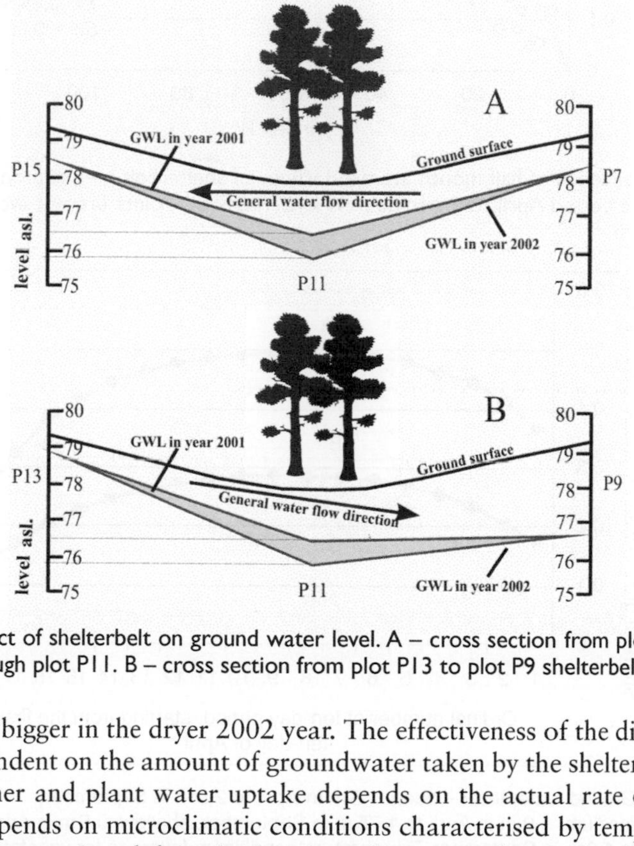

Figure 20.7 Impact of shelterbelt on ground water level. A – cross section from plot P15 to plot P7 through plot P11. B – cross section from plot P13 to plot P9 shelterbelt (See Fig. 20.3).

years, but was bigger in the dryer 2002 year. The effectiveness of the diffuse pollution change is dependent on the amount of groundwater taken by the shelterbelt. The ratio between weather and plant water uptake depends on the actual rate of evaporation *ETR* which depends on microclimatic conditions characterised by temperature, wind speed, vapour saturation deficits in the air and on groundwater levels. The index is given by the equation: $W = ETR/DGWL$.

The equation enabled the value of *p*-ratio for each ten-day period of the whole vegetation season (Fig. 20.9) to be calculated as well as the average value for p-ratio for the whole period under different thermal conditions and different *DGWL* (Table 20.3). At the beginning of the vegetation season the shelterbelt uptakes only 18% of the water from the deep saturated zone in cold weather while from shallow saturated zone under warm conditions the uptake amounts to 37% (Fig. 20.9). When, the temperature and evapotranspiration increases and water available in the aeration layer of the soil decreases, the uptake of water from the saturated zone increases by up to 30% from the deep saturated zone during cold weather and up to 50% from the shallow saturated

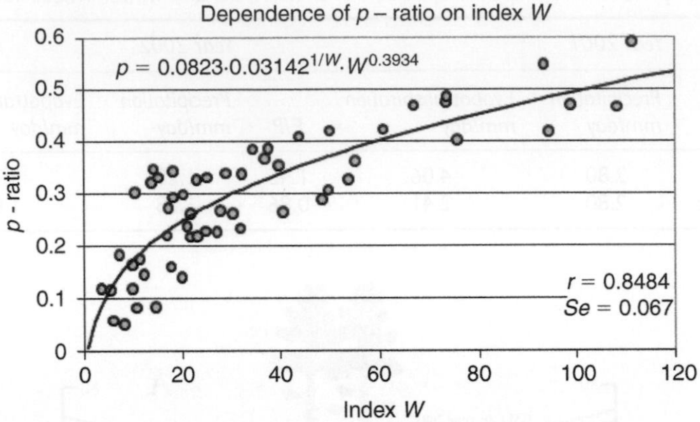

Figure 20.8 Dependence of half month averaged values of shelterbelt p – ratio on values of index W in the period April to September. $W = ETR/DGWL$. Points present average value for half month.

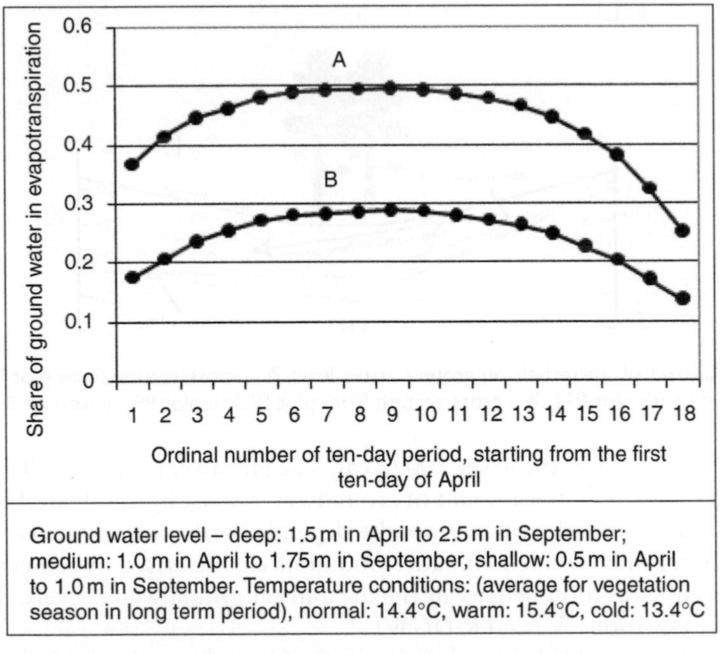

Figure 20.9 Seasonal values of p-ratio in warm year and shallow GWL (A), and in cold year and deep GWL (B).

zone when it is warm. In July the decreasing trend of water uptake from the saturated zone shows the higher rate under warm conditions (Fig. 20.9).

Thus, assuming that absorption of chemical compounds such as nitrate by plants is proportional to the rate of water uptake it can be concluded that shelterbelts limit the spreading of nitrate leached from cultivated fields better under warm weather

Table 20.5 Share of water taken up by shelterbelt from saturation zone in total evapotranspiration under different thermal conditions and different depth of groundwater level (Average value for period: April–September).

Depth of ground water level	*Thermal conditions* warm: 15.4°C	normal: 14.4°C	cold: 13.4°C
Deep: 1.5 m (April) – 2.5 m (September)	0.281	0.264	0.244
Medium: 1.0 m (April) – 1.75 m (September)	0.341	0.322	0.299
Shallow: 0.5 m (April) – 1.0 (September)	0.439	0.421	0.397

conditions with a shallow groundwater table than during cold weather and a deep groundwater level. On average the shelterbelt can use 24% of water from the saturated zone in a cold and dry year and 44% in a warm and wet year (Table 20.5).

20.5 CONCLUSION

Shelterbelts can reduce the concentration of chemicals transported by groundwater and their effectiveness depends on the intensity of evaporation, the depth to groundwater level, and prevailing weather conditions. The warmer and windier the day and shallower the groundwater the greater the effect of the trees in reducing pollutant quantities in groundwater.

REFERENCES

Kędziora A., Olejnik J. (2002) Water balance in agricultural landscape and options for its management by change in plant cover structure of landscape. In: *Landscape ecology in agroecosystems management*. Ed. L. Ryszkowski. CRC Press, Boca Raton, pp. 57–110.

Kędziora A., Ryszkowski L. (2004) Management of water resources in agricultural landscapes. In: *Ways to promote the ideas behind the CBS's ecosystem approach in Central and Easter Europe*. Eds. Horst Korn, Rainer Schilep, Jutta Stadler. BFN, Federal Agency for Nature Conservation. Bonn. Germany: 106–115.

Kędziora A. (2008) *Fundation of Agrometeorology* (In Polish). PWRiL, Poznan

Kessler J., Oosterbaan R.J. (1980) Determining hydraulic conductivity of soils. *Drainage principles and applications*, Publication 16 – Vol. III, Surveys and investigations, International Institute for Land Reclamation and Improvement, Wageningen, pp. 113–152.

Millennium Ecosystem Assessment (2005) Ecosystems and human well-being (Synthesis) Island Pres, Washington DC. 137 pp.

Mocek A., Drzymała S., Maszner P. (1997) *Genesis, analysis and classification of soils*. (In Polish). Wydawnictwo Akademii Rolniczej im A. Cieszkowskiego, Poznań. 416 pp.

Penman N.A. (1949) The dependence of transpiration on weather and soil conditions. *J. Soil Sci* 1: 74–89

Ryszkowski L., Szajdak L., Bartoszewicz A., Życzyńska-Bałoniak I (2002) Control of diffuse pollution by mid-field shelterbelts and meadow strips. In: *Landscape ecology in agroecosystems management*. Ed. L. Ryszkowski. CRC Press. Boca Raton, pp. 111–143.

Ryszkowski L., Bartoszewicz A. (1989) Impact of agricultural landscape structure on cycling of inorganic nutrients In: *Ecology of arable land*. Eds. M. Clarholm & L. Bergström. Kluwer Academic Publishers, pp. 241–246.

Ryszkowski L., Bartoszewicz A., Kędziora A. (1997) The potential role of mid-field forest as buffer zones. In: *Buffer zones: their processes and potential in water protection*. Ed. N. E. Haycock, T. P. Burt, K. W. T. Goulding & G. Pinay. Quest Environmental, Harpenden U. K., pp. 171–191.

Ryszkowski L., Bartoszewicz A., Kędziora A. (1999) Managements of matter fluxes by biogeochemical barriers at the agricultural landscape level. *Landscape Ecology* 14:479–492.

Ryszkowski L., Kędziora A. 2006. Modyfication of water flows and nitrogen fluxes by shelterbelts. *Ecol. Eng*. 29, 388–400.

van der Meer K., Messemaeckers van de Graaff R.H. (1980) Hydropedological survey. *Drainage principles and applications*, Publication 16 – Vol. III, Surveys and investigations, International Institute for Land Reclamation and Improvement, Wageningen, pp. 113–152.

Woś (1994) *Climate of Poland*. (In Polish). Wydawnictwo Naukowe PWN. Warszawa. 301 pp.

Part 4

Regional groundwater problems

Chapter 21

Hydrogeological characterisation of the heterogeneity of aquitards from a multilayered system

Olivier Cabaret, Alain Dupuy & François Larroque
Institut EGID, Université Bordeaux-3, Pessac, France

ABSTRACT

The characterisation of aquitards is obligatory to quantify the amount of leakage contributing to the recharge of aquifers in a multilayered system. This characterisation involves determining their hydrogeological heterogeneity. In the area of Bordeaux (south west France), the heterogeneity of the aquitards was studied at two different scales. The use of well logs allowed clarification of the vertical system and the geology of the aquitards. A high resolution approach consisted of drilling a well directly in an aquitard and studying the evolution of pressures across it. In addition, physical measurements on cores provided the hydraulic properties. These two methods gave information about the geological nature of the aquitards and their permeabilities. The results were used to update the hydrogeological model in order to assist in enhancing freshwater exploitation.

21.1 INTRODUCTION

In sedimentary basins, the vertical organisation of geological deposits leads to the existence of interbedded aquifers and aquitards. This alternation of hydrogeological units forms a complex multilayered aquifer system. The aquitards, also called leaky confining layers, are low permeability units generally composed of clayey materials. They can have very high storage capacities but they cannot transmit water at fast enough rates to supply wells. Nevertheless, they can transmit water slowly from one aquifer to another leading to issues of quality. This exchange is generally known as "leakage". This phenomenon is important in long-term transient systems and is generally considered a significant component of the total inter-aquifer recharge. It is therefore necessary to assess these vertical fluxes as accurately as possible and integrate them in general groundwater flows in order to tackle the management of groundwater resources. Stratified rock sequences, such as those composed by sedimentary formations, lead to composite porous media with heterogeneous properties. The flow through these formations, moreover of weak conductivity, is much more complex than the classical 1D model used, as pointed out by previous works (Remenda, 2001; Eaton *et al.*, 2007). This could be explained by the multiple sedimentary layers of differing properties found in stratified rock sequences. In order to quantify as accurately as possible the fluxes flowing through aquitards, the characterisation of their architecture and their hydraulic properties remains a major objective.

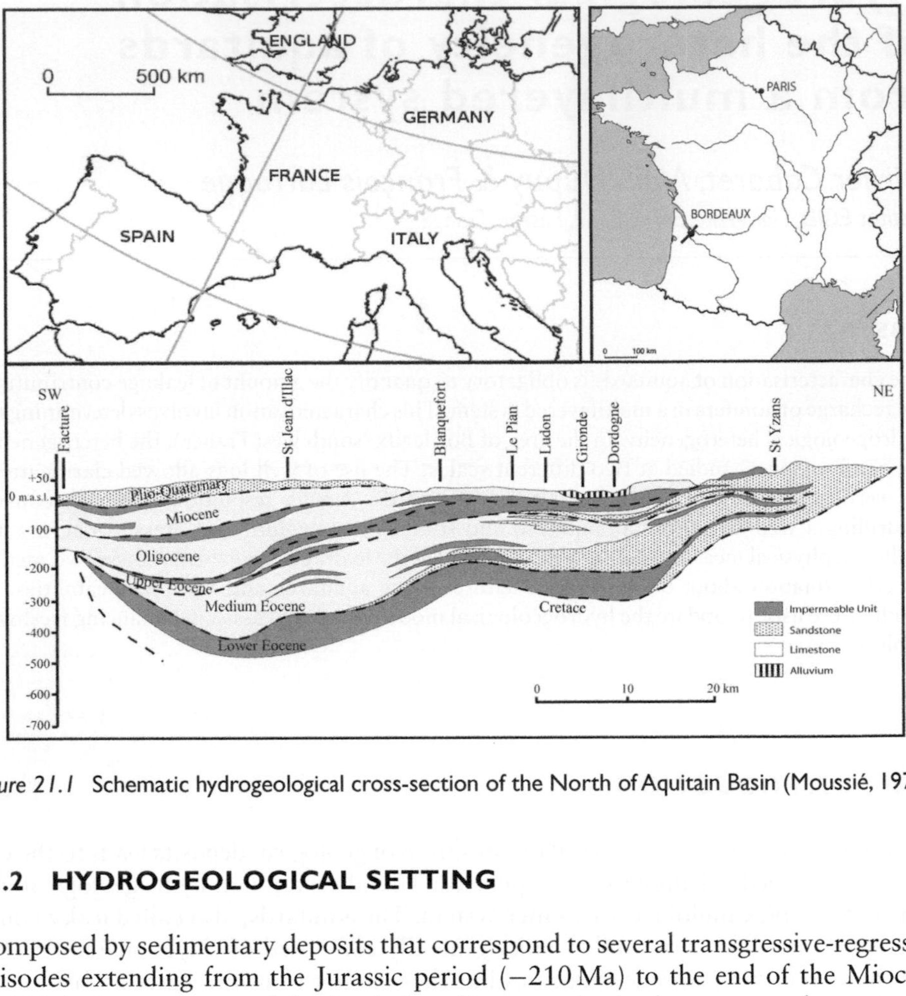

Figure 21.1 Schematic hydrogeological cross-section of the North of Aquitain Basin (Moussié, 1972).

21.2 HYDROGEOLOGICAL SETTING

Composed by sedimentary deposits that correspond to several transgressive-regressive episodes extending from the Jurassic period (−210 Ma) to the end of the Miocene (−5 Ma), the north part of the Aquitain sedimentary basin shows a complex structure both vertically and horizontally. Therefore, this basin is a multilayered aquifer system (Fig. 21.1) within which 6 main aquifers (one unconfined aquifer and five confined aquifers) are separated by low-permeability units that are more or less continuous.

Currently, groundwater resources are exploited for different uses (freshwater supply, geothermal energy, thermal water, agricultural and industrial domains, etc.). The amount of water available from such a system is an essential issue for the long-term management of freshwater supplies. The anthropogenic influence manifests itself through important withdrawals concentrated on the three shallowest confined aquifers (Eocene, Oligocene and Miocene aquifers). Two recent hydrogeological models developed in this area incorporate the confining layers. Even if they both identify the presence of vertical flows through the aquitards, they cannot correctly reproduce the functioning of the groundwater system and the state of the water reserves. The lack of knowledge about the structure of aquitards and the associated hydrodynamic measurements leads to uncertainties. The use of direct or indirect measurements to redefine

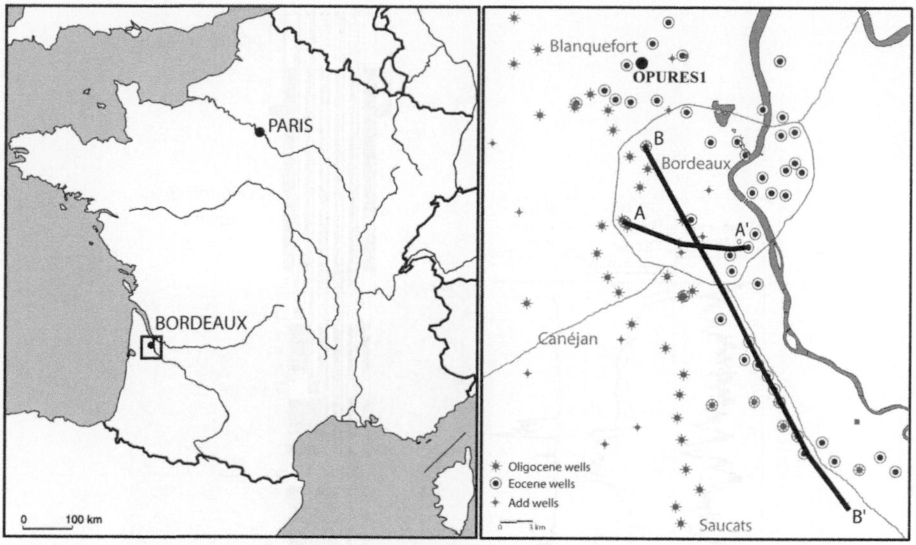

Figure 21.2 Map of the study area with boreholes and cross-sections A-A′ and B-B′ locations.

the vertical organisation of the multilayered system and the horizontal heterogeneity of the aquitards is a priority. These measurements are first based on the use of borehole logs. A 110 well database was constructed to organise the information on well completion, geological data and geophysical logs in the study area using Kingdom software (Fig. 21.2).

In addition, a well called OPURES1 (Fig. 21.3) was drilled into an aquitard and for the first time allowed direct long-term observation of the evolution of the hydraulic gradient, and assessment of hydraulic properties at different levels. The aquitard considered is the Oligocene/Upper Eocene Aquitard which is about 50 m thick. It underlies the Oligocene aquifer and confines the Upper Eocene and Medium Eocene aquifers. The latter is heavily pumped by a well 1 km away. The borehole was drilled using both rock coring and air-rotary methods, logged using downhole geophysics, and instrumented as a multi-level well. Five levels were selected for pressure monitoring inside the aquitard (Fig. 21.3).

21.3 BOREHOLE LOGGING CONTRIBUTION TO SPATIAL CHARACTERISATION

Borehole data allow re-examination of the aquitards and their heterogeneities to increase the hydrogeological understanding of the multilayered aquifer system prior to its numerical implementation. The measurements are based on the use of geophysical borehole logs such as Gamma-ray, normal-resistivity and flowmeters which, used as a complement to geological data, define facies boundaries with depth and assess the heterogeneity of the aquitards.

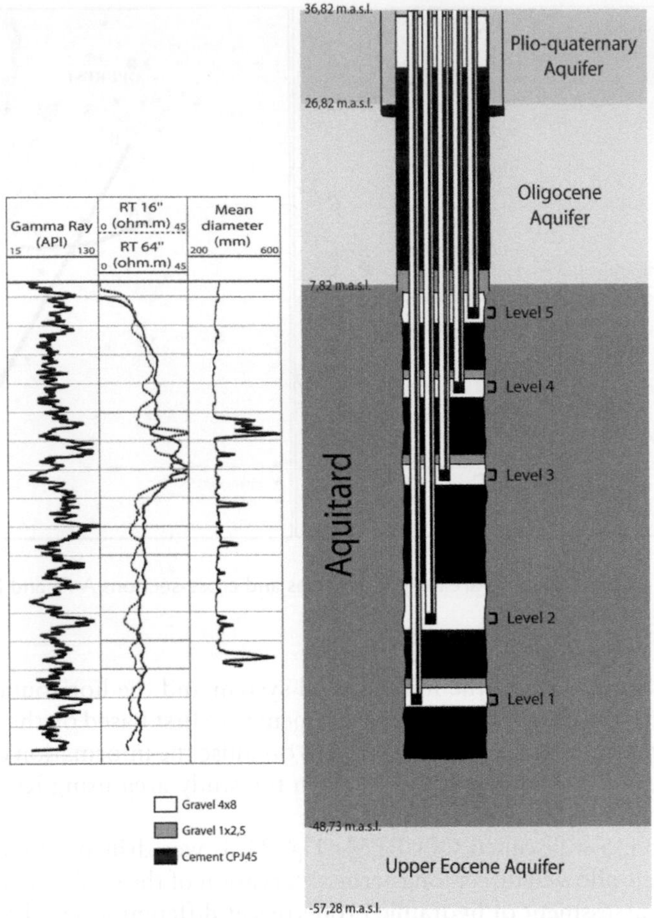

Figure 21.3 Borehole logs and completion of OPURES1.

Indeed, Gamma-ray logs allow for measurement of the total natural radioactivity of the formations intercepted by a well. This measurement is used to distinguish high radioactive shale beds, in our case corresponding to aquitards, from less radioactive sandstones and limestones. Normal-resistivity, enables the main conducting and non-conducting formations (aquitards and aquifers) to be located. Flowmeter log data further constrain the interpretation by locating the permeable zones of the aquifer's limits. Before correlating all these geophysical logs, raw gamma-ray data were pre-processed for import to various formats. Scaled gamma-ray logs were produced from raw gamma-ray data to eliminate calibration gaps induced by various logging probes. Gamma-ray data were scaled as follows (Miller *et al.*, 2000):

$$SCGR = \frac{(GR - Min)}{(Max - Min)} \times 100 \tag{21.1}$$

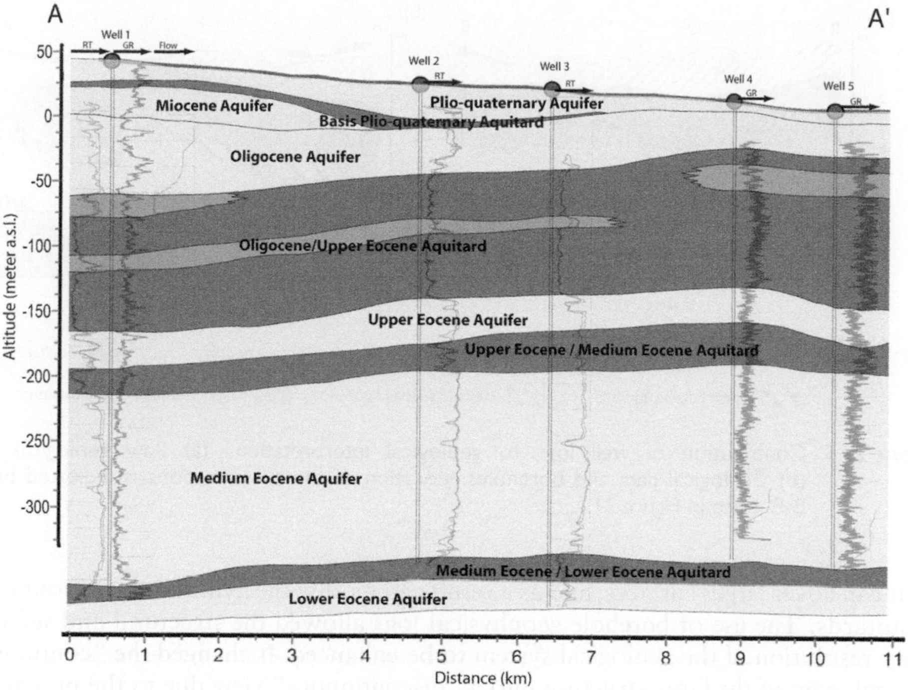

Figure 21.4 Example of interpretation from well logging data: RT = Normal-resistivity; GR = Gamma-Ray. Location of the cross-section is indicated by the A-A′ trace in Figure 21.2.

where *SCGR* is the percent of scaled gamma radiation, *GR* is the original gamma-ray value, *Max* is the base-line value for maximum gamma radiation and *Min* is the base-line value for minimum gamma radiation. The well-to-well correlations were constrained by using the original geological description to reduce the factors affecting the individual gamma-ray response, such as logging speed, and hole condition effects (tubing, casing, cement, diameter…) (Serra & Serra, 2004). This correlation provides the geology of the sedimentary basin in the area of Bordeaux (Fig. 21.4).

By comparing the two cross sections (Fig. 21.5), the contribution of borehole geophysical logs can be seen. This allows a more detailed description of the alternation of the lithologies and their distribution on a regional scale. The borehole logs revealed the presence of new clayey layers, which were not integrated in the generally accepted geological model, and which could play an important part in the recharge of aquifers and in their protection from pollution. They also allowed for the separation of an aquifer into subunits, such as for the Eocene aquifer which was considered homogeneous until now and was subsequently divided into three units. In contrast, the well logs showed that confining units do not exist everywhere between aquifers, allow possible hydraulic continuity, between the Oligocene and Miocene aquifers. The logs revealed their heterogeneous nature of the aquitards. These heterogeneities correspond to limestone and sandstone with variable clay contents. They generally form lenses of limited extension

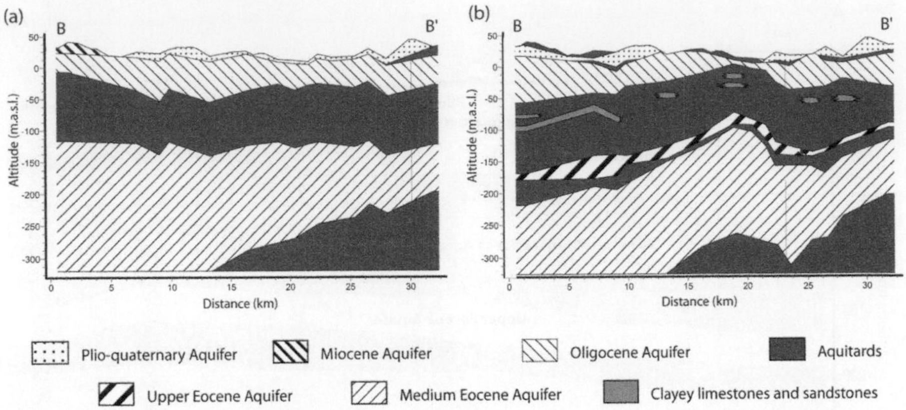

Figure 21.5 Contribution of well logs to geological interpretation. (a) Raw geological data. (b) Geological data and boreholes. Location of the cross-sections is indicated by the B-B′ trace in Figure 21.2.

but can cover larger surfaces, and as a result can modify the hydraulic behaviour of the aquitards. The use of borehole geophysical logs allowed the structural and sedimentary resolution of the geological system to be enhanced. It changed the "continuous" global view of the large structure into a "discontinuous" view due to the presence of clayey layers and their heterogeneity. Finally, it increases the accuracy of the existing numerical hydrogeological models.

21.4 HIGH RESOLUTION AQUITARD ASSESSMENT

The geological results based on borehole logs, cuttings and core descriptions confirm the heterogeneity of the aquitard. Within the clay matrix, some silty, sandy and calcareous layers appear. One of these, situated between 41 and 44 m below ground level shows a higher resistivity and corresponds to a silty and sandy layer. Moreover, the four-arm caliper and inspection of cores indicate the presence of fractures which could permit flows across the aquitard.

Hydraulic heads show a cyclic variation with different amplitudes between the Oligocene aquifer and the aquitard (Fig. 21.6). The hydraulic heads do not react as fast as in the Oligocene aquifer. Head values in the aquitard do not decrease uniformly with depth but there is a vertical downward flow. The distribution of hydraulic heads within the aquitard is not linear or monotonic. This non-linearity is likely due to contrasts between permeabilities across the aquitard.

Hydraulic properties of aquitard materials were measured on core samples. Permeability is usually linked to different properties of the pore space, such as porosity and pore size distribution (Xu *et al.*, 1997). These properties were determined by mercury intrusion porosimetry. This consists of injecting mercury at increasing pressures into a sample to force mercury into all accessible pores (Pellerin, 1980; Daïan, 2007). The total porosity, which corresponds to the interconnected porosity of pore volume

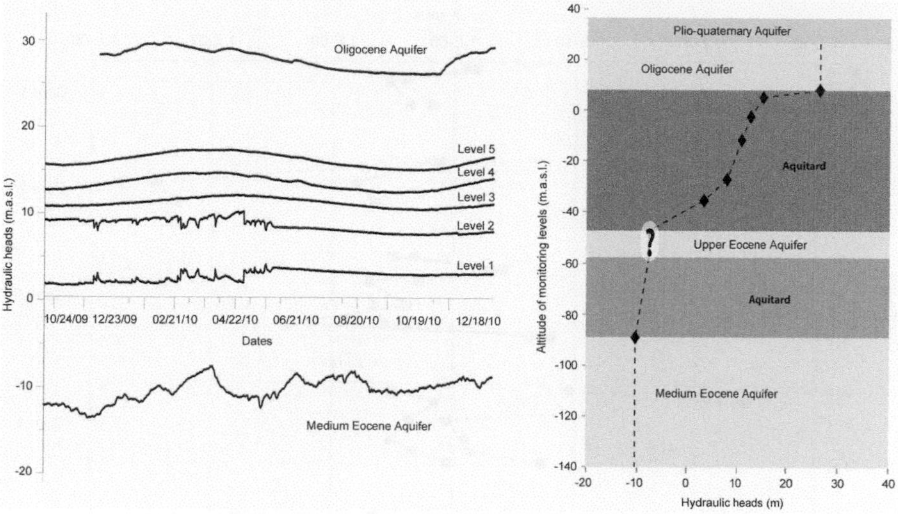

Figure 21.6 Evolution of hydraulic heads and hydraulic gradient.

accessible for mercury, is determined by the total volume intruded. The pore size distribution is determined by the volume injected at each pressure increment. The porous medium is assimilated into a bundle of cylindrical capillary tubes allowing for linking the pore throat size to capillary pressure (Washburn, 1921):

$$R_c = \left(\frac{2\sigma \cos\theta}{P_C}\right) \tag{21.2}$$

where R_c is the average pore-throat size, σ is the interfacial tension (=480 dynes $\cdot$ cm^{-1} for mercury), θ is the angle between the mercury meniscus and pore wall (=140° for mercury) and P_C is the capillary pressure (MPa). From the parameters determined with this method, different theories exist to link them to the permeability. Two models were used to determine the permeability of clays. The first one, called Katz-Thompson's model, is based on the Archie's law (1942) and uses the percolation theory to propose the following relationship for the permeability of rocks saturated with a single liquid phase (Katz & Thompson, 1986):

$$k = C \cdot l_c^2 \frac{1}{F} = C \cdot l_c^2 \frac{\sigma_0}{\sigma_w} \tag{21.3}$$

where k is the permeability (m^2) defined by the Darcy relation, C is a constant close to 1/226, σ_w is the conductivity of the rock saturated with a brine solution with conductivity σ_0 and l_C is the characteristic length of the pore space (m). It corresponds to a continuous path across the sample and is graphically defined as the inflection point of the rapidly rising portion of the curve determined with Washburn's equation.

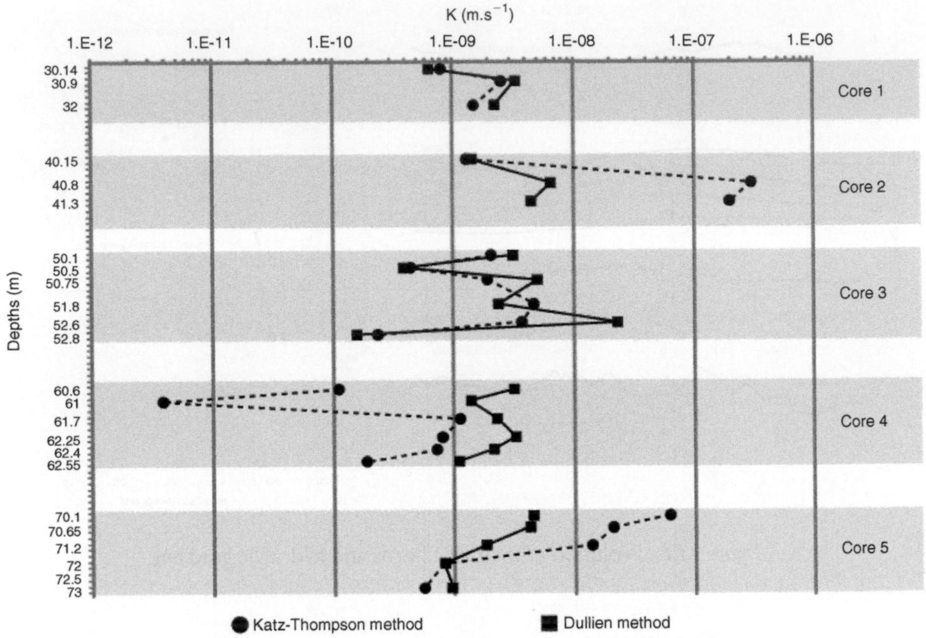

Figure 21.7 Permeabilities obtained by Katz-Thompson and Dullien methods.

The Dullien model (1992) is one of the numerous capillary models derived from the Hagen-Poiseuille equation (Garcia-Bengochea *et al.*, 1979; Juang & Holtz, 1986). The porous medium is considered as a distribution of parallel and circular capillaries with different diameters. The distribution of capillaries ($F(D)$) is linked to the permeability k (cm^2) by:

$$k = \frac{\omega}{96} \frac{\left[\int_0^\infty \frac{F(D)}{D^2} dD \right]^2}{\int_0^\infty \frac{F(D)}{D^6} dD} \tag{21.4}$$

where ω is the porosity (%) and D is the diameter of the capillary (m).

Although the application of these models remains debatable (Waxman & Smits, 1968; Clavier *et al.*, 1984; Lapierre *et al.*, 1990; Schneider, 2008), these models were used to estimate the permeability of the samples and assess the influence of the heterogeneities on the water flow across the aquitard.

The results (Fig. 21.7) from the two methods are quite similar, with a mean permeability close to $1 \cdot 10^{-16}\,m^2$ which corresponds to a hydraulic conductivity of $1 \cdot 10^{-9}\,m \cdot s^{-1}$ for water at 20°C (De Marsily, 1981). At the core scale, K values

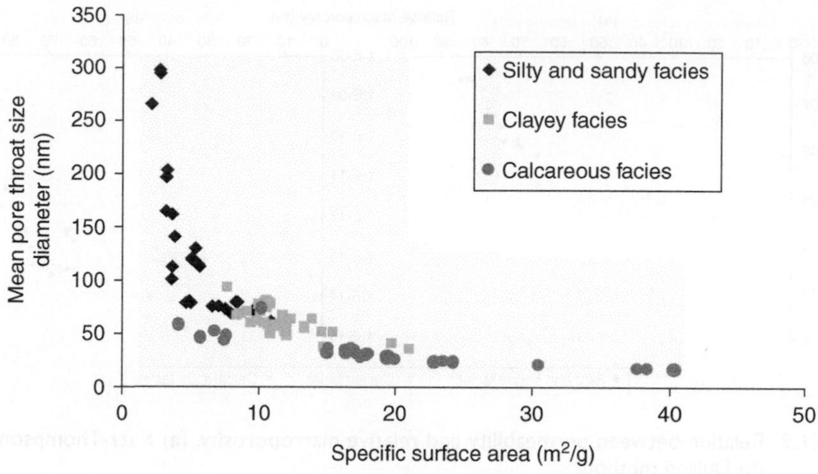

Figure 21.8 Relation between specific surface area and mean pore-throat size diameter.

describe the same patterns. Anyway, the Katz-Thompson model leads to values spreading over 5 orders of magnitude (from $4 \cdot 10^{-12}\,\text{m} \cdot \text{s}^{-1}$ to $4 \cdot 10^{-7}\,\text{m} \cdot \text{s}^{-1}$), while the Dullien model leads to more uniform values.

However, these results are relatively consistent with the variations of lithology observed in the cores. The silty and sandy layers show higher permeabilities than the clay and calcareous layers. These results can be explained by the different physical characteristics in the lithology. From the mercury intrusion porosimetry method, several parameters were determined such as macroporosity (pore-throat size $D_C > 100\,\text{nm}$) and mesoporosity ($D_C \leq 100\,\text{nm}$) (Cerepi *et al.*, 2000), the specific surface area, the mean pore-throat diameter and trapped porosity. This last parameter is inferred from the mercury volume that remains in the sample after the extrusion stage, and it is related to the complexity of the network geometry due to interconnections of large pores by a smaller pore throat (Wardlaw *et al.*, 1987; Sammartino *et al.*, 2003). The silty and sandy layers show the lowest specific surface areas (Fig. 21.8).

This characteristic indicates that these layers could have a higher amount of free water in their structure than the clayey and marly layers. In the same way, the macroporosity is well correlated with the permeability (Fig. 21.9). This one could therefore be essentially controlled by the macropores which correspond to the inter-aggregate pores of a clay.

Finally, the relative trapped porosity shows a positive correlation with the permeability, which is consistent with the proposal of Delage & Lefebvre (1984). They interpreted the trapped or constricted porosity as the inter-aggregate pore space, i.e. the large pores, while the free or non constricted porosity obtained with a second intrusion corresponds to the intra-aggregate porosity that is the smallest pores. The permeability results reflect some heterogeneities in the microstructural characteristics of clays (Fig. 21.10).

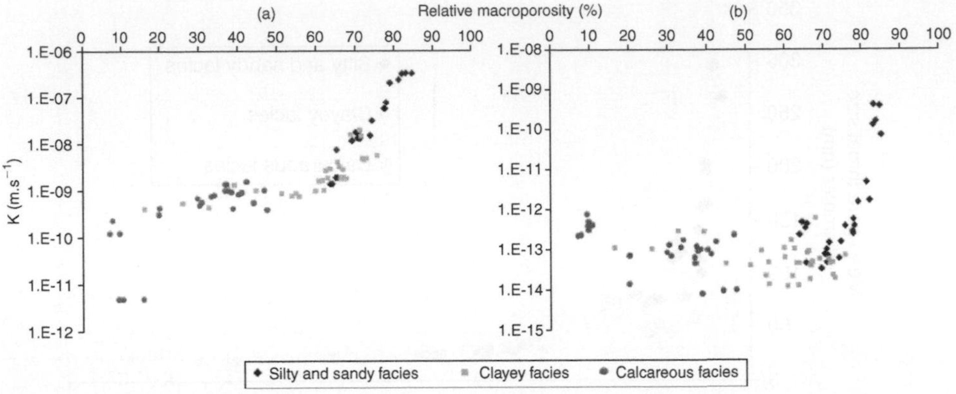

Figure 21.9 Relation between permeability and relative macroporosity. (a) Katz-Thompson method. (b) Dullien method.

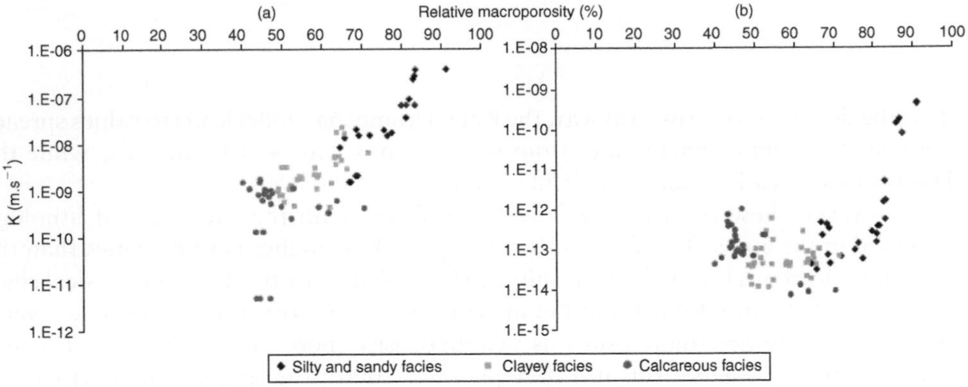

Figure 21.10 Relation between permeability and relative trapped porosity.

Because of the low macroporosity and high specific surface area, the clayey and marly layers are the least permeable. The presence of silts and sands modify the distribution of pore sizes. These layers show a relatively lower specific surface area and higher macroporosity leading to the possibility of free water which could be more easily available for groundwater flows.

21.5 SUMMARY AND CONCLUSION

The reinterpretation based on borehole logging combined with the original description of the geology provided a new view of the geometry of the complex multilayered aquifer system. It allowed the presence or absence of some clay layers between the aquifers to be identified and to define the characteristics of some heterogeneities within the aquitards. In addition, a high resolution description of an aquitard was proposed by means of petrophysical processes. The results gave information about the microstructural characteristics of the clays. As a first approach, these methods are useful to assess the hydraulic properties of the aquitard and to specify the role of some heterogeneities in terms of groundwater flows. Finally, the reinterpretation of the geology on a large scale, coupled with observations of the clay microstructure, allowed for the determination of the geological nature of rocks composing the aquitards in order to provide relative information on permeability values related to each facies.

The results provide an update to the regional model. The integration of the new geometry and the heterogeneities of the aquitards will allow their role in the groundwater flows to be assessed by groundwater modelling. These heterogeneities, even the ones of small thickness, could modify groundwater flows and travel times. This could lead to large changes of water quality coming from the aquitards and contributing to the recharge of the aquifers. The role of heterogeneities within an aquitard has to be specified in order to assess an accurate portrayal of leaky fluxes coming from these confining units in the scope of enhancing freshwater exploitation for complex multilayered aquifer systems.

REFERENCES

Cerepi A., Humbert L., Burlot R. (2000) Effets de la texture pétrographique sur les propriétés pétrophysiqucs d'un calcaire en zone de diagenèse météorique – calcaire oligocène d'Aquitaine, France (Effects of rock-fabric on the petrophysical properties of a carbonate-rock in meteoric diagenetic zone – Oligocene Aquitain limestone, France), *Bulletin de la Société Géologique de France*, no. 4:419–430 [in French].

Clavier C., Coates G., Dumanoir J. (1984) Theoritical and experimental bases for the Dual-Water model for interpretation of shaly sands, *Society of Petroleum Engineers Journal* 24, no. 2:153–168

Daïan J.F. (2007) Porométrie au mercure – Le modèle XDQ (Mercury porometry – XDQ model), 97p [in French]

De Marsily G. (1981) *Hydrologie quantitative* (Quantitative hydrogeology). Ed. Masson, Paris, 299p [in French].

Delage P., Lefebvre G. (1984) Study of the structure of a sensitive Champlain clay and of its evolution during consolidation, *Can. Geotech. Journal* 21, no. 1:21–35.

Dullien F.A.L. (1992) *Porous media: Fluid transport and pore structure.* Second edition, Academic press, 574p.

Eaton T.T., Anderson M.P., Bradbury K.R. (2007) Fracture control of ground water flow and water chemistry in a rock aquitard. *Groundwater* 45, no. 5:601–615.

Garcia-Bengochea I., Lovell C.W., Altschaeffl A.G. (1979) Pore distribution and permeability of silty clays, *Journal of the geotechnical engineering division*, 105 (GT7):839–856.

Juang C.H., Holtz R.D. (1986) A probabilistic permeability model and the pore size density function, *International journal for numerical and analytical methods in geomechanics*, 10:543–553

Katz A.J., Thompson A.H. (1986) Quantitative prediction of permeability in porous rock, *Physical review B*. 34, no.11:8179–8181.

Lapierre C., Leroueil S., Locat J. (1990) Mercury intrusion and permeability of Louiseville clay, *Can. Geotech. J.*, 27:761–773

Miller R., Castle J., Temples T. (2000) Deterministic and stochastic modeling of aquifer stratigraphy, South Carolina. *Groundwater* 38, no. 2:284–295.

Moussié B. (1972) Le système aquifère de l'Éocène moyen et supérieur du bassin nord aquitain – Influence du cadre géologique sur les modalités de circulation (The Upper and Medium Eocene aquifers system of North Aquitain Basin – Influence of geological context on groundwater flows). PhD Thesis, University of Bordeaux-1, France [in French].

Pellerin J.F. (1980) La porosité mercure appliquée à l'étude géotechnique des sols et des roches (Mercury intrusion porosimetry applied to the geotechnical study of soils and rocks), Bulletin de liaison des laboratoires des ponts et chaussées, 106:115–116 [in French].

Remenda V. (2001) Preface – Theme issue on confining units. *Hydrogeology Journal* 9, no. 1:3–4.

Sammartino S., Bouchet A., Prêt D., Parneix J.C., Tevissen E. (2003) Spatial distribution of porosity and minerals in clay rocks from the Callovo-Oxfordian formation (Meuse/Haute-Marne, Eastern France)—implications on ionic species diffusion and rock sorption capability, *Applied Clay Science*, 23:157–166.

Schneider S. (2008) Estimation des paramètres hydrodynamiques des sols à partir d'une modélisation inverse de données d'infiltration et de résistivité électrique (Estimation of the unsaturated soil hydraulic properties from joint inversion of tension infiltrometer and electrical resistance measurements) PhD Thesis, University of Paris-Sud 11, 146p [in French].

Serra O., Serra L. (2004) *Well logging – Data acquisition and applications*. Technip Editions, 674p

Wardlaw N.C., Li Y., Forbes D. (1987) Pore-throat size correlation from capillary pressure curves, *Transp. Porous Media*, 2:597–614

Washburn E.W. (1921) Note on a method of determining the distribution of pore sizes in a porous material, *Proc. Nat. Acad. Sci.*, 7:115–116

Waxman M.H., Smits L.J.M. (1968) Electrical conductivities in oil-bearing shaly sands, *SPE Journal*, 8, no. 2:107–122

Xu K., Daïan J.F., Quenard D. (1997) Multiscale Structures to Describe Porous Media, Part I: *Theoretical Background and Invasion by Fluids, Transport in Porous Media*, 26:51–73

Chapter 22

Hydrogeological characterisation of two karst springs in Southern Spain by hydrochemical data and intrinsic natural fluorescence

Juan A. Barberá & Bartolomé Andreo

Department of Geology and Centre of Hydrogeology of University of Malaga, Spain

ABSTRACT

Karst springs provide information that is essential for characterising hydrogeological behaviour and the optimum management and exploitation of karst aquifers. Hydrochemical, hydrodynamic and intrinsic natural fluorescence (IF) data obtained from the karst spring waters of El Burgo and Fuensanta (southern Spain) illustrate the hydrogeological behaviour within the Sierra Blanquilla and Sierra Hidalga aquifers. Discharge variations at Fuensanta are faster, but less pronounced than at El Burgo. Electrical conductivity (EC) time series and hydrochemical monitoring suggest that there is a more highly developed karst network in the system drained by Fuensanta. EC at El Burgo varies according to TAC, Ca^{2+}, Cl^{-} and total organic carbon (TOC) contents, while SO_4^{2-} and Mg^{2+} vary inversely. Fuensanta presents wider variations in most chemical components and TOC, in accordance with its functioning as a conduit flow system. The hydrochemical data indicate different degrees of functional karstification. IF time variations at Fuensanta suggest a source of organic matter (higher TOC and IF peaks) of surficial origin (runoff infiltration) during high water periods. All these observations need to be confirmed by tracer tests and isotopic studies to reinforce the above hypotheses.

22.1 INTRODUCTION

The Mediterranean countries experience climatic conditions that periodically cause a scarcity of water resources, and so groundwater is often the only constant water supply for urban, industrial and agricultural activities, among others. Karst aquifers represent the most accessible source for groundwater supply. Current research in this area is mainly focused on identifying and characterising hydrogeological behaviour (how the system reacts and behaves in response to recharge events), which is a very significant aspect of water management in areas with scarce water resources.

The carbonate aquifers which underlie most karst terrains are differentiated from other types of permeable formations (detrital and fissured aquifers) by intrinsic characteristics (White, 2002; Ford & Williams, 2007; Goldscheider & Drew, 2007), such as marked spatial heterogeneity, hydraulic conductivity scale-effect, duality of recharge and porosity, and high temporal variability.

In the absence of direct information (from prospective boreholes, speleological exploration, etc.), hydrogeological research normally focuses on the natural responses at discharge points. Diverse methods may be applied, of general application or specific to this type of aquifer. In order to assess renewable water resources, hydraulic connections and potentiometric distribution in the aquifer (based on water level measurements), a hydraulic approach is taken, via the analysis of hydrodynamic responses (Mangin, 1975; Bonacci, 1993; Jeannin, 1996). Hydrochemical monitoring provides insights into water qualities, karst spring behaviour and the functioning of the entire aquifer (Bakalowicz, 1979; Lastennet & Mudry, 1997; Moore *et al.*, 2009; Barberá & Andreo, 2010; Mudarra & Andreo, 2011, Ravbar *et al.*, 2011). Groundwater thermal properties have been widely used as a marker of infiltration (Andrieux, 1978; Genthon *et al.*, 2005, Liñán *et al.*, 2009), water-rock interactions, mixing among water volumes, and to infer surface-groundwater relationships (O'Driscoll & DeWalle, 2006). Recently, intrinsic natural fluorescence (IF) has been applied to karst systems as a useful tool to characterise rapid infiltration (Baker *et al.*, 1997, 1999; Blondel, 2008; Cruz Jr., 2005; Mudarra & Andreo, 2010) via dissolved organic matter analysis. The fluorescence properties of natural components in the soil are associated with total organic carbon (TOC) (Cumberland & Baker, 2007) and total dissolved nitrogen contents (Wilson & Xenopoulos, 2008). Furthermore, several researchers have investigated the kinetics of organic matter degradation, using continuous monitoring of IF, TOC, NO_3^- and even faecal bacteria (Perrin *et al.*, 2001; Pronk *et al.*, 2006; Blondel *et al.*, 2010).

Many of the techniques are applied in the present study, in which we characterise the karst behaviour of aquifers in southern Spain. The main aim of this work is to integrate the results obtained from these techniques to improve the hydrogeological conceptual model of the two most important karst springs draining the carbonate aquifers beneath Sierra Blanquilla and Sierra Hidalga (Fig. 22.1), a highland area classified as a UNESCO Biosphere Reserve.

22.2 DESCRIPTION OF TEST SITE

Sierra Hidalga and Sierra Blanquilla are located in the Malaga province, southern Spain. These mountains cover a surface area of approximately 90 km^2 and present a varied, rugged relief, rising to altitudes exceeding 1500 m a.s.l. (the highest peak of Hidalga reaches 1505 m a.s.l.). The Turón River is the most important hydrologic feature, gathering groundwater and surface water flows from the catchment area. The climate is of continental Mediterranean type, strongly influenced by Atlantic winds. There are two significant rainy periods, during winter and spring. Average precipitation values and mean annual temperatures are 650 mm and 15°C, respectively (Jiménez *et al.*, 2007). During the study period (August 2007–February 2009), the average annual precipitation was over 800 mm, and the mean air temperature was 14°C. Therefore, this period was both wetter and colder than the historical average.

From a geological standpoint, three main lithological groups (Figs. 22.1, 22.2) have been identified (Cruz Sanjulián, 1974; Martín Algarra, 1987): limestone, dolostone and clay with evaporite (Triassic age) at the bottom; these are overlain by several hundred metres of carbonate rock (Jurassic limestone and dolostone); and finally, an upper formation of Cretaceous-Paleogene marl and marly-limestone.

The geological structure is composed of NE-SW oriented folds, plunging towards the NE (Martín Algarra, 1987). The folds are box-type, with flat hinges and subvertical

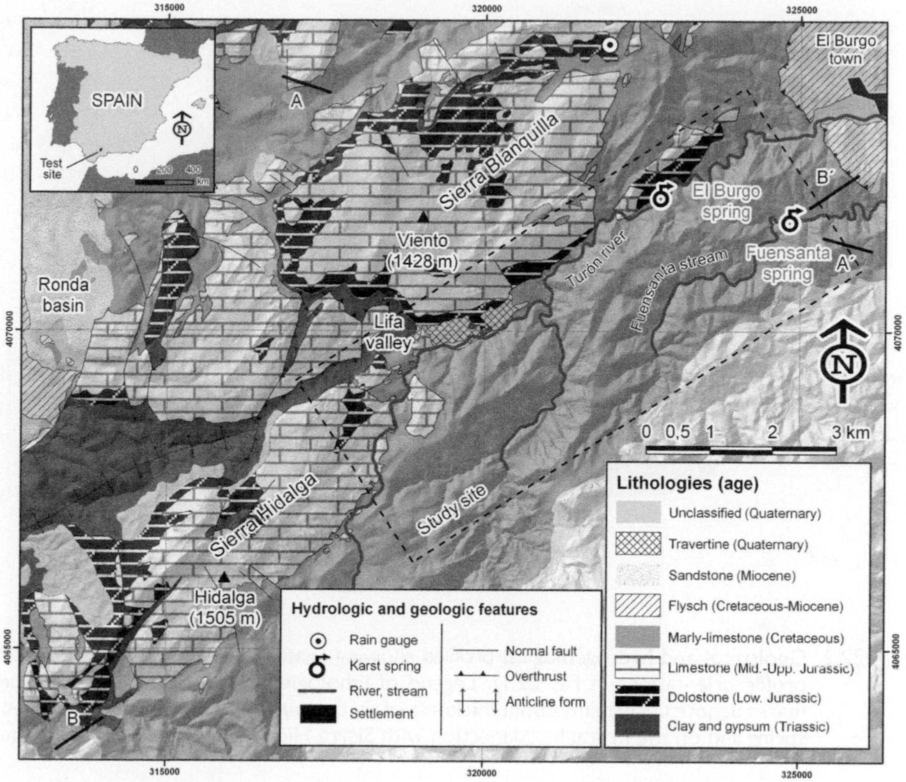

Figure 22.1 Location (dotted line rectangle, in detail) and geology of the study site. Direction of geological cross-sections (A–A′ and B–B′) of Fig. 22.2 is included.

flanks (Fig. 22.2; see both cross sections). All the fold structures are affected by fractures, which are preferentially oriented N50-70E and N150E (Fernández, 1981).

Sierra Hidalga and Sierra Blanquilla present significant karst features in the Jurassic formations, including large extensions of karrenfields, dolines, uvalas and karst sinkholes, mainly at the top of the massifs.

The geological settings of the springs (El Burgo and Fuensanta) are shown in Figure 22.2. The El Burgo spring is situated at the southern edge of Sierra Blanquilla aquifer, on the border between permeable rocks (Jurassic limestone) and impervious layers (Cretaceous marly-limestone). The Fuensanta spring emerges from marly-limestone, which is considered impermeable, but field observations and analysis of natural responses suggest there is a deep-lying hydraulic connection with the main karst aquifer. Furthermore, the upper limit of the Jurassic limestone cannot be very deep, perhaps just a few tens of metres. A final point of interest is the potential aquifer-river interaction, involving both outlets, due to the surficial drainage network configuration (Turón River and Fuensanta stream) in relation to the location of the springs.

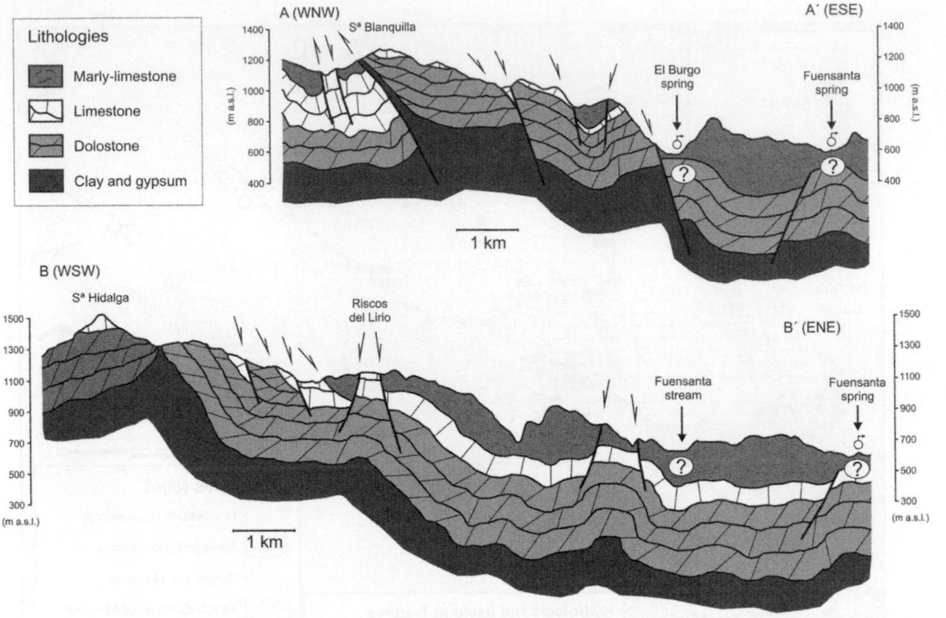

Figure 22.2 Geological and hydrogeological profiles showing spatial location of springs studied (see profile orientations in Fig. 22.1). Legend of lithologies is shown in Fig. 22.1. Question marks denote uncertainty about thickness of cretaceous marly-limestone below Fuensanta spring and on the hydraulic connection with Sierra Hidalga and Sierra Blanquilla aquifers.

22.3 SAMPLING PROCEDURE AND METHODS

Hydrodynamic, hydrothermal and hydrochemical parameters at El Burgo and Fuensanta springs were monitored from August 2007 to February 2009. Additionally IF-measurements were taken from July 2008 to the end of the study period. A total of 164 water samples were collected at both springs for chemical analysis under different hydroclimatic conditions, and *in situ* measurements were taken of electrical conductivity (EC), temperature, pH and flow rate. Sampling frequency was variable, from fortnightly (low waters) to daily (high waters). In addition, continuous (hourly) monitoring was taken of EC and temperature at both springs. Rainfall (15 min. time step) was recorded by an automatic weather station (see location in Fig. 22.1).

All the hydrochemical parameters considered in this work were analysed, within 24 hours of sampling, at the Centre of Hydrogeology at Malaga University. Total alkalinity (TAC) was determined by titration with 0.02 H_2SO_4 to reach pH 4.45. Major ions (Ca^{+2}, Mg^{+2}, Na^+, Cl^-, SO_4^{-2} and NO_3^-) were analysed by high pressure liquid chromatography (HPLC, Metrohm 792 IC Basic), and total organic carbon (TOC) using a SHIMADZU V-TOC carbon analyser, after HCl treatment to remove organic matter.

(IF) was measured with a Perkin-Elmer luminescence spectrometer (LS-55). The WinLab interface was used to obtain the excitation-emission spectrums (EEM)

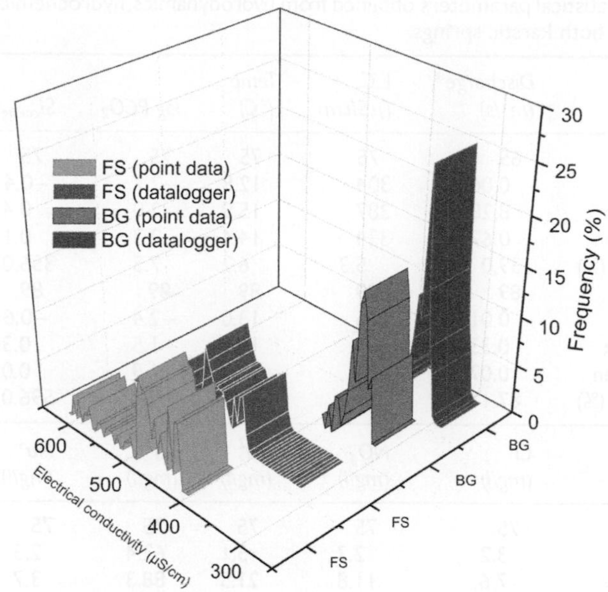

Figure 22.3 Frequency distribution curves of electrical conductivity (EC) data series at El Burgo (BG) and Fuensanta (FS) karst springs (EC intervals are of 5 µS/cm).

corresponding to each sample. In every case, the excitation wavelength (λ_{ex}) ranged from 200 nm to 350 nm with a 5 nm slit, while the emission wavelength (λ_{em}) ranged from 250 to 550 nm. The fluorescence values recorded from EEM are expressed as Intrinsic Fluorescence Units (IFU) measured in Uf · nm^2 (Baker *et al.*, 1997; Blondel, 2008).

22.4 RESULTS AND DISCUSSION

22.4.1 Spring water types

A preliminary analysis of water mineralisation comprised two types of electrical conductivity (EC) data sets (Fig. 22.3): continuous measurements (provided by dataloggers) and single measurements taken at the springs, to test the proper functioning of the instruments.

A methodology to classify karst aquifer types (fissured and conduit), based on the frequency distribution of EC values, was proposed by Bakalowicz (1979). EC in El Burgo spring ranged from 288–383 µS/cm (Fig. 22.3; datalogger data) and the shape of the frequency curve revealed a distribution with a single clearly-defined mode (unimodal response), around 320 µS/cm, with a maximum frequency of 26%. Thus, El Burgo spring response (unimodal distribution) is characterised by a less marked karst behaviour, although its range of EC variation is relatively high (over 100 µS/cm), according to Bakalowicz (1979) and Mudry (1987).

Table 22.1 Main statistical parameters obtained from hydrodynamics, hydrochemical and hydrothermal data in both karstic springs.

		*Discharge** *(m^3/s)*	*E.C.* *($\mu S/cm$)*	*Temp* *(°C)*	*log PCO_2*	*$SI_{calcite}$*	*TOC* *(mg/l)*	*TAC* *(mg/l)*
El Burgo	n	65	75	75	75	75	75	75
	Min	0.06	304	12.8	−2.7	−0.4	0.19	186.7
	Max	8.28	387	15.9	−2.0	0.4	1.44	245.0
	Mean	0.67	334	14.4	−2.3	0.1	0.58	210.4
	CV (%)	139.0	5.3	6.2	7.5	356.0	49.2	7.4
Fuensanta	n	89	89	89	89	89	89	89
	Min	0.01	416	13.0	−2.4	−0.6	0.26	232.8
	Max	0.33	609	16.7	−1.5	0.3	2.21	332.1
	Mean	0.07	477	14.3	−1.9	0.0	0.86	277.3
	CV (%)	87.1	10.2	6.2	9.2	596.0	49.1	10.1
		Cl^- *(mg/l)*	*NO_3^-* *(mg/l)*	*SO_4^{2-}* *(mg/l)*	*Ca^{2+}* *(mg/l)*	*Na^+* *(mg/l)*	*K^+* *(mg/l)*	*Mg^{2+}* *(mg/l)*
El Burgo	n	75	75	75	75	75	75	75
	Min	3.2	2.7	5.1	62.4	2.3	0.2	4.6
	Max	7.6	11.8	21.5	88.3	3.7	0.9	9.7
	Mean	5.0	5.0	10.7	71.6	2.9	0.5	6.6
	CV (%)	23.1	36.9	45.0	9.5	12.5	29.9	23.5
Fuensanta	n	89	89	89	89	89	89	89
	Min	4.4	0.0	11.2	81.0	3.6	0.3	4.7
	Max	11.7	5.2	146.5	121.5	5.6	2.7	27.8
	Mean	7.8	1.1	49.1	99.8	4.5	0.6	12.6
	CV (%)	27.8	96.3	75.5	10.4	12.2	45.9	50.0

*Discharge from El Burgo spring was calculated by subtracting the upstream from the downstream flow in the Turón River.

The EC record for Fuensanta spring produced a wider range of variation (336–600 μS/cm, analysing hourly measurements). Thus, the frequency distribution presented multiple modes, with low frequency values (in no case exceeding 10%). The EC data distribution values for the two springs show that water mineralisation presents greater variability at Fuensanta than at El Burgo. This wide variation of data distribution reveals considerable internal hierarchization within a conduit flow system. In Fuensanta spring, this may be due to the interference of rapid flow paths, from different origins.

22.4.2 Variability of spring hydrographs and chemographs

Table 22.1 summarises the physico-chemical data recorded at El Burgo and Fuensanta spring waters. The chemographs for El Burgo spring (Fig. 22.4) are shown with the hydrograph (single discharge measurements record) for the Turón River, which together constitute the main drainage axis for the Sierra Blanquilla watershed.

The Turón River hydrograph mainly gathers groundwater discharge from Sierra Blanquilla (the main outlet, located on the southern border) and Sierra Hidalga

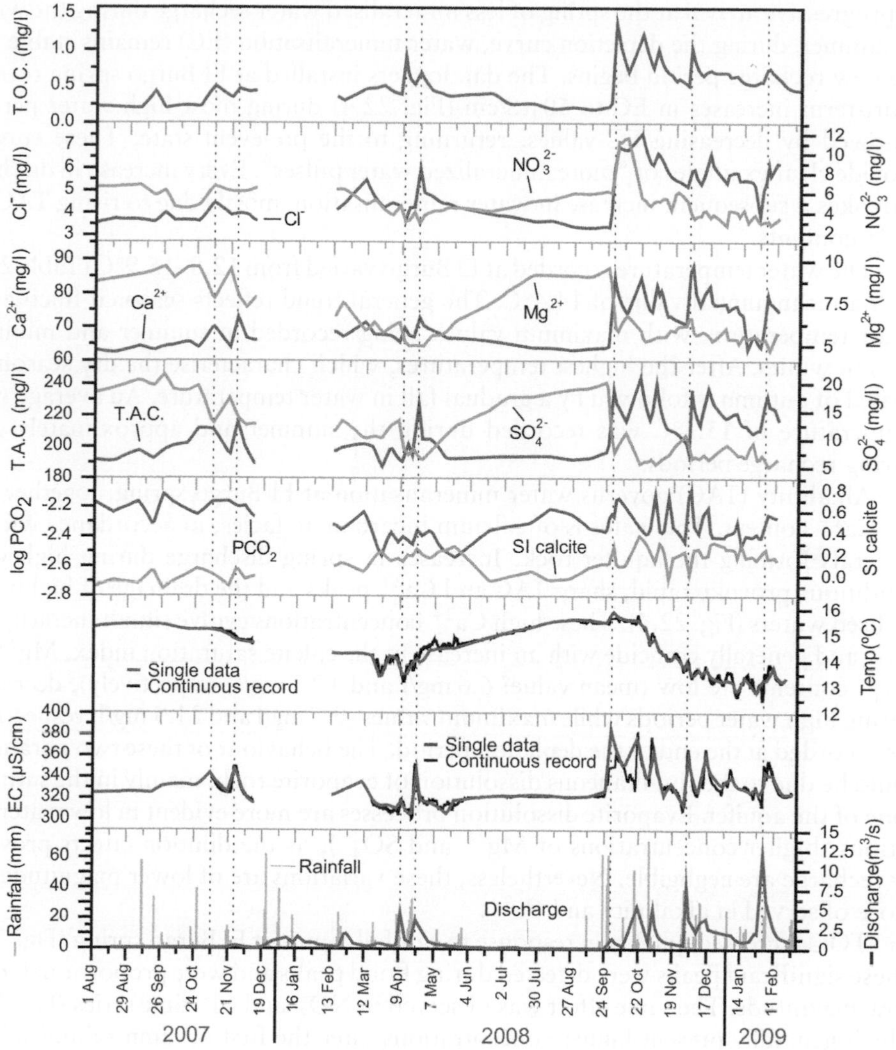

Figure 22.4 Chemographs of El Burgo spring, together with the Turón River hydrograph. Guidelines mark the most interesting flood events.

aquifers; apart from a small proportion of runoff from the south. After a relatively dry year, above-average rainfall during 2008/09 provoked a maximum discharge of 13.1 m^3/s on 2 February 2009 and, consequently, a higher flow rate drained by the spring. Mean annual groundwater discharge from Sierra Blanquilla aquifer toward the Turón River outlet exceeds 0.67 m^3/s, with flood peaks up to 8 m^3/s and minimum discharge values around 0.06 m^3/s during the dry season. The discharge from El Burgo spring increases in response to rainfall, after approximately one day (Fig. 22.4). The hydrograph shape suggests high variability in flow conditions, including rapid, sharp peaks, most of which rise to maximum values in response to significant rain inputs.

EC, temperature and most chemical components decrease significantly in response to progressive arrival at the spring of less mineralised water recharge during the winter. In summer, during the depletion curve, water mineralisation (EC) remains stable until the new recharge period begins. The dataloggers installed at El Burgo spring revealed short-term increases in EC to 50 μS/cm (Fig. 22.4) during most high-water periods, followed by decreasing EC values, returning to the pre-event state. These episodes provide clear examples of "more mineralized water pulses". Every increase in discharge provokes a subsequent increase in water mineralisation, mostly due to rising TAC and Ca^{+2} contents.

The water temperature recorded at El Burgo varied from 12.8–15.9°C (Table 22.1), with a mean annual value of 14.4°C. The general trend reflects seasonal fluctuations in air temperature, with maximum values being recorded in summer and minimum ones in winter. After the highest temperatures, which characterise the dry season, the arrival of autumn is followed by a gradual fall in water temperature. An average water temperature of 15.5°C was recorded during the summer and approximately 13°C during recharge periods.

Alkalinity (TAC) governs water mineralisation at El Burgo spring, together with the Ca^{2+} content. The water is of calcium bicarbonate facies, in accordance with the minerals forming the aquifer rock. Increases in spring discharge during high water conditions provoke rapid, sharp TAC and Ca^{2+} peaks and the drainage of highly mineralised waters (Fig. 22.4). These high Ca^{2+} concentrations evolve simultaneously with TAC and generally coincide with an increase in the calcite saturation index. Mg^{2+} and SO_4^{2-} contents are low (mean value: 6.6 mg/l and 10.7 mg/l, respectively), decreasing during high water periods while maximum values (9.7 mg/l and 21.5 mg/l, respectively) are recorded at the end of the depletion period. The behaviour of these two parameters could be due to the simultaneous dissolution of evaporite rock, mainly in the saturated zone of the aquifer. Evaporite dissolution processes are more evident in low water conditions (higher concentrations of Mg^{2+} and SO_4^{2-}), as the dilution effects provoked by recharge are negligible. Nevertheless, these variations are of lower magnitude than those observed in alkalinity and Ca^{2+}.

TOC increased quickly in response to rainfall inputs at El Burgo spring (Fig. 22.4). These significant peaks were detected during flood peaks and were proportional to outflow magnitude. The same effect was observed in NO_3^- and Cl^- time series (Fig. 22.4), which generally present higher concentrations after the first autumn rainfalls. Thus, after the stormy rains of October 2008, maximum concentrations (11.8 and 7.6 mg/l, respectively; see Table 22.1) were recorded. Natural tracers, such as TOC and NO_3^- are derived from the soil; these are considered indicators of transit time and water infiltration through the soil and the unsaturated zone.

Therefore, chemographs obtained during the high water period highlight the existence of rapid flows (guidelines in Fig. 22.4), which characterise the transit of recharge waters from the soil to the saturated zone, also draining the epikarst reservoir. Rapid responses are more marked, especially at the beginning of the hydrologic year, although rapid infiltration flows contribute to spring discharge during the whole of the wet season. The availability of TOC, NO_3^- and Cl^- in the soil layer is maximum at the end of summer because of biological activity and intense reconcentration by soil water evaporation. After that, autumn rains favour the progressive washing of the soil cover. This effect, together with the dilution of other parameters, such as SO_4^{2-} and Mg^{2+},

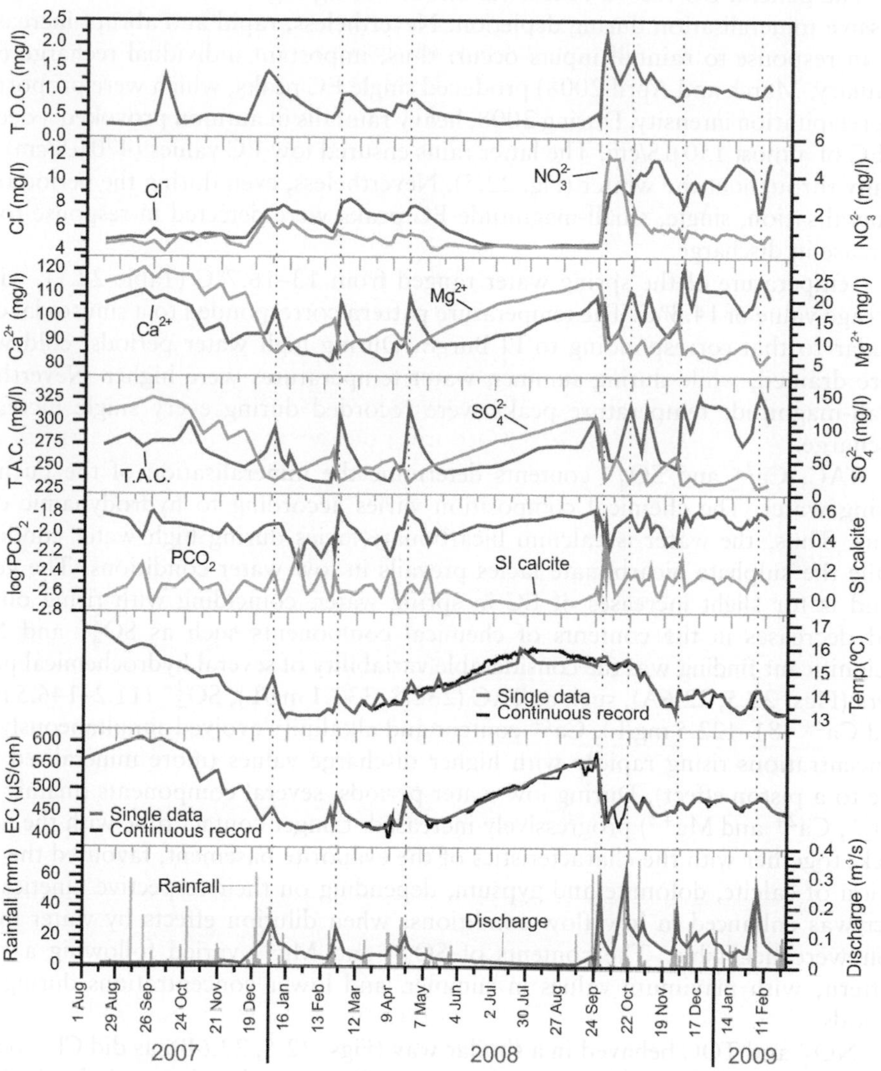

Figure 22.5 Temporal evolution of dissolved chemical components and discharge variations recorded at Fuensanta spring. Vertical dotted lines indicate main flood events which display significant chemical variations.

which are tracers of the saturated zone, suggests that the water stored in shallow parts of the aquifer becomes progressively mixed with the recharge water, which contains a lower concentration of natural soil tracers.

On the other hand, the mean discharge at Fuensanta spring was 0.071 m^3/s during the study period, while the maximum flood peak was 0.325 m^3/s. During the dry season, the discharge fell to scarcely 0.01 m^3/s, but the spring never became completely dry. Time response to rainfall was very short, less than one day, as demonstrated by the lag between rainfall and the outflow peaks (hydrograph in Fig. 22.5) and the chemographs.

The general EC record reflects dilutions during high water conditions and progressive mineralisation during depletion. Nevertheless, rapid and abrupt increases of EC in response to rainfall inputs occur: thus, important individual recharge events (January, March and April 2008) produced single EC peaks, which were proportional to precipitation intensity. During 2008, heavy rainfalls in autumn provoked a decrease in EC of almost 150 μS/cm. The latter rains ensured low EC values (470 μS/cm) practically throughout the winter (Fig. 22.5). Nevertheless, even during the period of low mineralisation, single, small-magnitude EC peaks were detected in response to each increase in discharge.

Temperature of the spring water ranged from 13–16.7°C (Table 22.1), with an average value of 14.3°C. The temperature pattern corresponded to a sinusoidal curve, similar to that corresponding to El Burgo. During high water periods, cold waters were drained, while during summer, water temperatures were higher. Nevertheless, small-magnitude temperature peaks were recorded during every single increase in discharge.

TAC, Ca^{2+} and SO_4^{2-} contents determine the mineralisation of the Fuensanta spring water. The chemical composition varies according to hydrodynamic conditions. Thus, the water is calcium bicarbonate facies during high water conditions, while the sulphate bicarbonate facies prevails in low water conditions. The general trend is for slight increases of EC in spring water, coinciding with rising outflow, and decreases in the contents of chemical components such as SO_4^{2-} and Mg^{2+}. A significant finding was the considerable variability of several hydrochemical parameters (Figs. 22.5, 22.6A), such as TAC (232.8–332.1 mg/L), SO_4^{2-} (11.2–146.5 mg/L), and Ca^{2+} (81–122.5 mg/L). Ca^{2+} content and alkalinity evolved simultaneously, with concentrations rising rapidly with higher discharge values (more mineralised water due to a piston effect). During low water periods, several components (mainly TAC, SO_4^{2-}, Ca^{2+} and Mg^{+2}) progressively increased. Longer contact time with the aquifer rock, together with the characteristics of the evaporite basement, favoured the dissolution of calcite, dolomite and gypsum, depending on their respective kinetics. This fact was enhanced in low flow conditions, when dilution effects by water infiltration were negligible. The contents of SO_4^{2-} and Mg^{2+} varied following a similar pattern, with maximum values in summer, and lower concentrations during rainy periods.

NO_3^- and TOC behaved in a similar way (Figs. 22.5, 22.6B), as did Cl^- contents, but conversely to EC and water temperature. During high water periods, the former parameters related to the soil presented rapid, but slight increases, almost always associated with increases in spring discharge. NO_3^- and TOC tended to be higher at the beginning of the hydrologic year (in the autumn), when the first rains fell, and evolved depending on water availability in the soil and the transit time of recharge waters through the aquifer.

22.4.3 Time series of intrinsic natural fluorescence (IF)

To complete the analysis of natural tracer responses, especially those originating in shallow parts of the aquifer (soil and epikarst), the IF peaks are represented in Figure 22.7, together with TOC and flow data for the springs. A perfect relationship is observed between TOC and the A and C peaks. The latter parameters evolve

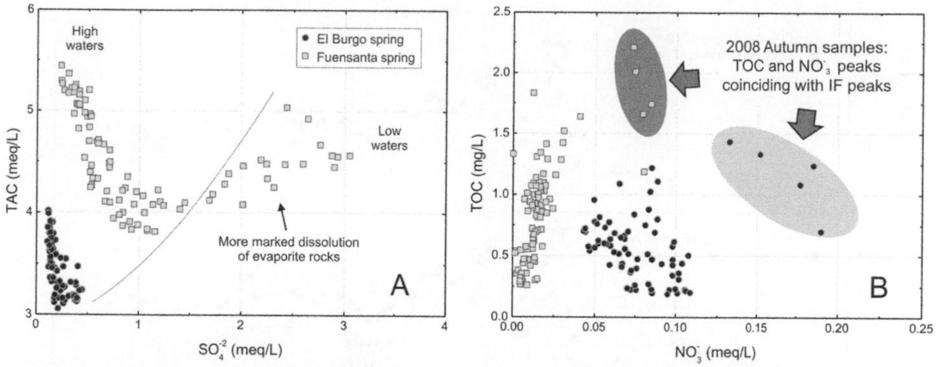

Figure 22.6 Binary plots showing TAC-SO_4^{2-} (A) and TOC-NO_3^- (B) relationships. Note the wide dispersion of TAC and SO_4^{2-} values at the Fuensanta spring, reflecting the dual water quality (changes in hydrochemical facies) depending on the hydrodynamic stage. This fact was not observed in the chemical variability of El Burgo spring waters. The autumn 2008 flood peak samples for both springs are shown within grey ellipses in figure B, which shows maximum values in TOC, NO_3^- and peaks A and C (IF).

synchronously, which means that TOC depends on organic acids, basically humic and fulvic-like substances (Cumberland & Baker, 2007). Therefore, variations in IF intensity are associated with differences in vegetation cover, soil type and humification (Baker & Genty, 1999).

The differences in IF peaks (34% in peak A and 15% in peak C) at the two springs suggest an additional source of organic content, rich in humic and fulvic acids, producing the higher values at Fuensanta spring. Allogenic recharge may be considered, via direct infiltration from surface waters, through fractured marl outcrops in the river bed. Concentrated recharge through swallow-holes was not observed, but Fuensanta stream loses part of its flow, upstream from the spring. During high water conditions, soil components can be washed by surficial waters, rich in TOC and IF, along nearby streams (i.e. Fuensanta stream; Fig. 22.1), and consequently may mix with karst groundwater. Additionally, the location of the spring, in the middle of a well fractured marly-limestone outcrop in a reforested area, together with a probable hydraulic connection between the stream river bed and the spring, could explain this modality of recharge, affecting the IF of the spring water. Evolution of the IF peaks was more sensitive to autumn rainfall and to isolated rainfall inputs (see October and November precipitations and sporadic storms in January 2009; Fig. 22.7), when the differences between IF intensity of A and C peaks were greater. High levels of organic matter production in the warmer seasons (spring and summer) favour the enrichment of soils by organic components developed in marly-limestone. Consequently, an ephemeral hydrodynamic regime in Fuensanta stream prevails after stormy periods, and flash floods may induce a significant proportion of allogenic infiltration toward the aquifer, provoking high TOC and IF values, as shown by the star symbols in Fig. 22.7, for the Fuensanta spring.

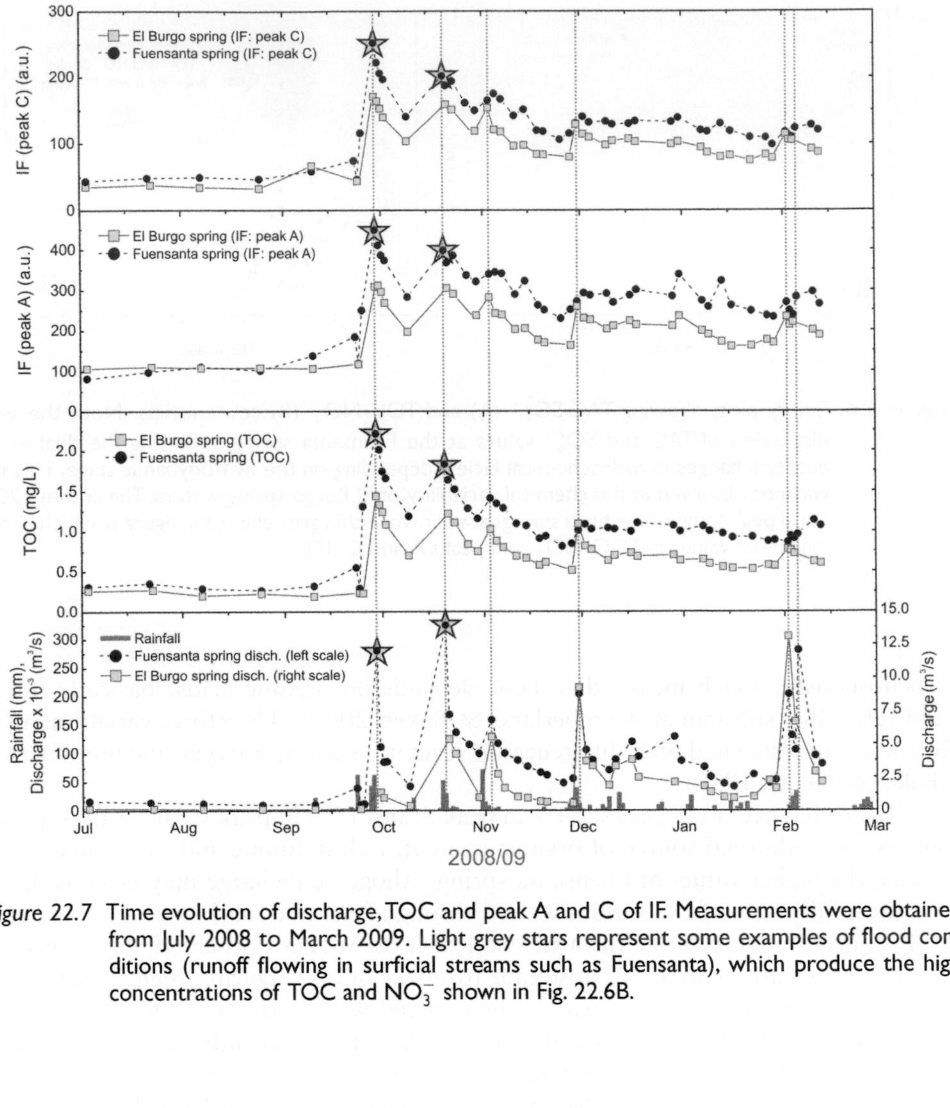

Figure 22.7 Time evolution of discharge, TOC and peak A and C of IF. Measurements were obtained from July 2008 to March 2009. Light grey stars represent some examples of flood conditions (runoff flowing in surficial streams such as Fuensanta), which produce the high concentrations of TOC and NO_3^- shown in Fig. 22.6B.

22.5 CONCLUSIONS

The hydrogeological tools used to analyse the data series in this study enabled the hydrochemical characterisation of the two springs, El Burgo and Fuensanta, which drain Sierra Blanquilla and Sierra Hidalga carbonate aquifers. Additional information was derived from spectrofluorometry techniques, which proved to be a useful complementary tool for characterising rapid-infiltration waters.

Discharge variations were faster at Fuensanta, although of lesser magnitude than at El Burgo. The latter, therefore, must drain a larger limestone/dolostone catchment area. In order to validate the permanent flow at Fuensanta spring, a groundwater contribution from the carbonate Jurassic aquifer must be assumed. Moreover, the IF characteristics of Fuensanta spring are produced by an allogenic source of water with an enhanced fluorescence signature, probably due to runoff of surficial origin.

The EC values recorded at the two springs presented differences in the degree of functional karstification; thus, Fuensanta spring drains a more karstified sector of the aquifer than does El Burgo.

Hydrochemical variations at the two springs differ slightly, indicating subtle differences in water quality and hydrogeological behaviour according to the sectors drained. El Burgo spring shows rapid flows and less hydrochemical variability with respect to Ca^{2+}, SO_4^{2-}, Mg^{2+} and TOC, which is due to a lower development of functional karstification. For every discharge increase, a single EC peak occurred (piston flow), involving some of the chemical parameters (EC, TAC, Ca^{2+} and NO_3^-, Cl^- and TOC at the very beginning of autumn). The remaining parameters (SO_4^{2-} and Mg^{2+}) evolved conversely, according to the dilution of water volume stored in the aquifer.

Fuensanta displayed considerable variations for most of its chemical components (TAC, Ca^{2+}, SO_4^{2-}, Mg^{2+}, NO_3^-, Cl^- and TOC), which is typical of karst springs and is indicative of a well developed network of internal drainage. The spring flow comprises flow paths from the Jurassic aquifer flow, draining evaporite rocks, and rapid infiltration waters that are rich in organic components (TOC and IF) and NO_3^-, coming from surface waters. Flow conditions are dominated by large dilutions of almost all chemical components during recharge periods (especially the autumn rains). The lowest EC values recorded during high water periods (around 470 μS/cm) determine the maximum dilution within the aquifer, where the rapid infiltration of less mineralised waters prevails. In low water periods, the influence of evaporite is significant up to a maximum level of mineralisation (mainly due to SO_4^{2-}, Ca^{2+} and TAC values), which produces a sulphate-bicarbonate chemical type of spring water. On the other hand, allogenic recharge occurs coinciding with the flood peaks generated by intense, isolated rainfall inputs. This normally occurs as a consequence of early precipitations (autumn rains) and stormy periods in summer and autumn.

The results obtained from this study show that the simultaneous use of different hydrogeological techniques is a valuable tool for studying complex karst systems. To complement the conclusions reached, more reliable techniques such as dye tracers and isotopic investigations are now being applied, to obtain new information from aquifer dynamics (geometry, recharge area and hydraulic connectivity), data which are essential for the suitable planning of groundwater quality monitoring, zoning protection and the further management of spring waters.

ACKNOWLEDGEMENTS

This work is a contribution to the projects CGL2008-06158, HP2008-47 and DE2009-0060 of DGICYT, and to the Research Group RNM-308 of the Junta de Andalucía.

REFERENCES

Andrieux C. (1978) Les enseignements apportés para la thermique dans le karst. Colloque de Tarbes, Le karst: son originalité physique, son importance économique. Association des Géologues du SudOuest (AGSO), France, 48–63.

Bakalowicz M. (1979) Contribution de la géochimie des eaux à la connaissance de l'aquifère karstique et de la karstification. *Thèse Doct Sci Nat, Univ P et M Curie, París-VI, Géol Dyn*; 269 p.

Baker A., Barnes W.L., Smart P.L. (1997) Variations in the discharge and organic matter content of stalagmite drip waters in Lower Cave, Bristol. *Hydrological Processes* 11: 541–555.

Baker A., Genty D. (1999) Fluorescence wavelength and intensity variations of cave waters. *J Hydrology* 217: 19–34.

Barberá J.A., Andreo B. (2010) Duality in the functioning in a karst system under Mediterranean climate conditions, deduced from hydrochemical characterization. In: *Advances in Research in Karst Media* 189–194.

Batiot C., Emblanch C., Blavoux B. (2003) Total Organic Carbon (TOC) and magnesium (Mg^{2+}): two complementary tracers of residence time in karstic systems. *Comptes Rendus Geoscience* 335: 205–214.

Blondel T. (2008) Traçage spatial et temporel des eaux souterraines dans les hydrosystèmes karstiques par les matières organiques dissoutes. PhD Thesis Académie d'Aix-Marseille. Université d'Avignon et des Pays de Vaucluse, p. 190.

Blondel T., Emblanch C., Dudal Y., Batiot-Guilhe C., Travi Y., Gaffet S. (2010) Transit time environmental tracing from dissolved organic matter fluorescence properties in karstic aquifers. Application to different flows of Fontaine de Vaucluse experimental basin (SE France). In: *Advances in Research in Karst Media*, 189–194.

Bonnacci O. (1993) Karst spring hydrographs as indicators of karst aquifers. *Journal of Hydrological Sciences* 38: 51–62.

Cruz Jr F.W., Karmann I., Magdaleno G.B., Coichev N., Viana Jr O. (2005) Influence of hydrological and climatic parameters on spatial-temporal variability of fluorescence intensity and DOC of karst percolation waters in the Santana Cave System, Southeastern Brazil. *J Hydrology* 302: 1–12.

Cruz Sanjulián J. (1974) Estudio geológico del sector Cañete la Real-Teba-Osuna (Cordillera Bética, región occidental). Doctoral Thesis, Univ of Granada, 431 p.

Cumberland A., Baker A. (2007) The freshwater dissolved organic matter fluorescence-total organic carbon relationship. *Hydrological Processes* 21: 2093–2099

Emblanch C., Blavoux B., Puig J.M., Mudry J. (1998) Dissolved organic carbon of infiltration within the autogenic karst hydrosystem. *Geophysical Research Letters* 25: 1459–1462.

Emblanch C., Zuppi G.M., Mudry J., Blavoux B., Batiot C. (2003) Carbon 13 of TDIC to quantify the role of the unsaturated zone: the example of the Vaucluse karst systems (Southeastern France). *J Hydrology* 279(1–4): 262–274.

Fernández R., Pulido-Bosch A. and FernándezRubio R. 1981 Bosquejo hidrogeológico de tres sistemas acuíferos kársticos al norte de Ronda (Málaga). *I Simp. Agua en Andalucía, II*: 643–658. Granada.

Ford D., Williams P. (2007) Karst Hydrogeology and Geomorphology. Edit Wiley Chichester (UK) 562 p.

Genthon P., Bataille A., Fromant A., D'Hulst D., Bourges F. (2005) Temperature as a marker for karstic waters hydrodynamics. Inferences from 1 year recording at La Peyrére cave (Ariège, France). *J Hydrology* 311: 157–171.

Goldscheider N., Drew D.P. (2007) *Methods in Karst Hydrogeology*. Taylor & Francis, London, United Kingdom. 262 p.

Jeannin P.Y. (1996) Structure et comportment hydraulique des aquifères karstiques. PhD thesis, Université de Neuchatel, 237 p.

Jiménez P., Fernández R., Jiménez Fernández P. (2007) Sierras Hidalga-Merinos-Blanquilla (MAS 060.043). Atlas hidrogeológico de la provincia de Málaga, 2: 49–58. Diputación de Málaga-IGME-UMA.

Lastennet R., Mudry J. (1997) Role of karstification and rainfall in the behaviour of a heterogeneous karst system. *Environmental Geology* 32: 114–123.

Liñán C. Andreo B., Mudry J., Carrasco F. (2009) Groundwater temperature and electrical conductivity as tools to characterize flow patterns in carbonate aquifers: The Sierra de las Nieves karst aquifer, southern Spain. *Hydrogeology J* 17-4: 843–853.

Mangin A. (1975) Contribution à l'etude hydrodynamique des aquifères karstiques. Ann. Spéleol 29 (3): 283–332 (4): 495–601 30 (1): 21–124.

Martín Algarra A. (1987) Evolución geológica Alpina del contacto entre las Zonas Internas y las Zonas Externas de la Cordillera Bética (Sector Occidental). Doctoral Thesis, Univ of Granada, 1171 p.

Moore P., Martin J., Screaton E. (2009) Evidence of recharge, mixing, and controls on spring discharge in an eogenetic karst system. *J Hydrology* 376: 443–455.

Mudarra M., Andreo B. (2010) Hydrogeological functioning of a karst aquifer deduced from hydrochemical components and natural organic tracers present in spring waters. The case of Yedra Spring (Southern Spain). *Acta Carsologica* 39/2.

Mudarra M., Andreo B. (2011) Relative importance of the saturated and the unsaturated zones in the hydrogeological functioning of karst aquifers: the case of Alta Cadena (Southern Spain). *J Hydrology* 397: 263–280.

Mudry J. (1987) Apport du traçage physico-chimique naturel à la connaissance hydrocinématique des aquifères carbonatés. Thèse Sciences Naturelles, Université de Franche-Comté, Besançon, 378 p.

O'Driscoll M., DeWalle D. (2006) Stream-air temperature relations to classify stream-groundwater interactions in a karst setting, central Pennsylvania, USA. *J Hydrology* 329 140–153.

Perrin J., Tomasi N., Aragno M., Rossi P. (2001) Evolution of inorganic (nitrates) and faecal bacteria contaminants in the water of a karstic aquifer, Milandre test site (Swiss Jura). *Proceedings of the 7th Conference on Limestone Hydrology and Fissured Media, Besançon (France)*, 273–276.

Pronk M., Goldscheider N., Zopfi J. (2006) Dynamics and interaction of organic carbon, turbidity and bacteria in a karst aquifer system. *Hydrogeology J* 14: 473–484.

Pronk M., Goldscheider N., Zopfi J., Zwahlen F. (2009) Percolation and Particle Transport in the Unsaturated Zone of a Karst Aquifer. *Ground Water* 47(3): 361–369.

Ravbar N., Engelhardt I., Goldscheider N. (2011) Anomalous behaviour of specific electrical conductivity at a karst spring induced by variable catchment boundaries: the case of the Podstenjsek spring, Slovenia. *Hydrological processes* DOI: 10.1002/hyp.7966.

White W.B. (2002) Karst hydrology: recent developments and open questions. Ed Elsevier. *Engineering Geology* 2016.

Wilson H., Xenopoulos M. (2008) Effects of agricultural land use on the composition of fluvial dissolved organic matter. *Nature Geoscience Letters*, vol. 2, 37–41.

Part 5

Data processing & modelling

Chapter 23

Integration of environmental tracer information with groundwater modelling

Fritz Stauffer & Wolfgang Kinzelbach
ETH Zurich, Institute of Environmental Engineering, Zurich, Switzerland

ABSTRACT

Numerical groundwater models, which are calibrated with the help of head data are often non-unique. A fit to head data alone is necessary but in general not sufficient. The basic problem in flow modelling consists of the fact that hydraulic conductivity is in general poorly known. Since boundary fluxes, recharge rates, and fluxes to or from rivers cannot be measured directly, they have usually to be estimated independently. Both types of information are uncertain, and therefore, the resulting flow model suffers from uncertainty. Model calibration cannot remove this uncertainty as there is usually no unique solution to the parameter estimation problem. Environmental tracers can be of help in this situation. Environmental tracers can for example indicate that there is recent recharge, they allow estimating the age of groundwater or its residence times, they can yield streamline information, ratios of fluxes including recharge rates, and effective porosity values. Indeed, examples from the literature show that various environmental tracers were successfully used in the past to check or to improve flow models, to select among several alternative scenarios or conceptual models, and to determine solute transport parameters. Expectations should, however, be modest concerning their ability to improve the characterisation of hydraulic conductivity or the accuracy of fluxes. This review suggests that environmental tracer information can primarily be used for improving ratios of parameters and constraining the conceptual model, but not necessarily for getting higher numerical accuracy. In the Bayesian sense, the integration of environmental tracers in groundwater models can reduce the range of possible alternative interpretations consistent with all observations.

23.1 INTRODUCTION

Numerical groundwater models, which are calibrated with the help of head data only, are often non-unique. A fit to head data is necessary but in general not sufficient to identify the required parameters. The basic problem in flow modelling consists of the fact that hydraulic conductivity of the aquifer is in general more or less poorly known. Consider the schematic cross section in Figure 23.1. The discharge Q through the cross section under steady-state conditions is given by Darcy's law, integrated over the cross section A:

$$Q = \int_A v_{\mathrm{Darcy,n}} dA = \int_A K(\mathbf{x}) I_{\mathrm{n}}(\mathbf{x}) dA \tag{23.1}$$

where K is the hydraulic conductivity, and I_{n} is the absolute value of the normal component of the hydraulic gradient, which can be approximated by interpolating the

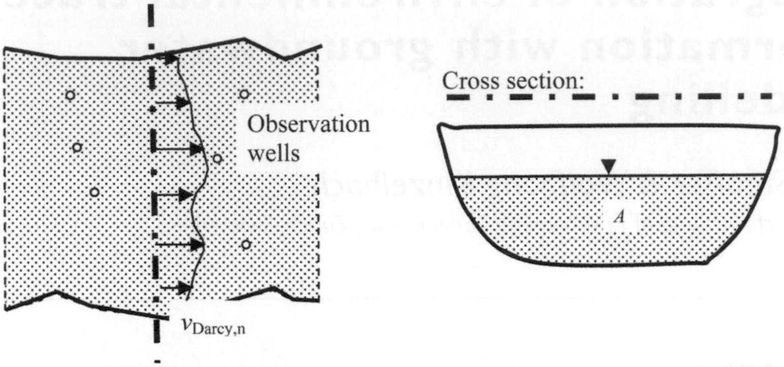

Figure 23.1 Flow through a cross section of an aquifer; plan view (left) and vertical cross section (right).

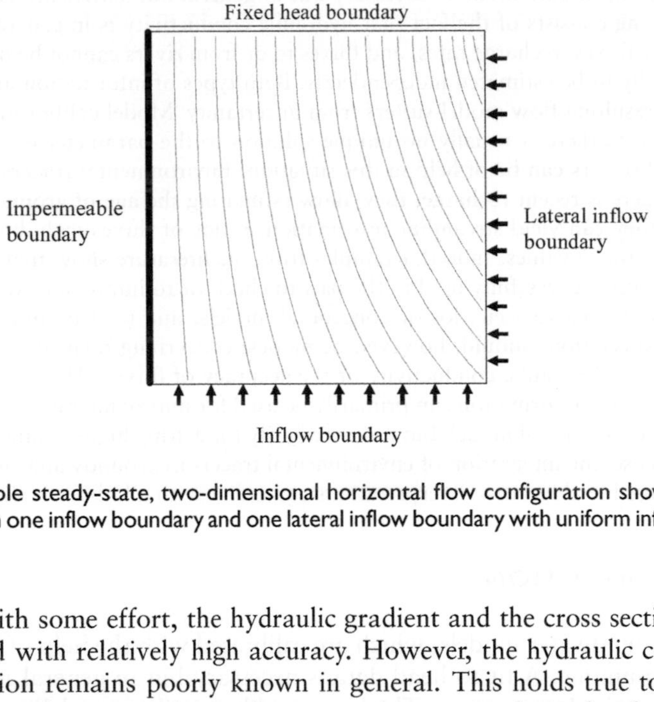

Figure 23.2 Simple steady-state, two-dimensional horizontal flow configuration showing streamlines with one inflow boundary and one lateral inflow boundary with uniform inflow distribution.

head field. With some effort, the hydraulic gradient and the cross sectional area can be determined with relatively high accuracy. However, the hydraulic conductivity in the cross section remains poorly known in general. This holds true to some degree even in the case where several pumping test or slug test results in the observation wells are available. The uncertainty is mainly due to considerable spatial variability of this parameter.

Another problem is posed by uncertain boundary water fluxes. Consider the steady-state flow field in Figure 23.2. Note that in the numerical solution the finite difference cell in the north-eastern corner belongs to the lateral inflow boundary in order to avoid inconsistencies.

The Figure shows that the amount of the lateral inflow rate can exhibit a large influence on the streamlines and therefore on the flow field. Since boundary fluxes, recharge rates, and fluxes to or from rivers cannot be measured directly, they have usually to be estimated independently. In numerical modelling this is often done by model calibration using head data. The resulting fluxes are non-unique, and therefore, the resulting flow model is uncertain. The prediction of plume propagation in the case of Figure 23.2 is poor and depends very sensitively on the ratio of the southern and eastern inflows, while heads vary only little in response to relative flux changes.

As a consequence of both problems, the resulting flow field is often too inaccurate for solute transport modelling. Environmental tracers offer a possibility to estimate or at least improve the estimates of recharge and boundary fluxes.

23.2 USE OF ENVIRONMENTAL TRACERS

Tracers often used in water resources management (Leibundgut *et al.*, 2009) are dating tracers and streamline tracers like ^3H, ^3He, the ratio ^3H to ^3He, ^{85}Kr, chlorofluorohydrocarbons, ^{14}C, ^{18}O, or ^2H. Evaporation tracers are ^{18}O-^2H, and chloride from sea salt aerosols.

Can environmental tracers help in improving flow and solute transport models in general? They can be useful at least to some degree. Environmental tracers can for example indicate that there is recent recharge, they allow estimating the age of groundwater or its residence times, they can yield streamline information, ratios of fluxes including recharge rates, effective porosity values, or evaporation rates in the case of chloride. All this information can be introduced in numerical groundwater modelling as additional information.

23.3 LIMITATIONS FOR THE USE OF ENVIRONMENTAL TRACERS IN GROUNDWATER MODELLING

The reason that environmental tracers are not a panacea for groundwater modelling is the fact that each tracer requires a number of new parameters concerning input functions, unsaturated zone properties, aquifer porosity, degassing, and adsorption parameters. So one is in a situation where the data increase but the number of parameters to be estimated also increases, and uncertainty needs to be reduced. Long time series could help to alleviate this problem, but while they are often available for heads they are usually not available for environmental tracer data in groundwater. This leads to limitations for the use of environmental tracers in groundwater modelling. There are more limitations. For example, even if at certain sampling locations long records of the input function exist, the local input function relevant for the aquifer is often not well known. The age window of a particular tracer is limited and while there are a number of tracers suitable for up to 50 year residence times, there is practically nothing suitable for the 50 to 300 years time scale. Environmental tracer data do not provide direct information on Darcy fluxes. They relate to pore velocities and therefore require an effective porosity in order to connect to Darcy fluxes. Estimated porosity values can still exhibit some uncertainty, especially in fractured media. Moreover, the effective porosity may not be constant on a given time-scale due to heterogeneity showing in dual porosity effects. Furthermore hydraulic heads show a momentary situation while tracer data integrate over longer time periods.

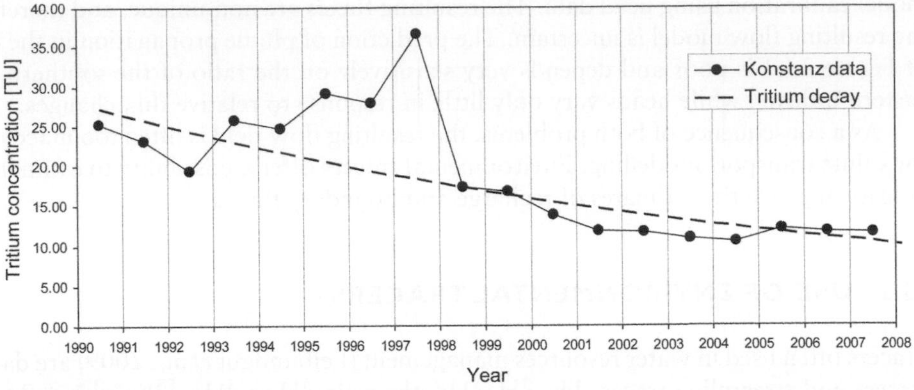

Figure 23.3 ^{3}H input function for the station Konstanz, Germany (Data IAEA, 2009), together with a ^{3}H decay function.

The residence time of tracers in the unsaturated zone can sometimes be larger or even much larger than in the saturated zone of the aquifer. This can increase the uncertainty since information on unsaturated flow conditions is usually very sparse. Dissolved gas tracers yield information which is different from that of solute markers of the water molecule. Nevertheless, the range of possible results from a calibrated flow model can sometimes be restricted considerably by using environmental tracer information in the modelling. This depends on the sensitivity of a particular parameter like effective porosity to simulated environmental tracer concentrations. In many cases transmissivity unfortunately is not sensitive.

The input function of environmental tracers may change over time. A prominent example is the ^{3}H input function, which exhibited strong peak values in the sixties of the last century of the order of 1000 TU (e.g., Leibundgut *et al.*, 2009). This property offered valuable possibilities for numerical evaluations using analytical and numerical models due to the large contrast of ^{3}H in the environment including groundwater. However, since then the input signal has decreased in general worldwide. On the southern hemisphere it has never been as pronounced as on the northern hemisphere. As an example the ^{3}H input concentration in precipitation is shown in Figure 23.3 for the station Konstanz (Germany, data from IAEA, 2009). The ^{3}H decay curve in Figure 23.3 suggests that, essentially, the input curve follows relatively closely the natural ^{3}H decay curve for the period after 1998. This property has consequences for the dating possibilities of ^{3}H, since it would severely hamper or even render infeasible the determination of the age of water samples. The application of such tracer data in numerical modeling would lead to insensitive results.

23.4 LESSONS FROM SIMPLE ANALYTICAL SOLUTIONS

The use of environmental data in modelling can be demonstrated with the help of simple analytical solutions. Although resulting from simplified models of flow and transport they can provide valuable insight into the possibilities of such methods.

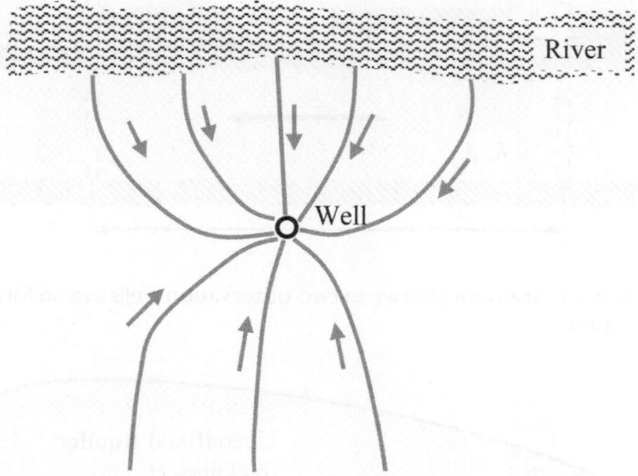

Figure 23.4 Flow from river and aquifer towards pumping well; schematic two-dimensional horizontal flow configuration.

Mixing of water from a river and an aquifer in a pumping well according to Figure 23.4 can be evaluated for steady-state conditions in a simplified manner using the water balance $Q_{Well} = Q_{River} + Q_{Land}$ and the tracer mass flux balance $c_{Well}Q_{Well} = c_{River}Q_{River} + c_{Land}Q_{Land}$ for given pumping rate Q_{Well} and measured tracer concentrations c_{Well}, c_{River}, and c_{Land}. Q_{River} and Q_{Land} are the discharge portions from the river and the aquifer into the well. The prerequisite of such an analysis is that all quantities are constant, and that the tracer concentrations are not spatially and temporally variable. Moreover, it is required that there is sufficient contrast in the tracer concentrations. For these conditions the ratio Q_{River}/Q_{Land} can be evaluated as follows:

$$\frac{Q_{River}}{Q_{Land}} = \frac{c_{Land} - c_{Well}}{c_{Well} - c_{River}} \qquad (23.2)$$

This example shows how environmental tracer information provides information about the ratio of two water fluxes. Only if one flux is given, the other one can be determined. In the bank filtration example this is the case as the pumping rate of the well is known.

One-dimensional steady-state tracer transport between two observation wells in a uniform flow field according to Figure 23.5 can be described using Darcy's law:

$$u = \frac{v_{Darcy}}{\phi_e} = \frac{K\Delta h}{L\phi_e} = \frac{L}{t_{Tracer}} \qquad (23.3)$$

where u is the mean pore velocity of an ideal tracer, ϕ_e is the effective porosity, K is the hydraulic conductivity of the aquifer, h is the head, L is the distance between the observation wells, and t_{Tracer} is the travel time between the piezometers, which can

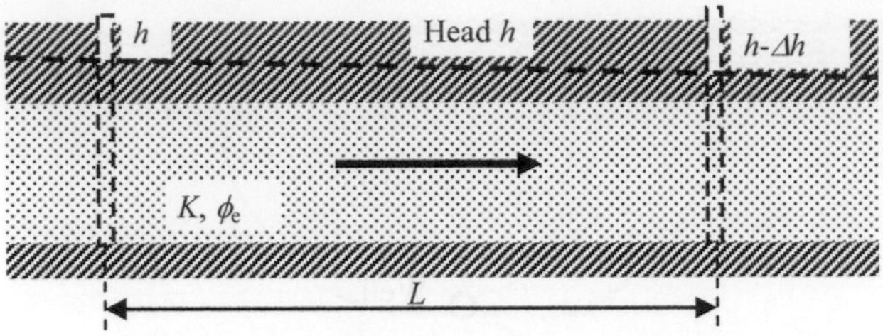

Figure 23.5 Flow and tracer transport between two observation wells in a uniform flow field; schematic cross section.

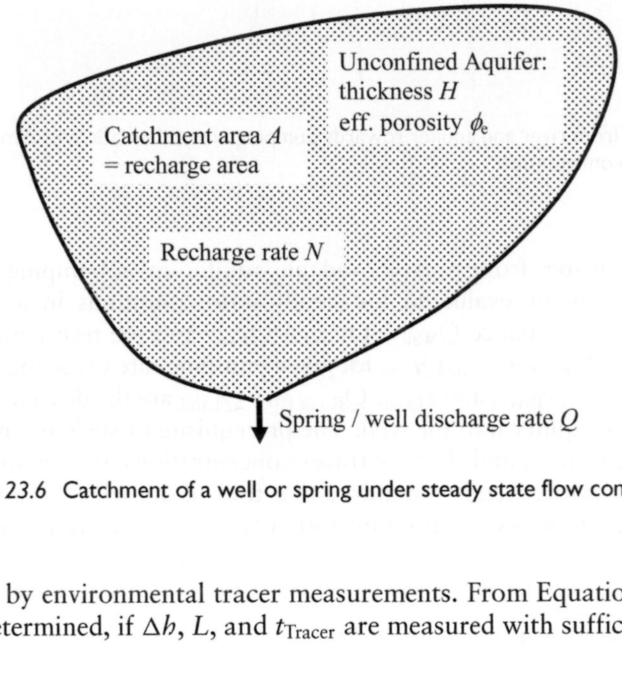

Figure 23.6 Catchment of a well or spring under steady state flow conditions.

be determined by environmental tracer measurements. From Equation 23.3 the ratio K/ϕ_e can be determined, if Δh, L, and t_{Tracer} are measured with sufficient accuracy:

$$\frac{K}{\phi_e} = \frac{L^2}{\Delta h \, t_{\text{Tracer}}} \tag{23.4}$$

The analysis yields again a ratio of two parameters. Whenever porosity can be accurately estimated the hydraulic conductivity can be deduced. However, porosity is in general also uncertain. In unconsolidated sedimentary aquifers the uncertainty may be relatively small (less than a factor of 2) but in fractured rock aquifers it may be much larger (more than a factor of 10).

The mean residence time of an ideal tracer in the saturated zone of a shallow unconfined aquifer within a spring or well catchment according to Figure 23.6 can be expressed for steady-state flow conditions as follows:

$$t_{\text{Tracer}} = \frac{AH\phi_e}{Q} = \frac{H\phi_e}{N} \tag{23.5}$$

where H is the mean aquifer thickness in the catchment, A is the catchment area, and N is the mean recharge rate of the catchment. Effective porosity is taken as constant. This consideration is based on purely hydrological reasoning. From estimates of the residence time of environmental tracers in the catchment, the ratio $H\,\phi_e/N$ can be evaluated from Equation 23.5. Note that the travel time is independent of transmissivity T. Of course the latter depends on aquifer thickness, which is determined via flow model calibration without using tracer information.

It can be concluded from the three simple cases that measured environmental tracer information can yield information on relationships between flow and/or transport parameters rather than the absolute parameter values directly. When applied in numerical flow and transport modelling these relationships can be used to test or to constrain and therefore to improve these models.

23.5 ANALYTICAL MODELS FOR GROUNDWATER DATING

Groundwater ages deduced from environmental tracer analyses represent valuable information for testing groundwater flow models, e.g., when using particle tracking in numerical flow models. Groundwater dating on the basis of simultaneous measurement of ^{3}H and tritiogenic ^{3}He was suggested by Tolstikhin & Kamensky (1969) and applied in the field of groundwater hydrology by a large number of investigators like Maloszewski and Zuber (1983), Poreda *et al.* (1988), Schlosser *et al.* (1988), Schlosser *et al.* (1989), Solomon *et al.* (1992), Szabo *et al.* (1996), Cook & Herczeg (1999), Holocher *et al.* (2001) and others. The method is based on the formula for the residence time as a function of the relative concentration of the two tracers in the saturated part of the aquifer:

$$t_{\mathrm{Residence}} = \frac{1}{\lambda}\ln\left(\frac{c_{^3\mathrm{He}}}{c_{^3\mathrm{H}}} + 1\right) \tag{23.6}$$

The coefficient λ is the decay constant ($\lambda = \ln(2)/t_{1/2}$, where $t_{1/2}$ is the half-life of ^{3}H). The formula is based on a moving small water parcel concept with negligible mixing and continuous accumulation of tritiogenic ^{3}He. Point sampling along a flow line allows estimates of the travel time. Age values are frequently used in numerical groundwater modeling (see below).

Recently, Stauffer *et al.* (2011) presented analytical solutions and discussed them for simplified groundwater systems with decaying environmental tracers such as ^{3}H including the formation of a decay product, such as tritiogenic ^{3}He. The developed solutions are applicable for shallow unconfined well or source catchments, which can be conceptually described by a steady state two-dimensional, semi-confined groundwater flow model with constant thickness, recharge rate and porosity. The prerequisite for the applicability of these solutions is that the pumping wells and observation wells at which tracer information is available are fully screened over the entire aquifer thickness. Sampling by pumping from such abstraction wells produces the complete mixing of water of different age and origin. Under these conditions the analytical solutions predict that the transient decaying and accumulating environmental tracer concentrations are homogeneously distributed in the model domain provided the assumptions are satisfied. The solutions furnish relations between aquifer thickness, porosity and recharge rate for given tracer concentrations. Moreover, for the ^{3}H-^{3}He system the residence time can be determined for given tracer concentrations.

Lumped-parameter or box models (e.g., Maloszewski and Zuber, 1983; overview in Leibundgut *et al.*, 2009) represent a further group of analytical solutions for simplified systems, which can be used for the interpretation of environmental tracer data.

23.6 USE OF ENVIRONMENTAL TRACERS IN NUMERICAL GROUNDWATER MODELLING

Environmental tracer information has been used in numerical groundwater modelling many times.

Wei *et al.* (1990) used ^{4}He and ^{14}C in their study of the Paris basin, France, to confirm their multi-aquifer flow model. The results of the simulations of the tracer transport strengthened the representativeness of their model. It allowed explaining the salinity distribution, the ^{4}He concentration distribution and the distribution of the residual ^{14}C activity.

Solomon & Sudicky (1991) used ^{3}H-^{3}He-age gradient evaluation near a groundwater table of an unconfined aquifer to estimate the recharge rate using simple one-and two-dimensional vertical flow models. Their simulations showed that groundwater ages determined using the ^{3}H-^{3}He ratio approximate the true groundwater travel times if the dispersive mixing between the water table (travel time zero) and the measurement location (measured travel time) is weak.

Smethie *et al.* (1992) used ^{85}Kr measurements in a two-dimensional vertical flow model (steady-state hydraulic potential and stream function model) for the Borden site (Ontario, Canada). Travel times calculated from a two-dimensional steady-state hydraulic potential and stream function model agreed well with the ^{85}Kr ages in the main recharge region of the aquifer.

Reilly *et al.* (1994) calculated groundwater recharge ages of shallow groundwater estimated from chlorofluorocarbon, ^{3}H, and other environmental tracers and combined them with numerical simulation techniques (flow and advective transport calculations using particle tracking). The combination of the methods enabled a coherent explanation of the flow paths and flow rates of their case study in Maryland, USA, but still indicated weaknesses in the understanding of the system.

Engesgaard *et al.* (1996) used ^{3}H data to estimate the range of large-scale dispersivity values of a sandy unconfined aquifer in Denmark. Numerical modelling comprised (a) the estimation of the ^{3}H content in the infiltration water, (b) the transport in the unsaturated zone, (c) the estimation of the flux-averaged ^{3}H concentration in the recharge water, and (d) the transport in the saturated zone. From the analysis it was not possible to identify a unique set of longitudinal and transversal dispersivity values, but to indicate a range of variation. In their analysis they could make use of the strong ^{3}H peak in the input function.

Sheets *et al.* (1998) used ^{3}H-^{3}He-ages to evaluate and to improve their flow model in a complex buried-valley aquifer. They combined the use of the ^{3}H-^{3}He dating technique and the particle tracking technique. The analysis led to a lower mean absolute error between measured and simulated heads. Discrepancies between simulated travel times and ^{3}H-^{3}He-ages were assumed to be due to improper conceptualisation or incorrect parametrization of the flow models. Selected modifications of the model resulted in an improved agreement.

Plümacher (1999) used a two-dimensional horizontal groundwater flow and transport model of a fractured carbonate aquifer in central Württemberg close to Stuttgart, Germany, to study the use of stable isotopes in model calibration and to define the capture zone of mineral springs. Stable isotopes allowed fixing the ratio between boundary inflow from the south and recharge in the west. The constant input of stable isotopes enabled time-averaged (steady-state) transport calculations independent of porosity.

Zoellmann *et al.* (2001) used SF6 and ^{3}H to calibrate porosity of an aquifer and field capacity of the overlying soil in Hessia, Germany. These tracers show different time scales due to their different transport mechanisms in the unsaturated zone. While solute tracers are moved advectively with the seepage water, gas tracers pass the unsaturated zone diffusively through the air phase. According to the properties of the unsaturated zone (hydraulic properties, thickness) the differences in behaviour were used to separate the subsurface transport process into unsaturated and saturated parts. The authors used a one-dimensional plug-flow model for the unsaturated zone combined with a detailed two-dimensional horizontal flow and transport model for the saturated zone. The simulations allowed for a consistent interpretation of the measured tracer concentrations within the aquifer. With this model it was possible to confirm measured nitrate concentrations at drinking water wells and predict times to effectiveness of remediation measures.

Mattle *et al.* (2001) used ^{3}H data to constrain their numerical three-dimensional flow model of a sandy gravel aquifer and to calibrate the leakage coefficient of the infiltrating river. The latter turned out to be highly sensitive with respect to the mean residence times in the aquifer. While the calibration of the flow model based on head measurements did not yield unique parameter values, the environmental tracer transport allowed to tune it much more sensitively.

Moltyaner *et al.* (2002) used ^{3}H and Cl^{-} data to determine the ^{3}H source function in a shallow sandy aquifer in Ontario, Canada, with the help of three-dimensional flow and advection-dispersion transport modelling. Dispersion parameters were estimated from the analysis of a controlled conventional natural-gradient tracer test.

Castro & Goblet (2003) tested different scenarios using ^{4}He data for their regional groundwater flow models. Four different calibrated groundwater flow scenarios were developed for the Carrizo aquifer in Texas, USA. All scenarios except one failed to reproduce coherent ^{4}He transport in the system.

Guell & Hunt (2003) studied the transport of radionuclides at an underground nuclear detonation test site in Nevada, USA. Transport simulation was used together with ^{3}H data to determine regional anisotropy in hydraulic conductivity and the reduced hydraulic conductivity in the cavity region. The calibrated model reproduced the ^{85}Kr breakthrough data when emplacement of ^{85}Kr by the upward migration of carbon dioxide was included.

Pint *et al.* (2003) analysed flowpaths using particle tracking in their case study in Allequash Basin, Wisconsin, USA. The analysis supported former conclusions from isotope and major ion chemistry.

Zuber *et al.* (2005) used various environmental tracers to check and to improve their three-dimensional flow and solute transport model of an aquifer in southern Poland. Time series of ^{3}H content in wells yielded quantitative information on age distribution and the total mean ages of flow through the unsaturated and saturated zones. Transport modelling of SF_6 using the MT3D code first showed large discrepancies to measured concentrations. Some discrepancies remained after calibration using the SF_6-data. ^{3}H simulation using this calibrated model yielded reasonable agreement in some wells and indicated the need for further investigations. The existence of distinct hydro-chemical zones turned out to be consistent with the tracer data.

Troldborg *et al.* (2007) used ^{3}H, ^{3}He, and chlorofluorocarbons to check four different conceptual models of their multi-aquifer system in Denmark. Each of the models was calibrated using head data and discharge measurements. Transport simulation results of the four models were compared with the tracer data. The models showed major differences in the prediction of the age of the groundwater and the environmental tracer concentrations. They concluded that it is crucial to take model conceptual uncertainty into account when making predictions beyond the calibration period.

Onnis (2007) investigated the potential of multiple ^{3}H, ^{3}He and ^{85}Kr tracer data from observation and pumping wells for constraining a groundwater model of a small shallow sandy gravel aquifer in northern Switzerland. Tracer transport in the unsaturated zone was modelled by a vertical advective model for ^{3}H and a gas-diffusion model for ^{85}Kr and ^{3}He. Transport in the saturated zone was simulated using the calculated tracer input from the unsaturated zone and based on an ensemble of equally-likely transmissivity realizations obtained by Monte-Carlo type inverse modelling. Due to the relatively thick and spatially variable unsaturated zone, transmissivity was not found to be an important source of uncertainty for tracer transport. Sensitivity analysis shows that tracer transport is instead dominated by unsaturated zone parameters (e.g., effective gas-diffusion coefficient) for ^{85}Kr and, to a smaller extent, by porosity. ^{3}H transport turned out to be insensitive to transmissivity and their contrasts. Moreover, not much sensitivity was found towards porosity.

Stichler *et al.* (2008) established a two-dimensional numerical water flow and tracer transport model using both hydraulic heads and stable isotope data in order to define the capture zone of drinking water wells.

Indeed, the examples from the literature show that various environmental tracers were successfully used in the past to check or to improve flow and transport models, to select among several scenarios or conceptual models, or to falsify conceptual models, and to determine solute transport parameters, and to some degree, also flow parameters.

From the above reviewed literature follows that parameter estimation using environmental tracers has usually been performed manually in the past. In principle automatic inverse procedures exist for tracer transport. However, the use of such procedures in connection with environmental tracers is not reported.

Why are environmental tracers not used even more intensively in flow and transport modelling? Besides the above mentioned problems, reasons might be the relatively high costs, and the lack of knowledge and experience about the sensitivity of a particular tracer. Moreover, too high expectations in the past may have led to frustration.

23.7 CONCLUSIONS

In conclusion, environmental tracer data can improve groundwater flow and transport models. Moreover, environmental tracers are the only available means to estimate effective porosity and travel times on the field scale. The combination of several environmental tracers with different properties (like gaseous tracers with different diffusion coefficients and water-bound tracers) is certainly more demanding, but can make the application more reliable. However, there are various limitations with respect to the use of environmental tracers.

Flow and transport models should be used to check the consistency of all data (even including proxy-data, which do not enter the model directly) and to confirm or to exclude hypotheses or scenarios. As a general suggestion, environmental tracer information should primarily be used for constraining or revising the conceptual model, and not necessarily for getting higher accuracy. The integration of environmental tracers in groundwater models can in general reduce the range of possible alternative interpretations, which are consistent with all observations.

ACKNOWLEDGEMENTS

The study was partially supported by the EU 7th Framework Programme GENESIS Project No. 226536.

REFERENCES

Castro M.C., Goblet P. (2003) Calibration of regional groundwater models: Working toward a better understanding of site-specific systems. *Water Resour Res* 39(6):1172, doi:10.1029/2002WR001653.

Cook, P., Herczeg, A. (1999) Environmental tracers in subsurface hydrology. Boston, USA, Kluwer Academic Press.

Engesgaard P., Jensen K.H., Molson J., Frind E.O., Olsen H. (1996) Large-scale dispersion in a sandy aquifer: Simulation of subsurface transport of environmental tritium. *Water Resour Res* 32(11): 3253–3266.

Guell M.A., Hunt J. (2003) Groundwater transport of tritium and krypton 85 from a nuclear detonation cavity. *Water Resour Res* 39(7): 1175, doi:1029/2001WR001249.

Holocher J., Matta V., Aeschbach-Hertig W., Beyerle U., Hofer M., Peeters F., Kipfer, R. (2001) Noble gas and major element constraints on the water dynamics in an alpine floodplain. *Ground Water* 39: 841–852.

IAEA (2009) International Atomic Energy Agency (IAEA, Vienna, Austria), http://www-naweb.iaea.org/napc/ih/IHS_resources_isohis.html.

Leibundgut C, Maloszewski P, Külls C (2009) Tracers in hydrology. Wiley-Blackwell, Chichester.

Maloszewski, P., A. Zuber (1983). Theoretical possibilities of the $_3$H-$_3$He method in investigations of groundwater systems. *Catena* 10(3), 189–198.

Mattle N., Kinzelbach W., Beyerle U., Huggenberger P., Loosli H.H. (2001) Exploring an aquifer system by integrating hydraulic, hydrogeologic and environmental tracer data in a threedimensional hydrodynamic transport model. *J Hydrol* 242: 183–196.

Moltyaner G.L., Klukas M.H., Takeda S., Krotzer T.G., Yamazaki L.S. (2002) Advection dispersion modelling of tritium and chloride migration in a shallow sandy aquifer at the Chalk

River Laboratories. In Use of Isotopes for Analyses of Flow and Transport Dynamics in Groundwater Systems. *IAEA*, 2002.

Onnis G.A. (2007) Interpreting Multiple environmental tracer data with a groundwater model in a perialpine catchment. PhD Thesis No 18003 ETH Zurich, 2007.

Pint C.D., Hunt R.J., Anderson M.P. (2003) Flowpath delineation and ground water age, Allequash basin, Wisconsin. *Ground Water* 41(7): 895–902.

Plümacher J. (1999) Kalibrierung eines regionalen Grundwasserströmungsmodells mit Hilfe von Umweltisotopinformationen. Schriftenreihe des Amtes für Umweltschutz Baden-Württemberg, Germany, Heft 1/1999.

Poreda R.J., Cerling T.E., Solomon D.K. (1988) Tritium and Helium isotopes as hydrologic tracers in a shallow unconfined aquifer. *J Hydrology* 103: 1–9.

Reilly T.E., Plummer L.N., Philips P.J., Busenberg E. (1994) The use of simulation and multiple environmental tracers to quantify groundwater flow in a shallow aquifer. *Water Resour Res* 39(2): 421–433.

Schlosser P., Stute M., Dörr H., Sonntag C., Münnich K.O. (1988) Tritium ^{3}He dating of shallow groundwater. *Earth and Planetary Science Letters* 89: 353–362.

Schlosser P., Stute M., Dörr H., Sonntag C., Münnich K.O. (1989) Tritiogenic ^{3}He in shallow groundwater. *Earth and Planetary Science Letters* 94: 245–256.

Sheets R.A., Bair E.S., Rowe G.L. (1998) Use of ^{3}H/^{3}He ages to evaluate and improve groundwater flow models in a complex buried-valley aquifer. *Water Resour Res* 34(5): 1077–1089.

Smethie W.M., Solomon D.K., Schiff S.L., Mathieu G.G. (1992) Tracing groundwater flow in the Borden aquifer using krypton-85. *J Hydrology* 130: 279–297.

Solomon D.K., Sudicky E.A. (1991) Tritium and Helium 3 isotope ratios for direct estimation of spatial variations in groundwater recharge. *Water Resour Res* 27(9): 2309–2319.

Solomon D.K., Poreda R.J., Schiff S.L., Cherry J.A. (1992) Tritium and Helium 3 as groundwater age tracers in the Borden aquifer. *Water Resour Res* 28(3): 741–755.

Stauffer F., Stoll S., Kipfer R., Kinzelbach W. (2011) Analytical model for environmental tracer transport in well catchments. *Water Resour Res* 47, W03525, doi:10.1029/2010WR009940.

Stichler W., Maloszewski P., Bertleff B., Watzel R. (2008) Use of environmental isotopes to define the capture zone of a drinking water supply situated near a dredge lake. *J Hydrology* 362, 220–233.

Szabo Z., Rice D.E., Plummer L.N., Busenberg E., Drenkard S., Schlosser P. (1996) Age dating of shallow groundwater with chlorofluorocarbons, tritium/helium 3, and flow path analysis, southern New Jersey coastal plain. *Water Resour Res* 32(4): 1023–1038.

Tolstikhin I., Kamensky I. (1969) Determination of groundwater ages by the T-^{3}He method. *Geochem. Int.*, 6: 810–811.

Troldborg L., Refsgaard J.C., Jensen K.H., Engesgaard P. (2007) The importance of alternative conceptual models for simulation of concentrations in a multi-aquifer system. *Hydrogeology J.* 15: 843–860.

Wei H.F., Ledoux E., de Marsily G. (1990) Regional modelling of groundwater flow and salt and environmental tracer transport in deep aquifers in the Paris basin. *J. Hydrology* 120: 341–358.

Zuber A., Witczak S., Różanski K., Sliwka I., Opoka M., Mochalski P., Kuc T., Karlikowska J., Kania J., Jackowicz-Korczynski M., Dulinski M. (2005) Groundwater dating with ^{3}H and SF_6 in relation to mixing pattern, transport modeling and hydrochemistry. *Hydrol. Proc.* 19: 2247–2275.

Zoellmann K., Kinzelbach W., Fulda C. (2001) Environmental tracer (3H and SF6) in the saturated and unsaturated zones and its use in nitrate pollution management. *J. Hydrology* 240: 187–205.

Chapter 24

Optimising groundwater monitoring networks using the particle swarm algorithm

Naser Ganji Khorramdel[1], *Saman Javadi*[2], *Kourosh Mohammadi*[3], *Ken Howard*[4] & *Mohammad J. Monem*[3]

[1]*Department of Water Engineering, Arak University, Arak, Iran*
[2]*Water Research Institute, Ministry of Energy, Tehran, Iran*
[3]*Tarbiat Modares University, Tehran, Iran*
[4]*Department of Physical and Environmental Sciences, University of Toronto Scarborough, Toronto, Canada*

ABSTRACT

Well designed monitoring networks are essential for the effective management of groundwater resources but the costs of monitoring well installations and sampling can prove prohibitive. The challenge is to obtain adequate water quality and quantity information with a minimum number of wells and sampling points, a task that can be approached objectively and effectively using numerical optimization methods. One recently developed optimization approach involves particle swarm optimization (PSO), a population based stochastic optimization technique that was inspired by the social behavior observed in bird flocks and schools of fish. The system is first initialized with a population of randomly generated particles (i.e. candidate solutions); thereafter, searches for optima are conducted iteratively. However, unlike genetic algorithms, PSO has no evolutionary operators (e.g. crossover and mutation) and instead, potential solutions "fly" through the problem space by following the current optimum particles. As a case study, the particle swarm algorithm technique was used to optimize an existing network of 57 monitoring wells located in the Astaneh aquifer in the north of Iran. The traveling sales person problem (TSP) analogy was used to initialize the problem and PSO was used to provide the optimal solution.

24.1 INTRODUCTION

Groundwater monitoring is relatively straightforward in concept but can be very difficult to perform efficiently. Monitoring is defined as the collection and analysis of data (chemical, physical, and/or biological) over a sufficient period of time and with adequate frequency to determine the status and/or trend in one or more environmental parameters or characteristics (EPA, 2004). Unfortunately, aquifer systems tend to be complex and monitoring can be very expensive, particularly when it requires the installation of a dedicated network of monitoring wells. In recent years, the challenge has been to design monitoring networks that are both efficient and cost effective.

Various methods have been tried to assist in the design of groundwater monitoring networks. Grabow *et al.* (1993) created a groundwater monitoring network using a "minimum well density" approach and Cieniawski *et al.* (1995) developed a long-term monitoring network using the Monte-Carlo simulation technique. Lee and Ellis (1996) examined a simple network location problem by applying eight heuristic algorithms used for solving nonlinear integer optimization problems, and were able to assess and compare their relative performance. More recently, Reed *et al.* (2000) applied a genetic algorithm to design a long-term groundwater monitoring network and identified robust parameter values using only three trial runs. They did, however, reveal that a major difficulty with the genetic algorithm approach to optimization lies in the large number of parameters that must be specified to control how the decision space is searched.

This chapter considers particle swarm optimization (PSO) as a potential means of improving the design of monitoring networks. PSO is a population based stochastic optimization technique (Kennedy and Eberhart, 1995) that is more commonly used for modelling social behaviour. It is an evolutionary algorithm (EA) which, despite its development primarily for the solution of optimization problems with continuous variables, has been successfully adapted in other contexts to problems with discrete variables (Montalvo *et al.*, 2008). PSO is an intelligence method developed to perform a heuristic direct search in a continuous parameter space without requiring any derivative estimation (Wang *et al.*, 2008). It has also been applied successfully in various water resource applications. Chau (2006) developed a PSO-based neural network approach to predict water levels in the Shing Mun River, Hong Kong, by adopting PSO to train multi-layer perceptrons. Similarly, a coupled simulation-optimization model was recently formulated using PSO and ant colony optimization (ACO), and applied to various groundwater management problems including maximisation of total pumping, minimisation of total pumping to contain contaminated water within a capture zone, and minimisation of the pumping cost to satisfy water demands for multiple management periods (Sedki and Ouazar, 2010). Both PSO and ACO were found to be promising methods for solving groundwater management problems and finding optimal or near-optimal solutions.

With regards to groundwater monitoring systems, where the challenge is to maximise the availability of good quality data while minimising the number of sampling sites and thereby limiting costs, optimization techniques clearly have a potentially valuable application. In turn, the task of minimising the number of sampling sites requires a close consideration of the distribution of the sampling sites and the ability of interpolation methods to predict reliable values in unsampled areas. A site in northern Iran is used to test the ability of PSO, when used in combination with Kriging, to lower the cost of a monitoring network by reducing the number of monitoring wells without compromising the quality of the interpolated data.

24.2 METHODS

Particle Swarm Optimization (PSO) was first introduced by Kennedy and Eberhart (1995) as an optimization method to solve nonlinear problems. PSO is a computational method that optimizes a problem by iteratively trying to improve candidate solutions with respect to a given measure of quality. Such methods are commonly known as metaheuristics as they make few or no assumptions about the problem being optimized and can find a solution to a problem in a large space of candidate solutions. PSO is easy to implement in computer codes and it uses few computational resources.

As population based stochastic optimization technique, PSO was inspired by the social behaviour observed in bird flocks and schools of fish. The algorithm can be best understood by imagining a group of birds that is randomly searching for food in an area. There is only one piece of food in the area being searched and, at the outset, none of the birds know where the food is exactly located. However, each bird does know the distance to the food source at every time step, and can find the food most efficiently by simply following the bird that is nearest to the food. PSO simulates this behaviour by searching for the best solution-vector in the search space. Candidate solutions are called particles. Each particle has a fitness/cost value that is evaluated by the function to be minimised, and each particle has a velocity that controls the "flight" of the particles. In effect, the particles "fly" through the search space in pursuit of the optimum particles.

PSO shares many similarities with evolutionary computation techniques such as genetic algorithms (GA). The system is initialised with a population of random solutions and searches for optima by updating generations. However, unlike GA, PSO has no evolution operators such as crossover and mutation. PSO optimizes a problem by taking a population of candidate solutions (the particles), and moving these particles around the search space according to the simple mathematical formulae. The movements of the particles are guided by the best found positions in the search space, which are updated incrementally as better positions are found by the particles. In GA, chromosomes share information with each other and, as a result, the whole population moves as a single swarm towards an optimal area.

If the search space is *D*-dimensional, then the *i*th particle of the population (or "swarm"), can be represented by a *D*-dimensional vector $X_I = (x_{i1}, x_{i2}, \ldots, x_{iD})$. The best previously visited position of the ith particle is denoted as $P_I = (p_{i1}, p_{i2}, \ldots, p_{iD})$. The best particle (solution) in the entire swarm is indexed as g. The velocity of this particle can be presented by another *D*-dimensional vector $V_I = (v_{i1}, v_{i2}, \ldots, v_{iD})$.

The swarm is manipulated according to Equations 24.1 and 24.2 (Oliveira, 2005):

$$v_{id} = w * v_{id} + c_1 * rand(\ ) * (p_{id} - x_{id}) + c_2 * Rand(\ ) * (p_{gd} - x_{id}) \tag{24.1}$$

$$x_{id} = x_{id} + v_{id} \tag{24.2}$$

where w is the inertia weight, c_1 and c_2 are positive acceleration coefficients, *rand*( ) and *Rand*( ) are random numbers uniformly distributed in the interval [0, 1], $i = 1, 2, \ldots, N$ and N is the size of swarm. The coefficient c_1 is the cognitive behaviour parameter and describes the extent to which a particle will follow its own best solution, while c_2 is the social behaviour parameter, and indicates how closely the particle will follow the swarm's best solution.

Equation 24.1 is used to adjust the velocity of each particle to follow two best solutions, the first being the best solution of the particle and the second being the best solution of the swarm. The particle is then moved to the new location using Equation 24.2. The performance of each particle is evaluated using a fitness function that is predefined for each problem.

Clerc (1999) recommends the use of a Constriction Factor, *K*, to guarantee convergence of the algorithm and improve the ability of the PSO to constrain and control velocities. The Constriction Factor is described as a function of c_1 and c_2 according to:

$$K = \frac{2}{|2 - \varphi - \sqrt{\varphi^2 - 4\varphi}|}, \quad \text{where } \varphi = c_1 + c_2 \quad \varphi > 4 \tag{24.3}$$

to produce:

$$v_{id} = K * [v_{id} + c_1 * rand(\ ) * (p_{id} - x_{id}) + c_2 * Rand(\ ) * (p_{gd} - x_{id})] \tag{24.4}$$

φ is >4 and is usually set to 4.1 such that K will be 0.729 and:

$$K * c_1 = K * c_2 = 0.729 \times 2.05 = 1.49445 \tag{24.5}$$

Most PSO algorithms are designed to search in continuous domains. However, there are a number of PSO variants that operate in discrete spaces and these can pose difficulties for obtaining solutions. One variant proposed for discrete domains is the binary particle swarm optimization algorithm (Kennedy and Eberhart, 1997). In this algorithm, the position of the particle is discrete but its velocity is continuous. The jth component of a particle's velocity is used to compute the probability with which the jth component of the particle's position vector takes a value of 1. Velocities are updated as in the standard PSO algorithm using Equation 24.6, but positions are updated using the following rule (Equation 24.7):

$$v_{i,j} = v_{i,j} + pw * rand(\ ) * (x^*_{i,j} - x_{i,j}) + nw * rand(\ ) * (x^*_{i',j} - x_{i,j}) \tag{24.6}$$

$$x_{i,j} = \begin{cases} 1 & \text{if} \quad rand(\ ) < S(v_{i,j}) \\ 0 & \text{otherwise} \end{cases} \tag{24.7}$$

where $x_{i,j}$ is the jth component of the position vector of particle p_i, $x^*_{i,j}$ is the best position visited by the particle. It is referred to as the neighbourhood best and is the best position ever located by any particle in the neighbourhood of particle p_i. w is the inertia weight, pw is the weight for the best position of the jth particle and nw is the weight for the velocity of the best position in the entire swarm. $S(x)$ is a sigmoid function as defined in Equation 24.8 and can constrain v_{id} to the interval [0,1], while $S(v_{id})$ can be considered as a probability (Shi *et al.*, 2007).

$$S(x) = \frac{1}{1 + e^{-x}} \tag{24.8}$$

24.3 ANALYSIS

24.3.1 Study area

For the purpose of testing the optimization approach, the monitoring network for the Astaneh aquifer in northern Iran was selected (Figure 24.1). The study area lies between 49°32′ and 50°05′ east and 37°07′ and 37°25′ north and has an area of 1100 km^2. The land surface in this region ranges from 25 m below sea level along its northern Caspian

Sea Coast to 2705 m above sea level inland. The aquifer comprises alluvial sediments deposited by the Sefid Rud River and is currently monitored by a network of 57 wells.

24.3.2 The traveling sales person problem

The traveling sales person problem (TSP) analogy was used to initialize the problem and PSO was used to provide the optimal solution. The TSP is one of the most intensively studied problems in computational mathematics and involves finding the shortest itinerary between a series of cities under the condition that each city may be visited only once. For our purposes, the cities can be substituted by monitoring wells. In TSP, the distance between the monitoring wells and the order they are visited are important considerations. However, in applying PSO to optimize the monitoring network, the order and distances are not directly related to the objective function and the imposed constraints. Distances between wells do, however, affect the accuracy of the water level interpolations.

24.3.3 Establishing the PSO-TSP algorithm

For this algorithm, the particle selects a well to visit based on the relative importance of that well compared to its neighbors. The objective function is to minimise the overall data loss in the optimized monitoring network. This can be quantified using the root mean square error (RMSE) according to:

$$\min Z = \sqrt{\frac{\sum_{i=1}^{m}\left(\frac{WT_{est,i} - WT_{act,i}}{\min(WT_{est,i}, WT_{act,i})}\right)^2}{m}} \tag{24.9}$$

$$m = S_{goal}$$

where m is the number of wells eliminated from the network, $WT_{act,i}$ is the actual groundwater level in the eliminated well i, $WT_{est,i}$ is the estimated groundwater level in the eliminated well i based on measurements in neighbouring wells and S_{goal} is the total number of wells that should be eliminated. S_{goal} is selected by the user.

To conduct the analysis, the Kriging algorithm was used to estimate groundwater levels at unsampled locations. Kriging is a geostatistical technique that is used to interpolate the value of a random field (e.g. water level) at an unobserved location based on observations of its value at nearby locations. It belongs to the family of linear least squares estimation algorithms. Different types of Kriging have been developed according to the stochastic properties of the random field. Classic types include simple, ordinary, universal, indicator, disjunctive and lognormal. In this study, ordinary Kriging was used as it provides the best fit between observed and estimated groundwater levels for the region. Optimization of the monitoring network involved the following series of steps:

Step 1 Initialization: The iteration counter is set to zero ($k=0$). n particles are randomly produced ($\{X_i^0, i=1,2,\ldots,n\}$). These particles have initial velocities of $V_i^0=[v_{i1}^0, v_{i2}^0,\ldots,v_{id}^0]$. The objective function of each particle, $f(X_i^0)$, is evaluated. If the constraints are met, then the best particle is stored as $PB_i^0 = X_i^0$. In addition, the particle having the best objective function value throughout the entire swarm is stored as the best global particle (GB^0). If the problem constraints are not met, the initialization is repeated.

Step 2 The iteration counter is updated.

Step 3 Velocity vector is updated using Equation 24.1.

Step 4 The position of the particle is calculated using the sigmoid function (Equations 24.7 and 24.8).

Step 5 The best particle position (PB_i^k) in iteration k is updated. If $f_i(X_i^k) < f_i(PB_i^{k-1})$ then $PB_i^k = X_i^k$, otherwise, $PB_i^k = PB_i^{k-1}$.

Step 6 The best particle position (GB^k) for the entire swarm is updated ($f(GB^k) = \min\{f_i(PB_i^k)\}$). If $f(GB^k) < f(GB^{k-1})$ then $GB^k = GB^k$, otherwise, $GB^k = GB^{k-1}$.

Step 7 Stop criterion: when the iteration reaches the set number or the best fitness reaches the prescribed value.

Both RMSE and relative estimation error (REE) were calculated to measure the fitness of solutions. Equations 24.10 and 24.11 show how these parameters are calculated.

$$RMSE = \sqrt{\frac{1}{n}\sum_{i=1}^{n}(WT_{est,i} - WT_{act,i})^2} \tag{24.10}$$

$$REE = \frac{|WT_{est,i} - WT_{act,i}|}{\min(WT_{est,i}, WT_{act,i})} \tag{24.11}$$

where $WT_{est,i}$ and $WT_{act,i}$ are estimated and actual water table levels, respectively.

24.4 RESULTS AND DISCUSSION

The PSO-TSP algorithm was applied to the Astaneh monitoring network with the objective of minimising the number of water level sampling points, while minimising the RMSE error between the full monitoring network and the optimized network. The algorithm required only 2850 objective function evaluations to reach an optimal solution for the network of just 42 monitoring points. By comparison, 2.206 × 1013 calculations would have been required to evaluate all possible permutations using a routine analytical approach.

A summary of the results for different monitoring network scenarios is shown in Table 24.1. For comparative purposes, RMSE values calculated using the genetic algorithm are also included in this table. Figure 24.1 shows how the RMSE and relative estimation error (REE) varies according to the number of monitoring wells selected, while Figures 24.2 to 24.4 provide comparisons of the water level maps for selected network scenarios.

The PSO algorithm successfully evaluated all feasible solutions. RMSE values shown in Table 24.1 and Figure 24.2 reveal that a reduction from 57 to 48 monitoring

Table 24.1 Results of the PSO and GA optimization algorithms for different monitoring networks.

Remaning Wells	*Wells eliminated using PSO-TSP*	*RMSE (PSO-TSP) m*	*RMSE (GA) m*
54	21, 23, 52	0.190	0.104
51	13, 21, 44, 47, 52, 53	0.159	0.122
48	1, 2, 7, 15, 21, 23, 40, 47, 53	0.236	0.177
45	3, 5, 13, 15, 21, 22, 36, 44, 47, 49, 52, 53	0.300	0.165
42	1, 5, 7, 13, 15, 21, 22, 23, 36, 40, 42, 44, 47, 52, 53	0.322	0.203

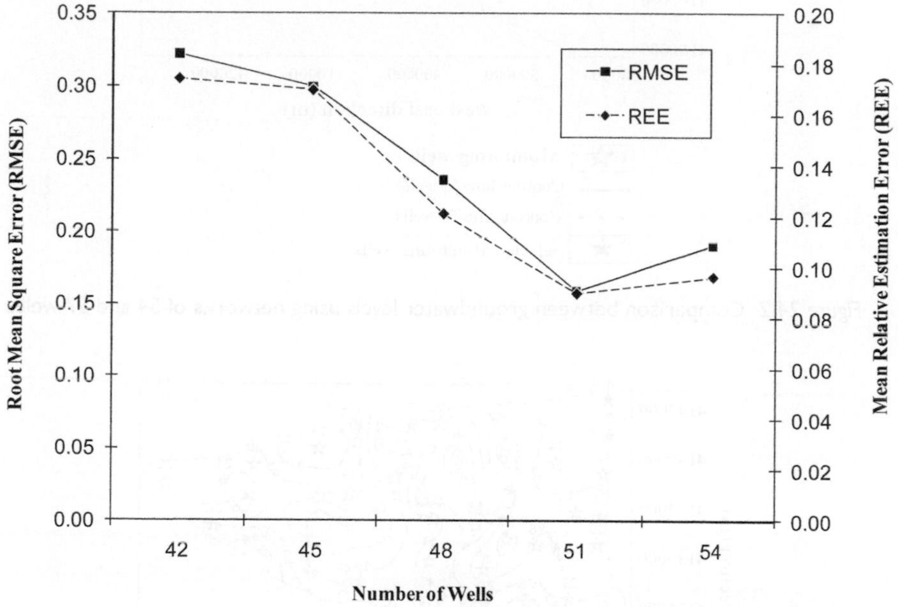

Figure 24.1 The variation in RMSE with the number of monitoring wells.

wells results in a relatively small error. Thereafter, however, the RMSE increases linearly as more wells are eliminated from the network. Figures 24.3–24.5 show a progressive deterioration of the match between the interpolated and the original water level contours, as more wells are removed. Even then, the discrepancy in the water level contours for a 26% reduction of wells (from 52 to 42 wells; Figure 24.5) is barely discernible. As expected, most of the wells eliminated by the analysis were located in areas that originally had a relatively high density of monitoring wells.

The proposed PSO method was also compared with the more conventional GA technique and showed good agreement. Although the GA technique showed better accuracy and lower RMSE, CPU times for the PSO-TSP algorithm were significantly smaller (Table 24.2). It is expected that for larger aquifers and more complex situations the speed performance of the PSO method will be even more pronounced.

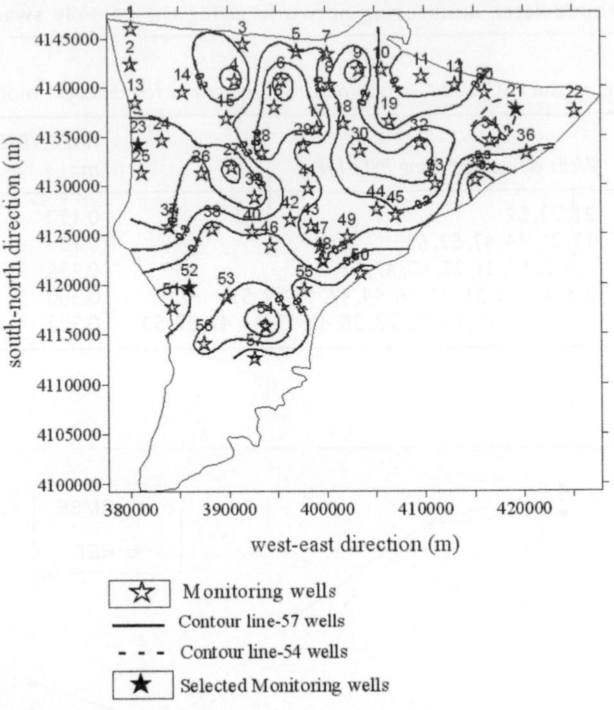

Figure 24.2 Comparison between groundwater levels using networks of 54 and 57 wells.

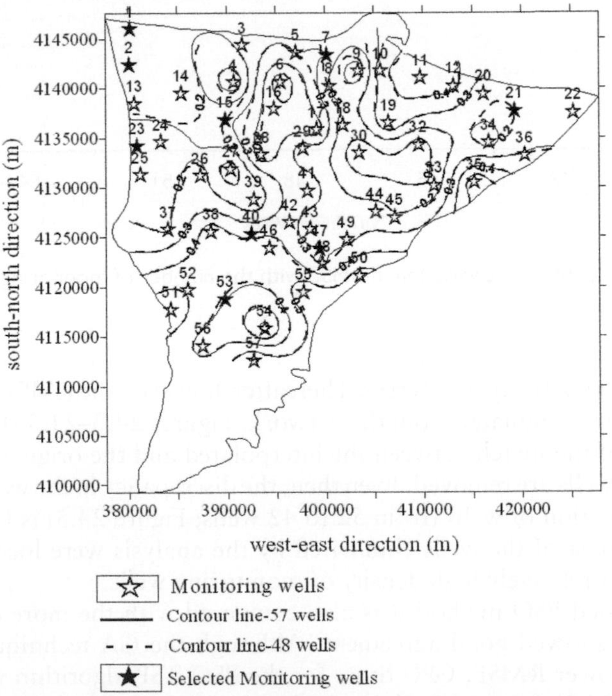

Figure 24.3 Comparison between groundwater levels using networks of 48 and 57 wells.

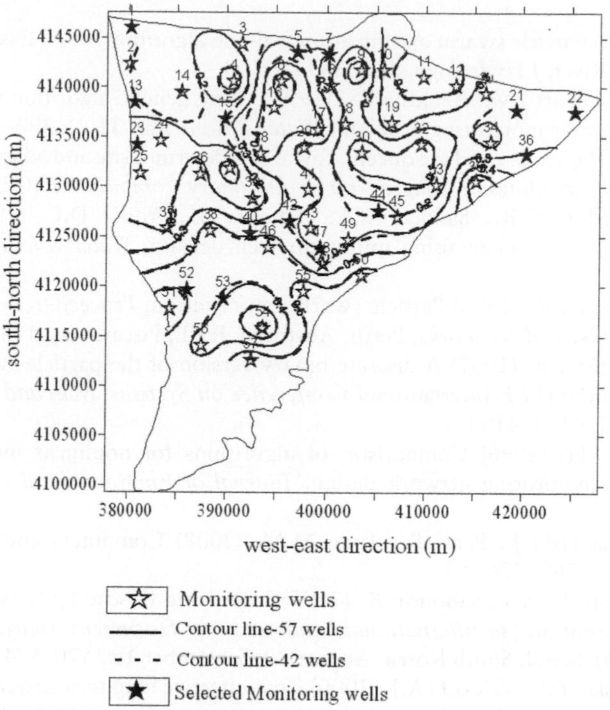

Figure 24.4 Comparison between groundwater levels using networks of 42 and 57 wells.

Table 24.2 CPU time of GA and PSO-TSP algorithms for the Astaneh aquifer.

Remaining Wells	*CPU Times of GA (sec)*	*CPU Times of PSO-TSP (sec)*
54	32.4219	17.2656
51	61.4844	32.9062
48	88.4219	47.5312
45	112.828	60.5312
42	136.6094	72.2344

24.5 CONCLUSIONS

In this chapter, the developed PSO-TSP algorithm would reduce the number of wells in a groundwater monitoring network without seriously compromising either the quality of the data collected or the reliability of the contour maps that are subsequently generated. The approach proved to be highly successful. The results of the optimization showed that the number of observation wells in the Astaneh aquifer monitoring network could be reduced by 26% from 57 to 42 without a significant loss of information. The root mean square error (RMSE) for the final optimized network was 0.322 m. A comparison of RMSE values determined using the PSO-TSP algorithm with those calculated using the more conventional genetic algorithm technique showed good agreement and provides strong support for the new, more efficient PSO-TSP approach.

REFERENCES

Chau K.W. (2006) Particle swarm optimization training algorithm for ANNs in stage prediction of Shing Mun River, *J Hydrology*, 329: 363–367.

Cieniawski S.E., Eheart J.W., Ranjithan S. (1995) Using genetic algorithm to solve a multiobjective groundwater monitoring problem. *Water Resour Res*, 31(2): 399–409.

Clerc M. (1999) The swarm and the queen: Towards a deterministic and adaptive particle swarm optimization. *Proceedings of Congress on Evolutionary Computation* (CEC'99), 1951–1957.

Grabow G.L., Mote C.R., Sanders W.L., Smoot J.L., Yoder D.C. (1993) Groundwater monitoring network design using minimum well density, *Water Science Technology*, 28: 327–338.

Kennedy J., Eberhart R. (1995) Particle swarm optimization, *Proceedings of the International Conference on Neural Networks*, Perth, Australia, IEEE, Piscataway, 1942–1948.

Kennedy J., Eberhart R. (1997) A discrete binary version of the particle swarm algorithm. In *Proceedings of the IEEE International Conference on Systems, Man and Cybernetics*, IEEE, Piscataway, NJ, 4104–4108.

Lee Y.M., Ellis J.H. (1996) Comparison of algorithms for nonlinear integer optimization: application to monitoring network design. *Journal of Environmental Engineering* 122(6): 524–531.

Montalvo I., Izquierdo J., Perez R., Tung M.M. (2008) Computers and Mathematics with Application, 56: 769–776.

Oliveira L.S., Brito Jr. A.S., Sabourin R. (2005) Improving Cascading Classifiers with Particle Swarm Optimization, *8th International Conference on Document Analysis and Recognition (ICDAR 2005)*, Seoul, South Korea, August 29–September 1st, 570–574.

Reed P.M., Minsker B.S., Valocchi A.J. (2000) Cost-effective long-term groundwater monitoring design using a genetic algorithm and global mass interpolation. *Water Resour Res* 36 (12): 3731–3741.

Sedki A., Ouazar D. (2010) Swarm intelligence for groundwater management optimization, *Journal of Hydroinformatics*, 163 (in press).

Shi X.H., Liang Y.C., Lee H.P., Lu C., Wang Q.X. (2007) Particle swarm optimization-based algorithms for TSP and generalized TSP, *Journal of Information Processing Letters*, 103: 169–176.

Wang K., Wang X., Wang J., Lv G., Jiang M., Kang C., Shen L. (2008) Particle swarm optimization for calibrating stream water quality model. *Second International Symposium on Intelligent Information Technology Application*, 20–22 December.

Chapter 25

A novel approach to groundwater model development

Thomas D. Krom[1] *& Richard Lane*[2]
[1] *Silkeborg, Denmark*
[2] *Christchurch, New Zealand*

ABSTRACT

The hydrogeologic conceptual model is a key source of uncertainty in predicting groundwater flow. The data used to develop the hydrogeologic conceptual model include geologic sample descriptions, interpretations of geophysical data and geochemical information. The proposed method is based on implicit modelling concept as there is a direct link between data – interpretation – model. The method uses Radial Basis Functions (RBF) to form grid free hydrogeological conceptual models. RBF models are developed for the key components in a hydrogeological conceptual model. This approach has significant advantages. Firstly, the models are consistent with the known data and can be automatically updated when new data comes to hand. Secondly, the models can be influenced by both the choice of high level input such as anisotropy while maintaining consistency with the data. Thirdly, the user can add manual interpretations (trends or a priori information) that are maintained separately from measurements, but are then merged in the model building process to produce a model consistent with both measured and interpreted data. Once created, the model (i.e. equations) can be gridded at any resolution or fitted to any mesh, a process that provides a flexible interface to flow simulators. The methodology is tested on data from Denmark.

25.1 INTRODUCTION

The hydrogeologic conceptual model is a key source of uncertainty in predictions of groundwater flow. This is however an area of uncertainty that is frequently unaddressed in model analysis, while say parameter uncertainty is explored. In part this has been due to the availability of tools, like PEST (Doherty, 2005), that ease parameter uncertainty analysis; combined with a paucity of methods that address geological uncertainty. Another problem that hinders the exploration of geological uncertainty is the time required to develop alternative conceptual models.

A method is presented that addresses the key issues that hinder the development of multiple competing conceptual models: the methodology facilitates the creation of multiple conceptual models; and by accessing modern IT methods it facilitates the model building process by increasing transparency and speed of development. Furthermore, the method allows the development of alternative numerical meshes or grids for each conceptual model in order to explore gridding effects.

Hydrogeological models can be created from the 3 and 4-dimensional data sets using Radial Basis Function (RBF) models. RBF models for the components in a hydrogeological model are: aquifers, aquitards and boundaries. This approach has significant advantages. Firstly, the models are consistent with the known data and can be automatically updated when new data comes to hand. Secondly, the models are influenced by both the choice of high level parameters such as anisotropy while maintaining consistency with the data. Thirdly, the user can add manual interpretations (trends or a priori information) that are maintained separate from measurements, but are then merged in the creating contact surfaces to produce a model consistent with both measured and interpreted data. Once created, the model (i.e. equations that describe the surfaces) can be gridded at any resolution or fitted to any mesh, a process that provides a flexible interface to flow simulators.

25.2 RADIAL BASIS FUNCTIONS

The equations used to represent hydrogeological system elements are developed by fitting RBF's to the data set. RBF's are well established set of methods used in scattered data interpolation, signal processing and artificial intelligence methods. RBFs are real valued functions defined as:

$$y(x) = \sum_{i=1}^{N} w_i \phi(\|x - c_i\|)$$

where, c is the ith center, x is locations in space (or space-time), w is a weight and function ϕ can be any one of a number of functions but typically is either Gaussian, quadratic, or a type of spline function. A polyharmonic (thin-plate) splines function for ϕ is applied. The key factor in the application of a RBF is the determining the weights w_i associated with centers c_i, through optimization methods.

RBF's have been used in hydrogeology for a number of years; and fuller descriptions of RBF's is found in Govindaraju & Rao (2000) and a comparison to Kriging in Chilés & Delfiner (1999).

25.3 DATA MODEL

Today, a huge amount and a wide variety of data are collected for hydrogeological problems. To develop a complete understanding of the system, it is necessary to integrate all of these data into the analysis. Figure 25.1 shows a simplified example of the data that can be used in Leapfrog Hydro (software implementation of our method) in the development of a conceptual model for a hydrogeological system.

An example of the type of analyses that Hydro can carry out is; an equation that describes geophysical data can be developed for a specific value (e.g. 20 ohm-m for a clay horizon) which can then be combined with well log data as well as manual interpretation to define a contact surface.

25.4 IMPLICIT MODELLING

A key advantage of the implicit modelling approach is that once data is within Leapfrog Hydro, there is well defined link between data – model – interpretation – flow mesh.

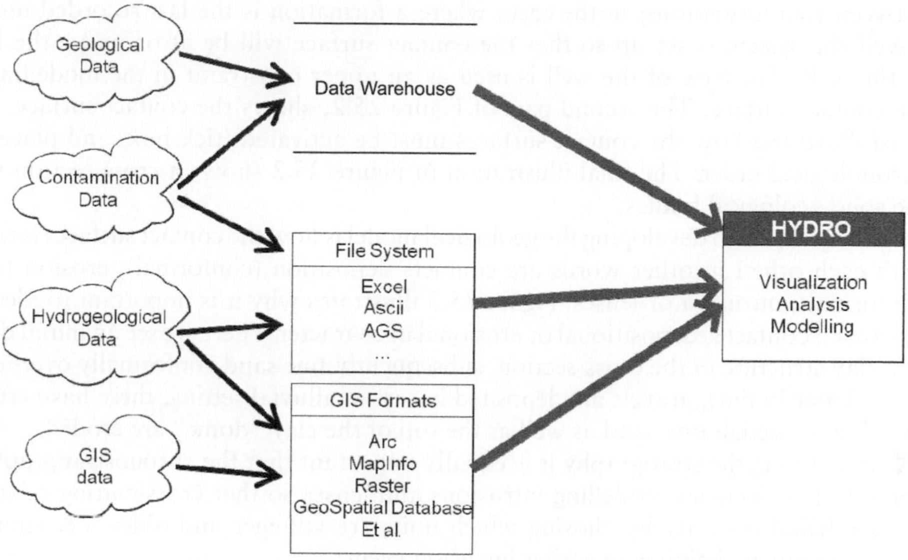

Figure 25.1 Data flow that lies behind the development of hydrogeological models in Hydro.

Furthermore, the results from flow modelling can be imported into Leapfrog Hydro. A link is defined as an object oriented relationship. This is most clearly described by example: The user has developed a geological model and a flow mesh for that model. So what happens if the user imports a new well; the geological model and the flow model will be automatically updated. This also applies to simple things like if one changes a colour for a lithology in the well data set; then the colours in the geological and flow model will also be updated.

25.5 CONSTRUCTING GEOLOGICAL MODELS

The data that can be used to develop hydrogeologic conceptual models include: geologic sample descriptions, interpretations of geophysical data, geochemical information and manual interpretations. There is never sufficient information to fully describe the hydrogeology without considerable interpretation from an expert. Existing methods for capturing this expert knowledge rely on solely manual interpretation of hydrogeological structures or the definition of statistical relations (e.g. in T-Progs (Carle, 1999)). Either procedure is time consuming, difficult to update or repeat. Certainly the manual interpretation approach makes it difficult to maintain alternative interpretations of the hydrogeology. The method proposed here is objective in nature; geological contacts are determined directly from the data (Figure 25.2), and the parameters behind the RBF are clearly exposed. Furthermore, the methodology documents in a clear manner the interaction of data versus manual interpretation; different types of data are maintained separately.

The first illustration in Figure 25.2 shows how the system identifies the contacts between two formations; in the cases where a formation is the last recorded unit in a well the system is set up so that the contact surface will be at or under the base of the well. The base of the well is used as an upper constraint in the modelling of the contact surface. The second part of Figure 25.2, shows the contact surface. The third illustrates how the contact surfaces must be activated (tick box) and placed in chronological order. The final illustration in Figure 25.2 shows a cross section with the solid geological bodies.

A key aspect to developing the geological model is how the contact surfaces interact with each other; in other words are contacts deposition (conformal), erosion (non-conformal), intrusion or lenses. Figure 25.3 illustrates why it is important to identify whether a contact is depositional or erosional in character. There we see an initial dome like clay structure in the cross-section, subsequently fine sand conformally overlie the clays. Later in time, gravels are deposited in say an alluvial setting, these have eroded the older material; fine sand as well as the top of the clay "dome" are eroded.

In defining the stratigraphy it is equally important that the chronostratigraphy is correct. This includes modelling intrusions and lenses so that cross-cutting relations are modelled correctly by chosing which units are younger and older (i.e. ignoring younger units in defining an earlier intrusion event).

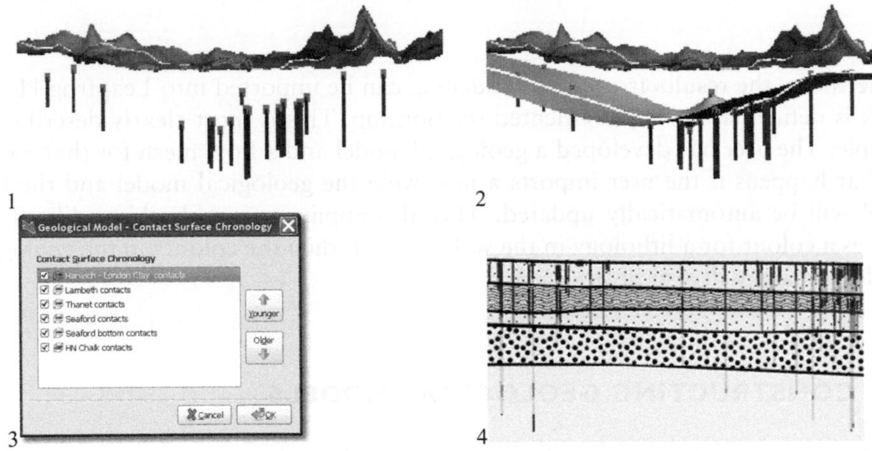

Figure 25.2 Process in developing a geological model (1 → 2 → 3 → 4). The user defines a set of contacts (Hydro can automatically extract these from well logs), a surface is fitted and put into a stratigraphic order, and finally a solid 3D description of each formation is created. The contact points are selected using database technology.

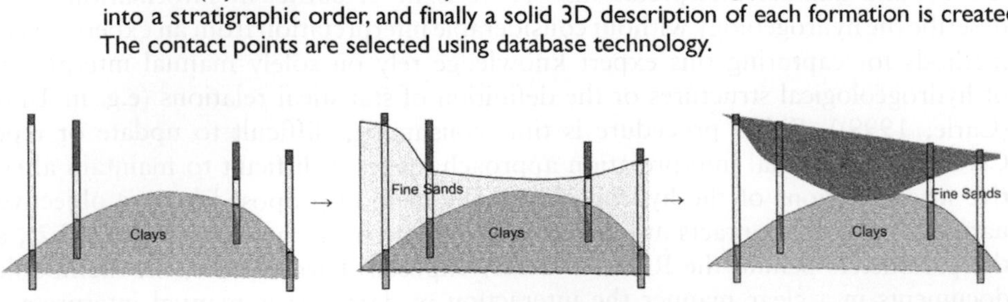

Figure 25.3 Evolution of a geological profile illustrating erosion versus deposit contact relations.

In this entire process RBF models are used to describe each contact surface. This approach has three significant advantages. Firstly, the models are consistent with the known data and can be automatically updated when new data comes to hand. Secondly, the models can be influenced by both the choice of high level parameters such as anisotropy while maintaining consistency with the data. Thirdly, the user can add manual interpretations (trends or a priori information) that are maintained separately from measurements, but are then merged in the model building process to produce a model consistent with both measured and interpreted data. Similarly, soft data such as the interpretation of geophysics data can be used in the definition of contact surfaces.

25.6 CASE STUDY: DJURSLAND AQUIFER SYSTEM

In Djursland, Denmark a series of glacial and postglacial sediments lie on top of an erosion surface which consists of chalk. The glacial sediments have been partially reworked by subsequent glaciations and Quaternary sedimentation. The result is a far more complex geological setting than is discussed in the geological model section. In this case study we have non-layered geology (till complex) resting on an erosion surface (Table 25.1). The main hydrogeologic issue in the region is potable water quality degradation due to nitrates.

The data for the development of the hydrogeologic framework are 3851 well logs of varying depth. In addition to the well logs there is a digital elevation model and groundwater chemistry data. It is very time consuming if one is to group formations by hand for 3851 well logs, though one can learn by inspecting cross-sections and from experience, which formations are hydrogeologically similar. The knowledge extraction process quickly resolves the hydrogeological formations at the site.

The hydrogeologic problem is to define sand and gravel aquifers and specifically identify where there is good groundwater protection, i.e. thick tills overlying sand and chalk. The upper 10–30 meters of the chalk is a fractured aquifer, regardless of the type of chalk. The geology is simplified into hydrogeological formations in the following manner: tills and silts and other fine grain sediments form one hydrogeologic formation; while sands and gravel form a second formation; finally chalk formations are placed into a 3rd formation. The surface between chalk and the other formations is defined as an erosion surface.

Figure 25.4 shows the conceptual problem that is solved in the Djursland hydrogeology; that is we have a non-layered system overlying a layered system. The proposed

Table 25.1 Hydrogeology for the Djursland aquifer system.

Formation	*Environment*	*Type*	*Chronology*
Postglacial sediments	**Various**	**Mainly aquitards**	**youngest**
Till	**Glacial**	**Aquitard**	**same age**
Sands and gravels	**Glacial**	**Aquifer**	
Erosion surface			
Chalk	**Marine**	**Aquifer**	**Oldest**

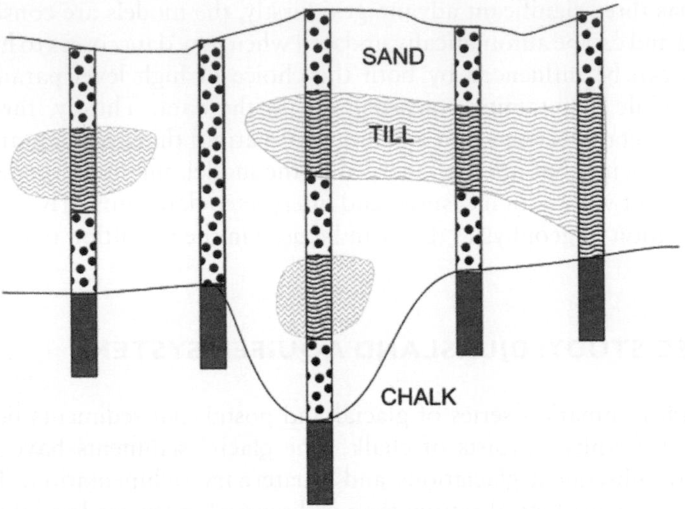

Figure 25.4 Conceptual problem in the Djurs hydrogeology.

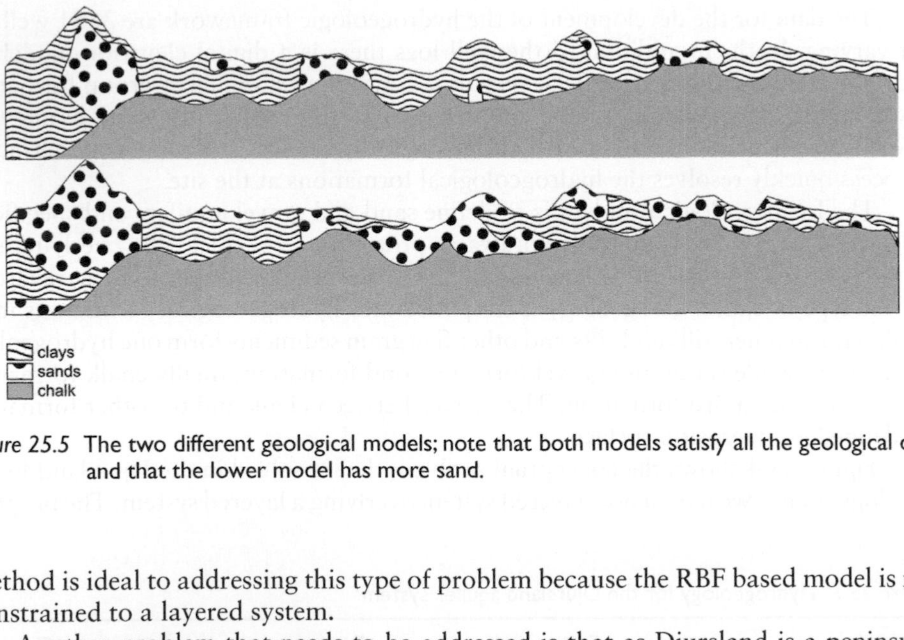

Figure 25.5 The two different geological models; note that both models satisfy all the geological data and that the lower model has more sand.

method is ideal to addressing this type of problem because the RBF based model is not constrained to a layered system.

Another problem that needs to be addressed is that as Djursland is a peninsula, there are no wells to the north and south. This causes a problem with the chalk surface as most interpolation routines will result in the chalk being exposed to the north and south of Djursland. However, the addition of dip points just off shore for the erosion surface "forces" the chalk surface deeper. Figure 25.5 shows the result of the modelling exercise for Djursland where there is a very heterogeneous system overlying the chalk surface.

Table 25.2 Flow models based on 2 grids and 2 conceptual models.

	Conceptual model A	*Conceptual model B*
Grid 1	modelA1	modelB1
Grid 2	modelA2	modelB2

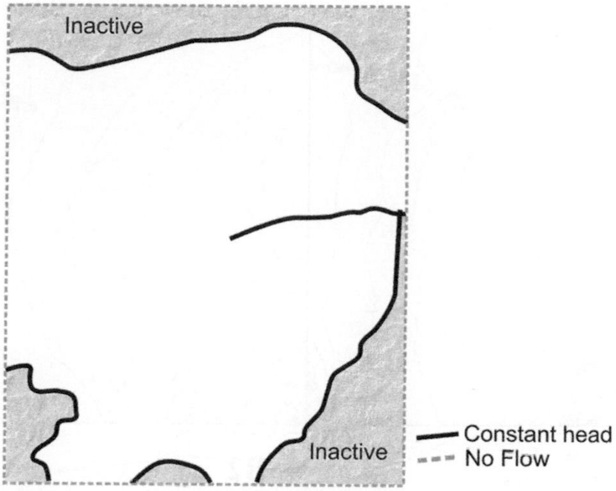

Figure 25.6 Boundary conditions for the flow models, constant head boundaries are a mean sea level.

Since an RBF is a function; once created, the model can be iso-surfaced or gridded at any arbitrary resolution or fitted to any mesh or grid, a process that provides a flexible interface to flow simulators. This does not require re-fitting the data and interpretations, as opposed to say a Kriging approach that would require re-solving the Kriging equations if new grid resolution was chosen.

Figure 25.5 shows a result of the modelling exercise for Djursland where there is a very heterogeneous system overlying the chalk surface. Hydrogeologic parameters are assigned to geological units and transferred to gridded models. There are two alternative conceptual models; one where sand dominants in cases where there is doubt as to the geology and another where fines dominate.

Hydrological model. The geological models are transferred to a hydrological model implemented in Modflow2000. This is done with two different grids to show the implications those choices have on the results of flow modelling (Table 25.2). The purpose is not necessarily show the value of using alternative conceptual models or alternative gridding; but to show that it is easy to carry out.

The outer flow boundary conditions for the models are the same in all four cases outlined in Table 25.2. Along the coast line a constant head boundary condition is assumed at mean sea level; elsewhere a no-flow boundary condition is assumed (Figure 25.6).

The two gridding concepts are: 1) a layer based approach where the top of the chalk is honoured and there are four numerical layers above the chalk and, 2) a voxel based approach with equally thick layers for the model cells where there are 15 layers.

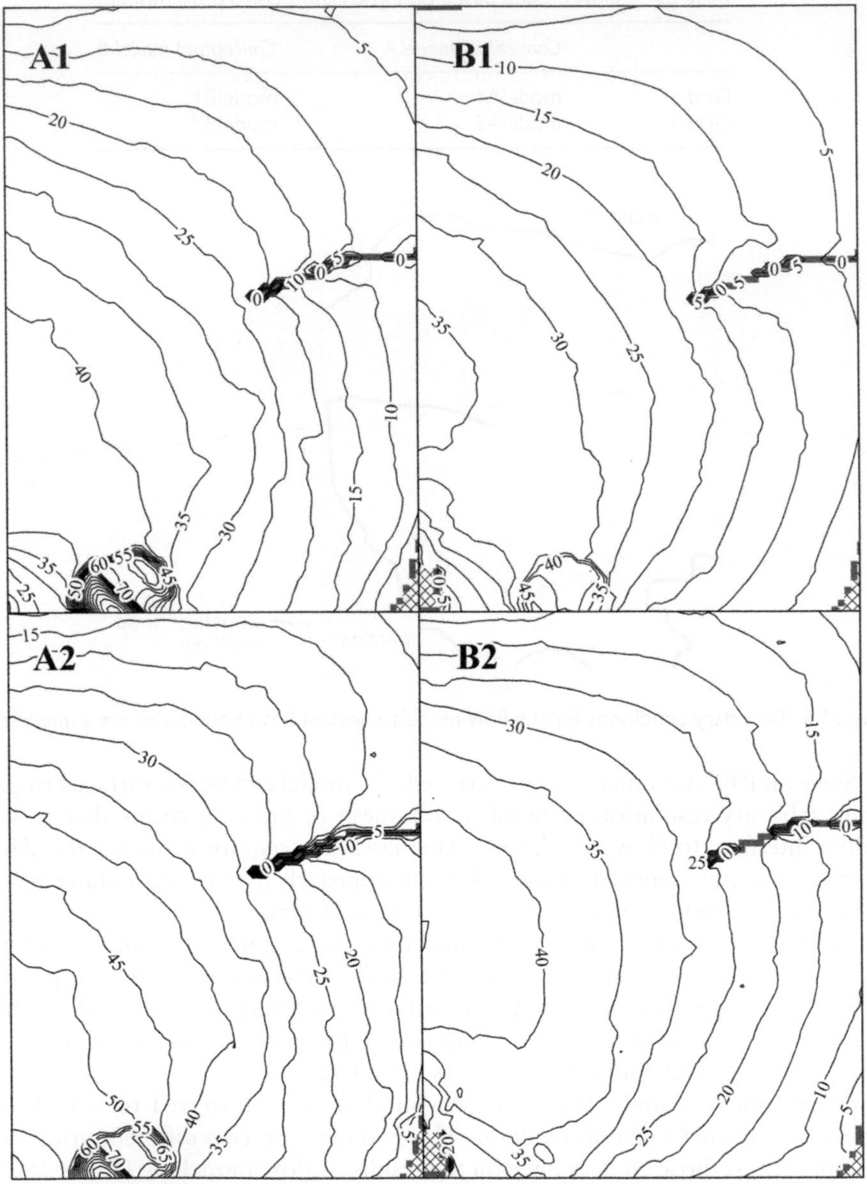

Figure 25.7 Groundwater table for 4 different combinations of the conceptual model and Modflow Grid.

The models are otherwise identical. Recharge is parameterised by being 50% more for sandy areas versus clayey areas. There is a constant head boundary along the coastal edge of the model. Figure 25.7 shows the water table for the four different models. There are substantial differences due to both the different gridding as well as the different conceptual models. Both conceptual models honour all the field data.

25.7 CONCLUSIONS

This is a methodology that can be applied to develop hydrogeological models in complex hydrogeological systems. The method shows advantages in extracting knowledge from the data and developing grid free hydrogeological models. The method also shows advantages in clearly indicating knowledge that comes from data versus expert opinion/a priori knowledge.

A significant part of the advantage with the method is how it is employed in the implementation and is experienced carrying out the work flow in developing a hydrogeological model as well as in the collaboration between the geologist and hydrogeologist. The implicit modelling approach makes the analysis robust, transparent and easily updatable. Also since the implicit approach will use all the data it is more objective than an approach based solely on drawing contact surfaces.

The method is designed to facilitate the creation of competing geological models for the hydrological problem, thus addressing geological uncertainty. The method also allows the testing of alternative meshing or gridding concepts to test there implication on the problem solution.

REFERENCES

Carle S.F. (1999). T-PROGS: Transition Probability Geostatistical Software version 2.1. Davis, California, USA: University of California.

Chilés J.P., Delfiner P. (1999). *Geostatistics: modeling spatial uncertainty*. Wiley.

Doherty J. (2005). *PEST: Model-Independent Parameter Estimation User Manual*: 5th Edition, Brisbane, Australia: Watermark Numerical Computing.

Govindaraju R.S., Rao A.R. (2000). Artificial neural networks in hydrology. Netherlands: Kluwer.

Chapter 26

Determining natural background values with probability plots

Thomas Walter[1], Antje Beer[2], D. Brose[3], Dörte Budziak[4], Patrick Clos[5], Thomas Dreher[6], Hans-Gerhard Fritsche[7], Matthias Hübschmann[8], Silke Marczinek[9], Anett Peters[10], Heidrun Poeser[8], Hans-Jörg Schuster[11], Bernhard Wagner[9], Frank Wagner[5], Günther Wirsing[12] & Rüdiger Wolter[13]

[1]*Landesamt für Umwelt- und Arbeitsschutz, Saarbrücken, Germany*
[2]*Landesamt für Geologie und Bergwesen Sachsen-Anhalt, Halle, Germany*
[3]*Landesamt für Bergbau, Geologie und Rohstoffe Brandenburg, Cottbus, Germany*
[4]*Landesamt für Bergbau, Energie und Geologie, Hannover, Germany*
[5]*Bundesanstalt für Geowissenschaften und Rohstoffe, Hannover, Germany*
[6]*Landesamt für Geologie und Bergbau Rheinland-Pfalz, Mainz, Germany*
[7]*Hess. Landesamt für Umwelt und Geologie, Abt. Geologie, Dezernat Hydrogeologie, Wiesbaden, Germany*
[8]*Sächsisches Landesamt für Umwelt, Landwirtschaft und Geologie, Dresden, Germany*
[9]*Bayerisches Landesamt für Umwelt, Hof, Germany*
[10]*Thüringer Landesanstalt für Umwelt und Geologie, Weimar, Germany*
[11]*Geologischer Dienst NRW, Krefeld, Germany*
[12]*Regierungspräsidium Freiburg, Landesamt für Geologie, Rohstoffe und Bergbau, Freiburg, Germany*
[13]*Umweltbundesamt (UBA), Dessau, Germany*

ABSTRACT

The EU Water Framework Directive demands the achievement of a good qualitative status for all groundwater bodies. Therefore, a reference for the natural background values in groundwater is needed. The German Geological Surveys have established such background values based on previously determined hydro-geochemical units, based on a method using probability nets. The probability net displays normal or lognormal distributions as straight lines and mixtures of distributions as sequences of straight line segments. Through an iterative process of excluding the present anomalies, a trend line can be fitted to the bulk of data, considered as background population. With the line's slope and intercept, which are equivalent to the underlying populations' standard deviation and mean, every percentile can be calculated, using the 90th or the 95th the as upper limit of the background population. In order to accelerate and facilitate the process of calculation, an Excel application has been developed with extensive graphical and statistical controls.

26.1 INTRODUCTION

One of the main objectives of the EU Water Framework Directive (EU-WFD) is the achievement of a good qualitative status of groundwater bodies. Hence, the derivation of groundwater background values is required to detect significant point source contamination, or to identify whole groundwater bodies at risk of failing to meet the required EU water quality standards. For the State Geological Surveys of Germany, a working group 'Groundwater Background Values' (WG GBV) has engaged in establishing criteria for defining groundwater background values since 2005. The main objective of the WG BGV was the derivation of nation-wide characteristic groundwater background values based on previously determined hydro-geochemical units (HGC) for a range of major, minor and trace elements. The outcome of this project is publicly available as Web Map Service on the internet at http://www.bgr.de/Service/grundwasser/. The following describes the approach and development of the project, and the statistical method to derive the groundwater background values from the original data set.

26.2 OBJECTIVE AND APPROACH

The physical and chemical properties of groundwater are mainly determined by the composition of the percolating water, its alterations during passage through the unsaturated zone, the lithology of the aquifer, and the travel time of the groundwater in the subsurface. The natural geogenic properties of groundwater result from a dynamic equilibrium of groundwater and rock surface chemistry. At this interface complex chemical, physical, and biological processes take place. Ancient and therefore mainly deep groundwater is primarily influenced by geogenic processes, while younger and shallower waters tend to show more surface related influences including anthropogenic factors. In such groundwater pure geogenic background values can often no longer be determined (Wagner *et al.*, 2003; Kunkel *et al.*, 2004). Therefore, the background values in this study represent a combination of the geogenic component and a component from ubiquitous and diffuse anthropogenic influences. A hypothetical combination of background population and anomaly is presented in Figure 26.1.

If the background values of groundwater are defined as the upper threshold of the typical regional groundwater composition, it might seem that under similar regional conditions groundwater should show a more or less uniform, normal or – in most of the cases – lognormal distribution of chemical parameters, especially in the case of trace elements (Koch & Link, 1970). In reality, though, the chemical composition of groundwater is the result of many chemical reactions and other natural and anthropogenic influences, which act on different temporal and spatial scales (Marczinek *et al.*, 2008). Anomalies can be caused by anthropogenic influences, but also by local natural phenomena such as ore mineralisation, coastal and inland salinisation (Grube *et al.*, 2000) or acidification of crystalline rock regions (Hinderer & Einsele, 1998). However, it is not possible to separate a diffuse ubiquitous component from long-term atmospheric and agricultural inputs (e.g. fertilisation, soil melioration or traffic related inputs), if these impacts have already been integrated into the normal population. Kunkel *et al.* (2004), therefore, established the term 'natural, ubiquitously influenced groundwater composition'.

The concentration distribution can be affected, with varying intensity for each parameter, by different initial concentrations, changes in the reaction mechanisms, varying transmission paths, or various external influences, e.g. vegetation,

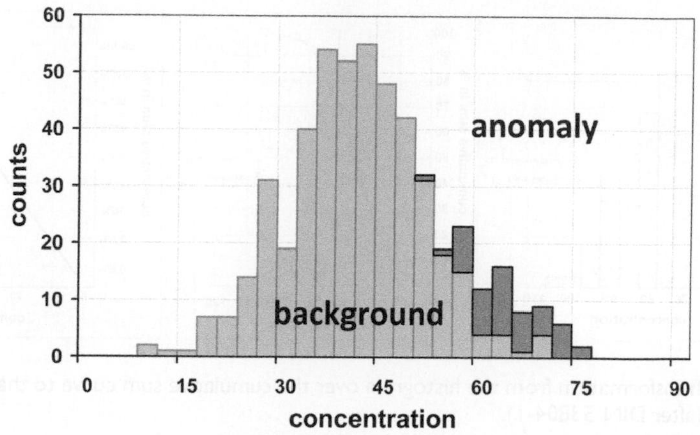

Figure 26.1 A hypothetical combination of background population and anomaly.

atmospheric, or anthropogenic influences. Anomalies and differing subpopulations will affect the parameter distributions with growing numbers of samples and increasing size of the investigated area. In a complex outset it is always difficult to decide, if multiple peaks or breaks in the cumulative curve or histogram of a data set should be interpreted as the existence of overlapping subpopulations. The interpretation depends largely on the experience of the evaluator and it can only be done with large numbers of analyses, when relying on histograms.

26.3 DERIVATION OF BACKGROUND VALUES WITH PROBABILITY PLOTS

For decades exploration geochemistry has successfully used statistical methods to identify and delineate anomalies resulting from ore mineralisation, clearly of geogenic origin. In this context, an anomaly is defined as a deviation from the characteristic regional distribution and it can so be found in the higher as well as the lower segments of the concentration range (Van den Boom, 1981). While in ore exploration normally the lower boundary of the anomaly is of interest for separating the high concentration of the anomaly from the less interesting lower concentration of the background, the upper boundary of the normal population is of much interest for hydrogeological investigations, as it can be interpreted as background value. Despite this slightly different viewpoint the statistical methods used in exploration geology can be transferred to groundwater data in order to identify local and regional anomalies against the normal population.

Lepeltier (1969) developed a method on the basis of probability plots using cumulative percentages, which is now an established standard in exploration geology. The method had to be slightly adjusted as the identification and calculation of distribution parameters for anomalies is somewhat different from the approach of determining groundwater background values, which are solely related to the normal population.

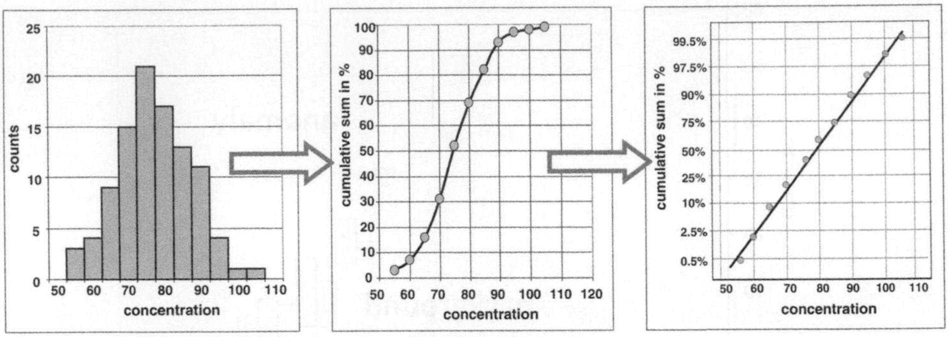

Figure 26.2 Transformation from the histogram over the cumulative sum curve to the probability plot (after DIN 53804-1).

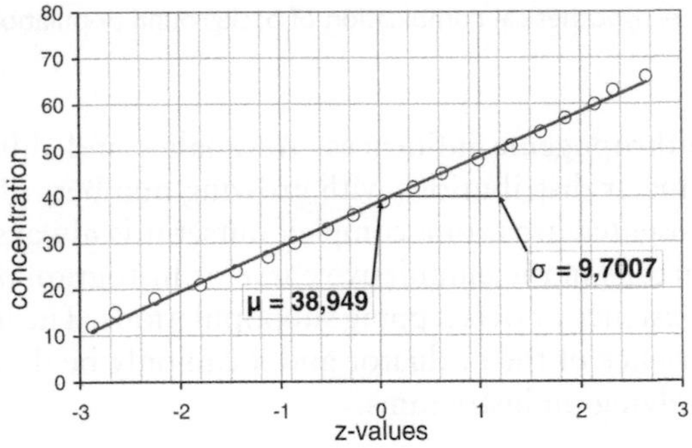

Figure 26.3 Probability net with a hypothetical distribution, randomly generated with mean = 40 and stdev = 10. Compared to Fig. 26.1, x- and y-axes are transposed, the slope allowing so a direct judgement of the standard deviation (steep = high, gentle = small standard deviation).

The probability plot is a simple graphical procedure to determine the distribution form of a random variable. It consists of a grid that is calculated linearly on one-axis and on the other axis determined by an integrated normal distribution. Normal distributions or normally distributed sections of a random variable are then represented by straight lines (Fig. 26.2 and 26.3), allowing a simple visual examination of the data set at hand and facilitating a quick distinction between normal and anomalous components. The great advantage of the probability plot is that the statistical distribution parameters of the underlying distribution can easily be determined: the median corresponds to the arithmetic mean of the population while the slope defines the standard deviation (Fig 26.4). Lognormal distributions also produce a straight line when the x-axis is scaled logarithmically. Using regular spreadsheet software, linearity is achieved by a simple z-transformation of the values, resulting in a standard normal distribution with a mean of $\mu = 0$ and a standard deviation $\sigma = 1$ and plotting the original values against their z-values:

$$z_i = \frac{x_i - \mu}{\sigma}$$

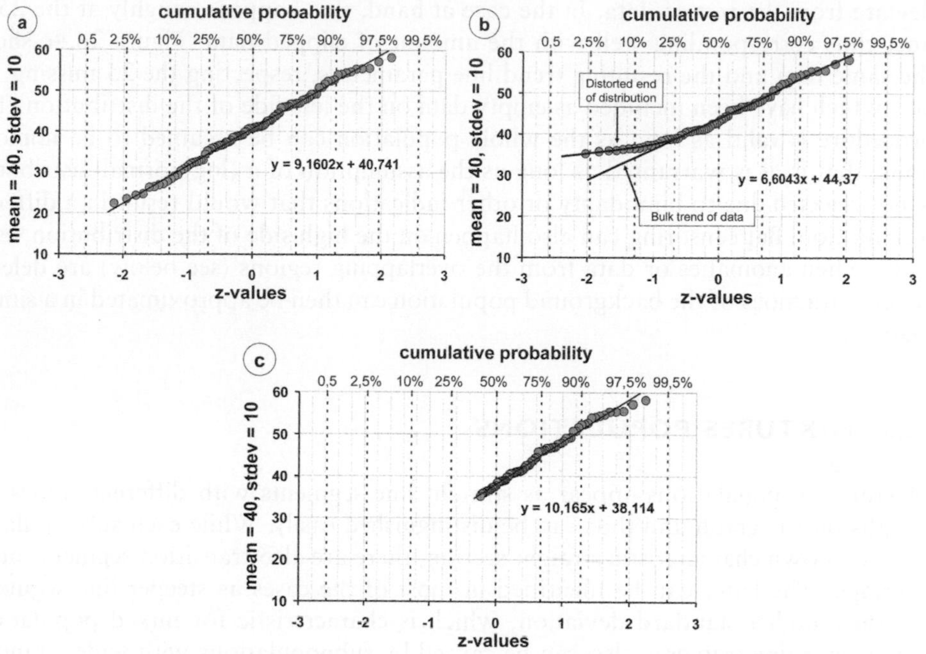

Figure 26.4 a) randomly generated data set of 100 values (mean = 40, stdev = 10); b) left censored data set after deletion of the 25 lowest data (note that distortion from linearity starts at the 25%-line); c) censored data set after addition of 25 empty data values at the left side.

For every z-value a cumulative probability can easily be calculated with any spreadsheet software, using built-in functions (in Excel, NORMSDIST(z) returns standard cumulative distribution function).

26.4 SINGLE DISTRIBUTIONS IN THE PROBABILITY NET

Whereas normal distributions plot straight in a probability plot when the full range of data is honored, censored data show a different behaviour. Censoring occurs when the values of an observation are only partially known. A typical case is data below the detection level, whose real value is not known. Excluding in this manner a range of low values generally results in higher means and smaller standard deviations compared to the statistics that would apply to the full range of values. To overcome this problem, often a work-around is used by including these values set to half the detection limit or following other substitution rules. But depending on the severity of censoring, the resulting distributional statistics may be highly questionable (Crogan & Egeghy, 2003). In the probability net, the behaviour of the statistics at first is similar: the mean rises and the standard deviation drops compared to their real means (Fig. 26.4a and b), because less values are distributed over a narrower but in general higher range. But additionally, the normally straight line is distorted at the clipped end and tends to flatten out. After Sinclair (1976), the amount of missing data is roughly equivalent to the cumulative percentage at the point where the distorted end of the curve starts to deviate from the rest of data. In the case at hand, this happens roughly at the 25%-line, thus corresponding well with the amount of clipped data. Figure 26.4c shows the same data and the resulting trend line parameters, respecting the 25 missing values, which have been included as empty data on the left side of the distribution. This procedure is valid as long as the whole population can be assumed to be homogenous, that is, it is warranted as long as the assumption of a (log)normal distribution is not rejected due to bimodality or other indications that would result in a different distribution. But censoring can also happen on the high side of the distribution, especially when anomalies or data from the overlapping regions (see below) are deleted. Deleted fractions of the background population can then be approximated in a similar manner.

26.5 MIXTURES POPULATIONS

Mixtures of populations appear as straight line segments with different slopes and lengths on the graph and thus can be distinguished easily. While each subpopulation shows its own characteristic straight section, there are also transition segments at the overlaps. The latter can be identified in most of the cases as steeper line segments, i.e. their higher standard deviation, which is characteristic for mixed populations, but steeper line segments also can be caused by subpopulations with wider standard deviations. Sinclair (1976) and Van den Boom (1981) describe the most frequent combinations of straight lines in the probability plot and their causes. Probability plots, thus, give an overview of data heterogeneity at first glance. Therefore, it is in any case advisable to examine the complete data set before starting to exclude data.

In Figures 26.5 and 26.6, two randomly generated populations (background with mean = 40 and standard deviation = 10, anomaly with mean = 65 and standard deviation = 5) have been mixed and plotted in a probability net. The usual pattern of mixed distribution is clearly visible: background population and anomaly are represented by more or less straight line segments, interconnected by a steeper segment representing the interval, where both populations overlap. The point of inflexion indicates roughly the percentage splitting of both fractions (Sinclair, 1976), in this case 80% background and 20% anomalous population. Figure 26.6 shows the effect of varying percentages of the anomaly. Here, different realisations of hypothetical background and anomaly populations – each with their respective trend lines – have been randomly generated and mixed in different proportions. Even relatively small anomalous fractions of 2.5% are already distinguishable as deviations from straight linearity. With increasing percentages of the anomalous fraction, the upper branch of the curve approximates the line representing the ideal anomalous component, while the lower branch moves slowly away from the bottom line given by the trend line for the background values. But while the proportion of anomaly does not exceed 25%, this deviation is so small that the distributional statistics derived from slope and intercept are virtually identical to those of the original background population. However, proportions of anomalous data of 25% or higher always indicate a strong heterogeneity of the data set, so that in these cases it is advisable to classify the data in a more effective way.

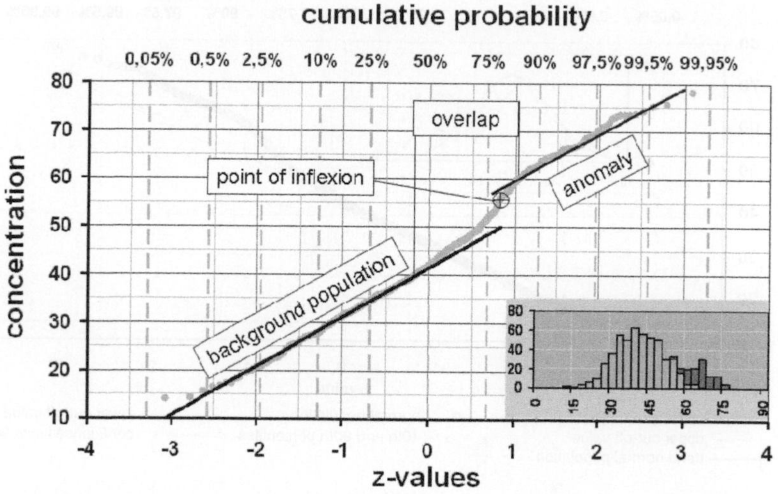

Figure 26.5 Randomly generated mixed distribution (background population with mean = 40 and standard deviation = 10, with mean = 65 and standard deviation = 5). The cumulative probability of the point of inflexion roughly indicates the ratio of background to anomaly, in this case 80% background population and 20% anomaly.

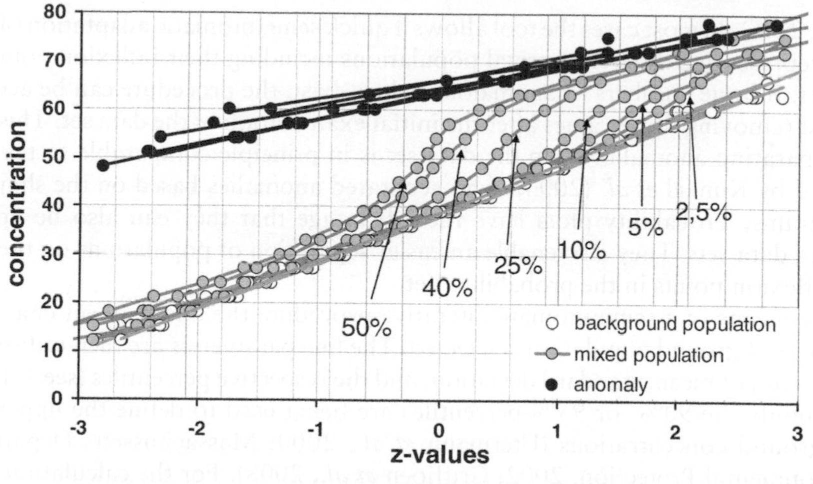

Figure 26.6 Overlay of several randomly generated mixed distributions with identical distributional statistics (background with mean = 40 and standard deviation = 10, anomaly with mean = 65 and standard deviation = 5) and varying proportions between 2.5 and 50% of the anomalous fraction.

26.6 SOFTWARE TOOL

For an efficient evaluation, a software tool was developed in Microsoft Excel (Walter, 2005), which iteratively fits a trend line to the main population of the distribution, thereby excluding anomalies on both sides of the distribution (see Table 26.1 and

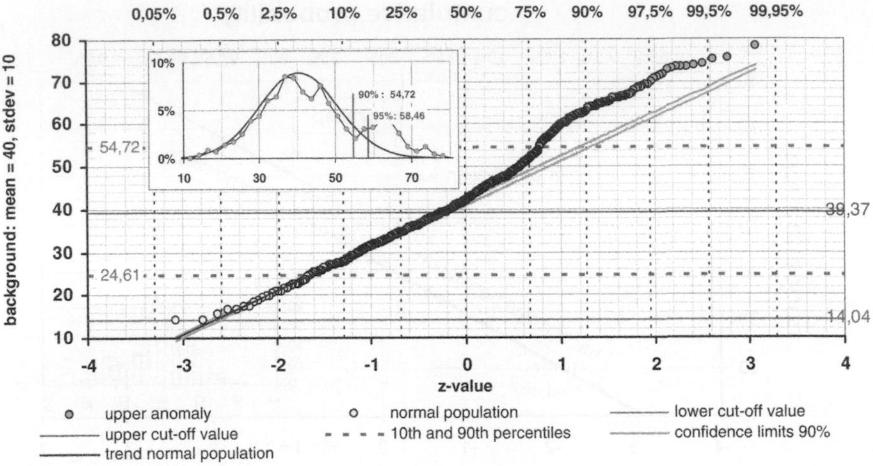

Figure 26.7 Data set from Figure 26.6 after fitting procedure. The graph shows the data excluded as anomaly, the trend line fitted to the remaining data interpreted as background population, the cut-off value at 39.37 as well as the 10th and the 90th percentile. The inset represents the theoretical distribution curve fitted to the original data in a histogram, together with the values for the 90th and the 95th percentile for comparison purposes.

Figure 26.6). In most cases the tool allows a quick semiautomatic adaptation of straight lines representing different normal populations including their inflexion points to the dataset. If large numbers of anomalous values exist, the procedure can be accelerated by first removing those values after the initial examination of the data set. This method for separating anomalies from the data set is in principle comparable to the method devised by Kunkel *et al.* (2004), who separated anomalies based on the shape of the histograms. Probability plots have the advantage that they can also be applied to smaller data sets. They also enable an easier separation of populations on the basis of the inflexion points in the probability net.

By means of a semiautomatic, iterative procedure the line segment characteristic for the background population is selected. The line parameters are then utilised for the calculation of mean, standard deviation, and the respective percentiles (see Table 26.1). Commonly the 90%- or 95%-percentiles are being used to define the upper limit of background concentrations (Utermann *et al.*, 2000; Massachussetts Department of Environmental Protection, 2002; Griffioen *et al.*, 2008). For the calculation of background values for the hydrogeochemical units of Germany (Wagner *et al.*, 2010), the 90th percentile was chosen as the upper limit of the background population, in compliance with the generalised usage of this percentile for identification of anomalies. But considering that the anomalous fraction already had been separated previously, the 95th percentile might be even more justified.

The results of the procedure can be controlled visually by comparing the calculated normal populations to a histogram of the data (Fig. 26.7). Goodness of fit is also tested by a correlation coefficient adapted to the probability plot (Ryan & Joiner, 1976) and the d'Agostino-Pearson test for normality (Sheskin, 2007), both given for normal and lognormal distribution. These tests allow an easy verification of the concordance of

Table 26.1 Statistical metadata to the analysis presented in Figure 26.7. Only after exclusion of 58% the data, convergence of the iteration process is achieved (the preset value for alpha is successively replaced with the resulting percentage of excluded data). The d'Agostino test does not support the assumption of normality, in contrast to the Ryan-Joiner test, based on the correlation coefficient. However, using random number generators does not necessarily produce real random numbers, especially when using Excel. The correlation coefficient (r) is given and in the original application its significance is signalised by a colour coding (red = not significant, green = significant, bold green = highly significant).

file:	Synthetic distributions.xls				
dataset:	background, mean = 40, stdev = 10 background, mean = 40, stdev = 10				
parameter:	20% anomaly				
tails:	1-tailed		distribution:	normal	
number of data below detection limit: 0					
alpha:	58.00%	entire data set		normal population	
		normal	lognormal	normal	lognormal
number of data		600		252	277
maximum		78.18		39.37	40.92
median		42.07		33.20	34.07
minimum		14.04		14.04	14.04
	−standard deviation	30.86	30.67	31.24	30.55
mean		43.77	41.81	41.53	44.08
	+standard deviation	56.68	57.01	51.82	63.60
d'Agostino-Pearson-K2-Test	K2 =	14.17	20.95	13.00	12.12
	p =	0.0008	0.0000	0.0015	0.0023
	normal distribution can	not be assumed!	not be assumed!	not be assumed!	not be assumed!
goodness of fit	r =	0.9915	0.9908	0.9980	0.9928
quantiles	5.0%	22.54	25.11	24.61	24.12
	10.0%	27.23	28.10	28.35	27.55
	25.0%	35.06	33.92	34.59	34.42
	50.0%	43.77	41.81	41.53	44.08
	75.0%	52.48	51.53	48.47	56.44
	90.0%	60.31	62.21	54.72	70.52
	95.0%	65.00	69.62	58.46	80.56
excluded values	high low	318	351	348	323
excluded (%)				58.00%	53.83%

assumed and actual distributions, and their results also help to determine if the data are best represented by a normal or lognormal distribution. However, the plausibility of the data has also to be assessed in consideration of the regional hydrogeology. Based on standard deviation and mean – calculated using the parameters of the trend line fitted to the portion of data identified as belonging to the background population – percentiles for the theoretical distribution of the background population are given. A minimum of 10 measured values in the normal population (i.e. without data below detection limit or excluded anomalies) is required to guarantee statistical reliability. When dealing with missing data, as a general rule for major elements it was established that at least 60% of all data should be used for the calculation of the background value. This criterion could not be maintained for minor and trace elements as it would have resulted in an exclusion of too many results due to high percentages of data below the detection limit.

Figure 26.7 and Table 26.1 give good, albeit slightly overestimated values for the theoretical distributional parameters (mean = 41.53 and standard deviation = 10.19), but using synthetic data sets produced by random number generators is likely to produce such deviations. The d'Agostino-Pearson test does not support the basic assumption of normality for the section of values identified as background population, in contrast to the Ryan-Joiner test. The test does not always give coherent results, what in this case also might be related to the "not-randomness" of random number generators.

REFERENCES

DIN 53804-1(2002): Statistische Auswertungen, Teil 1: Kontinuierliche Merkmale., Beuth-Verlag, Berlin.

Croghan C., Egeghy P. (2003) Methods of Dealing with Values Below the Limit of Detection using SAS, *11th Annual SouthEast SAS Users Group Conference*, September 22–24, 2003, St. Pete Beach, FL, http://analytics.ncsu.edu/sesug/2003/SD08-Croghan.pdf, cited August 2011.

Griffioen J., Passier H., Klein J. (2008) Comparison of Selection Methods To Deduce Natural Background Levels for Groundwater Units, *Environ. Sci. Technol.* 42: 4863–4869.

Grube A., Wichmann K., Hahn J., Nachtigall K. (2000) Geogene Grundwasserversalzung in den Porengrundwasserleitern Norddeutschlands und ihre Bedeutung für die Wasserwirtschaft. – *Veröff. aus dem Technologiezentrum Wasser Karlsruhe*, vol. 9, Hamburg.

Hinderer M., Einsele G. (1998) *Grundwasserversauerung in Baden-Württemberg.* 210 pp., Landesanstalt für Umweltschutz Baden-Württemberg Karlsruhe.

Koch G.S., Link R.F. (1980) Statistical Analysis of Geological Data, Dover Inc., New York.

Kunkel R., Wendland F., Voigt H.J., Hannappel S. (2004) Die natürliche, ubiquitär überprägte Grundwasserbeschaffenheit in Deutschland. *Schriften des Forschungszentrums Jülich*, Reihe Umwelt, 47, 204 pp.

Lepeltier C. (1969) A simplified statistical treatment of geochemical data by graphical representation. *Econ Geol* 64: 538–550.

Marczinek S., Beer A., Budziak D., Dreher T., Elbracht J., Fritsche H.G., Hotzan G., Hübschmann M., Panteleit B., Peters A., Poeser H., Schuster H., Wagner B., Walter T., Wirsing G., Witthöft M., Wolter R. (2008) Natürliche Hintergrundwerte im Grundwasser – Erste Ergebnisse. – in: Sauter, M., Ptak, T.H., Kaufmann-Knoke, R., Lodemann, M., Van den Kerk-Hof, A. (Hg.): *Grundwasserressourcen: Kurzfassungen der Vorträge und Poster*. Tagung der Fachsektion Hydrogeologie in der DGG, Göttingen, 21. bis 25. Mai 2008, Hannover.

Massachusetts Department of Environmental Protection (2002) Background Levels of Polycyclic Aromatic Hydrocarbons and Metals in Soil, http://www.mass.gov/dep/cleanup/laws/backtu.doc, cited September 2011.

Ryan T.A., Joiner B.L. (1976) Normal Probability Plots and Tests for Normality. http://www.minitab.com/uploadedFiles/Shared_Resources/Documents/Articles/normal_probability_plots.pdf cited February 2011.

Sheskin D.J. (2007) *Handbook of parametric and nonparametric statistical procedures.* Chapman & Hall, Boca Raton.

Sinclair A.J. (1976) Applications of probability graphs in mineral exploration. *Assoc. Explor. Geochem.*, spec. vol. 4, 95 pp, Rexdale, Ontario, Canada.

Utermann J., Fuchs M., Düwel O. (2008) Flächenrepräsentative Hintergrundwerte für As, Sb, Be, Mo, Co, Se, Tl, U und V in Böden Deutschlands aus länderübergreifender Sicht, http://www.bgr.bund.de/DE/Themen/Boden/Produkte/Schriften/Downloads/Hintergrundwerte.html, cited September 2011.

Van den Boom G. (1981): Geochemische Prospektionsmethoden. in: Bender, F.: *Angewandte Geowissenschaften*, Band 1, 327–357, Stuttgart.

Wagner B., Töpfner C., Lischeid G., Scholz M., Klinger R., Klaas P. (2003) Hydrogeochemische Hintergrundwerte der Grundwässer Bayerns. *GLA-Fachberichte*, 21, 250 pp., München (Bayer. Geol. L.-Amt).

Wagner B., Beer A., Brose D., Budziak D., Clos P., Dreher T., Fritsche H.-G., Hübschmann M., Marczinek S., Peters A., Poeser H., Schuster H., Wagner F., Walter T., Wirsing G., Wolter R. (2010) A Web Map Service of Groundwater Background Values in Germany. p. 714–715, *Proceedings of the IAH-conference*, 12.–17. September 2010, Krakow, Poland.

Walter T. (2005) An automated Excel-tool to determine geogenic background values using a probability net, BGR Workshop on Groundwater Bodies in Europe and adjacent Countries, Berlin; http://www.bgr.bund.de/cln_101/nn334348/EN/Themen/Wasser/Veranstaltungen/workshop_gwbodies/Poster_06_Germany_LUA_WAlter_pdf, cited February 2011.

Author index

Subject index

SERIES IAH-Selected Papers

Volume 1–4 Out of Print

5. Nitrates in Groundwater
Edited by: Lidia Razowska-Jaworek and Andrzej Sadurski
2005, ISBN Hb: 90-5809-664-5

6. Groundwater and Human Development
Edited by: Emilia Bocanegra, Mario Hérnandez and Eduardo Usunoff
2005, ISBN Hb: 0-415-36443-4

7. Groundwater Intensive Use
Edited by: A. Sahuquillo, J. Capilla, L. Martinez-Cortina and X. Sánchez-Vila
2005, ISBN Hb: 0-415-36444-2

8. Urban Groundwater – Meeting the Challenge
Edited by: Ken F.W. Howard
2007, ISBN Hb: 978-0-415-40745-8

9. Groundwater in Fractured Rocks
Edited by: J. Krásný and John M. Sharp
2007, ISBN Hb: 978-0-415-41442-5

10. Aquifer Systems Management: Darcy's Legacy in a World of impending Water Shortage
Edited by: Laurence Chery and Ghislaine de Marsily
2007, ISBN Hb: 978-0-415-44355-5

11. Groundwater Vulnerability Assessment and Mapping
Edited by: Andrzej J. Witkowski, Andrzej Kowalczyk and Jaroslav Vrba
2007, ISBN Hb: 978-0-415-44561-0

12. Groundwater Flow Understanding – From Local to Regional Scale
Edited by: J. Joel Carrillo R. and M. Adrian Ortega G.
2008, ISBN Hb: 978-0-415-43678-6

13. Applied Groundwater Studies in Africa
Edited by: Segun Adelana, Alan MacDonald, Tamiru Alemayehu and Callist Tindimugaya
2008, ISBN Hb: 978-0-415-45273-1

14. Advances in Subsurface Pollution of Porous Media: Indicators, Processes and Modelling
Edited by: Lucila Candela, Iñaki Vadillo and Francisco Javier Elorza
2008, ISBN Hb: 978-0-415-47690-4

15. Groundwater Governance in the Indo-Gangetic and Yellow River Basins – Realities and Challenges
Edited by: Aditi Mukherji, Karen G. Villholth, Bharat R. Sharma and Jinxia Wang
2009, ISBN Hb: 978-0-415-46580-9

16. Groundwater Response to Changing Climate
Edited by: Makoto Taniguchi and Ian P. Holman
2010, ISBN Hb: 978-0-415-54493-1

Printed and bound by CPI Group (UK) Ltd, Croydon, CR0 4YY
24/10/2024
01778286-0006